Problems & Solutions of
ENGINEERING MECHANICS

Problems & Solutions of

ENGINEERING MECHANICS

By Experienced Teachers

CBS

CBS Publishers & Distributors Pvt. Ltd.

New Delhi • Bengaluru • Chennai • Kochi • Kolkata • Mumbai

Hyderabad • Nagpur • Patna • Pune • Vijayawada

Dedicated to the

Engineers In the Making

ISBN: 81-239-0666-8

First Edition: 2000
Reprint: 2002, 2004, 2012, 2016

Published by:
Satish Kumar Jain for CBS Publishers & Distributors Pvt. Ltd.,
4819/XI Prahlad Street, 24 Ansari Road, Daryaganj, New Delhi - 110002
delhi@cbspd.com, cbspubs@airtelmail.in • www.cbspd.com
Ph.: 23289259, 23266861, 23266867 • Fax: 011-23243014

Corporate Office: 204 FIE, Industrial Area, Patparganj, Delhi - 110 092
Ph: 49344934 • Fax: 011-49344935
E-mail: publishing@cbspd.com • publicity@cbspd.com

Branches:
- ***Bengaluru:*** 2975, 17th Cross, K.R. Road, Bansankari 2nd Stage, Bengaluru - 70 • Ph: +91-80-26771678/79 • Fax: +91-80-26771680 E-mail: cbsbng@gmail.com, bangalore@cbspd.com
- ***Chennai:*** No. 7, Subbaraya Street, Shenoy Nagar, Chennai - 600030 Ph: +91-44-26681266, 26680620 • Fax: +91-44-42032115 E-mail: chennai@cbspd.com
- ***Kochi:*** Ashana House, 39/1904, A.M. Thomas Road, Valanjambalam, Ernakulum, Kochi • Ph: +91-484-4059061-65 Fax: +91-484-4059065 • E-mail: cochin@cbspd.com
- ***Kolkata:*** 6-B, Ground Floor, Rameshwar Shaw Road, Kolkata - 700014 Ph: +91-33-22891126/7/8 • E-mail: kolkata@cbspd.com
- ***Mumbai:*** 83-C, Dr. E. Moses Road, Worli, Mumbai - 400018 Ph: +91-9833017933, 022-24902340/41 • E-mail: mumbai@cbspd.com

Representatives:
- Hyderabad: 0-9885175004
- Patna: 0-9334159340
- Vijayawada: 0-9000660880
- Nagpur: 0-9021734563
- Pune: 0-9623451994

Printed at:
J.S. Offset Printers, Delhi (India)

PREFACE

The engineer of the coming decades, in which the world will be more like a global village connected through *infobahns*, is supposed to be more diversified in his knowledge than the conventional engineers of the past used to be. He is expected to be highly versatile in his compatibility, whether it be in the field of *financial engineering* or in the making of *Jurassic Park*.

With this thought in our mind, we have made this book to which we don't claim as solution to something but an approach to make engineering easy to the students, so as to provide them with enough time and scope to broaden their horizon.

Publishers

CONTENTS

CHAPTER ONE

RESULTANTS OF FORCE SYSTEMS

The subject of mechanics occupies a unique position in the physical science because it is fundamental to so many field of studies. Mechanics may be defined as the science which describes and predicts the condition of rest or motion of bodies under the action of forces. To engineers the important areas of mechanics are the mechanics of solids and of fluids.

- Engineering Mechanics
  - Mechanics of solids
    - Rigid bodies
      - Statics
      - Dynamics
        - Kinematics
        - Kinetics
    - Deformable bodies
      - Strength of Materials, Theory of elasticity, Theory of plasticity
  - Mechanics of fluids

TYPES OF FORCE SYSTEMS :

(1) Coplaner force system : In this force system, all the forces lie in one plane.

(2) Spatial force system : In this force system, all the forces do not lie in a single plane but are in space.

(3) Concurrent force system : In this system, all the forces intersect at a single common point.

(4) Non–concurrent force system : In this system, all the forces do not intersect at a single common point.

If all the forces are parallel then it is known as parallel force system and if all the forces are not parallel then it is known as Non–parallel force system.

RESULTANT OF FORCE SYSTEM :

If a number of forces, acts on a body, it is possible to find the single force which will replace or which will produce the same effect as produced by all the given forces. This single force is called *Resultant force.*

The effect of forces on a body is usually expressed in terms of its resultant. Since the value of this resultant determines the motion of the body, as we shall see, if the resultant is zero the body will be in equilibrium and will not change its original state of motion. But if the resultant of force system is not zero then the body will have varying state of motion, thereby creating problem in its dynamics.

RESOLUTION OF FORCES :

As a number of coplaner and concurrent forces can be combined into a single force, similarly a single force can be split up into a number of forces. This process of splitting up the given the force into a number of components, without changing its effect on the body is called Resolution of force. A force is generally resolved along two mutually perpendicular directions.

The force F can be resolved in two components along the axes as follows:

$$F_x = F \cdot \cos\theta \ \& \ F_y = F \cdot \sin\theta$$

$$\text{Resultant } F = [F_x^2 + F_y^2]^{1/2}$$

and $$\tan\theta = \frac{F_y}{F_x}$$

Sign convention to be followed : Forces which are acting upwards and rightward are to be treated as positive whereas those acting downward and leftward are to be treated as negative.

X is horizontal and positive rightwards and Y is vertical and positive upwards.

We denote it like

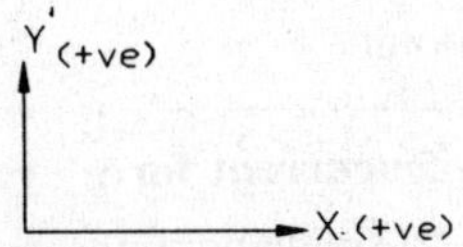

MOMENT OF FORCE :

The moment of a force about a point is defined as the product of the magnitude of the force by the perpendicular distance from the point to the action line of the force.

COUPLE :

A couple is defined as a set of two equal and opposite parallel forces, whose lines of action are different. The perpendicular distance 'd' between the lines of action of the two equal and opposite parallel forces is known as *arm of the couple.* The moment of a couple is the product of the force (i.e. one of the forces of the two equal and opposite parallel forces.) and the arm of the couple.

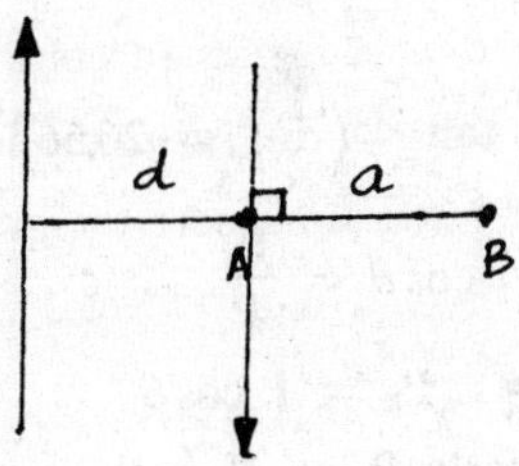

The vector sum of these two forces is zero but their moment sum is not zero. The couple dose'nt produce any translatory motion (i.e. the motion in a straight line) but produces rotation in the body on which its acts. The only effect of a couple on a body is a tendency to rotate the body about an axis perpendicular to the plane of the couple.

While attempting these problems, sometimes the direction of force is given in terms of slope of the line along the direction of force, in that case we must first find the direction of force in terms of angle.

Problem 1. : Resolve the force P in X and Y direction.

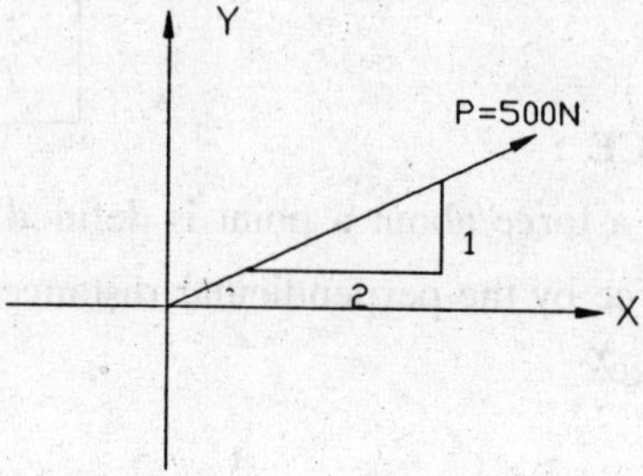

Solution : Let the angle made by P with positive X–axis be θ, then

$$\tan\theta = \left(\frac{1}{2}\right);$$

$$\theta = \tan^{-1}\left(\frac{1}{2}\right) = 26.56^{\circ}$$

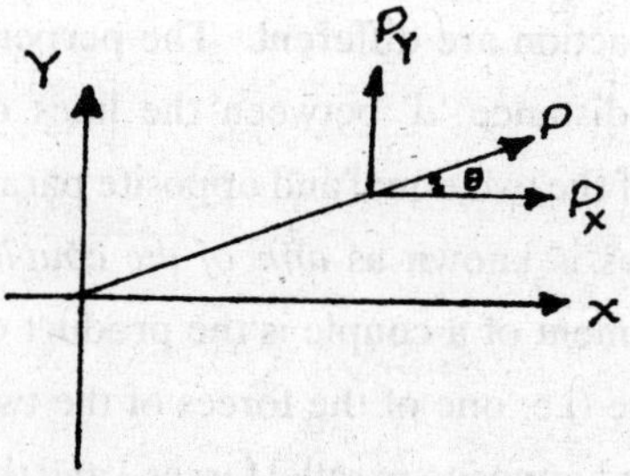

$$\text{Now } \cos\theta = \frac{P_x}{P}$$

$$\Rightarrow \quad P_x = P\cos\theta$$

$$\text{Similarly } P_y = P\sin\theta$$

$$\therefore \quad P_x = 500\cos 26.56$$

$$= 447.23 \text{ N}$$

$$P_y = 500\sin 26.56$$

$$= 223.56 \text{ N}$$

Problem 2. : Determine the X and Y components of each of the forces shown.

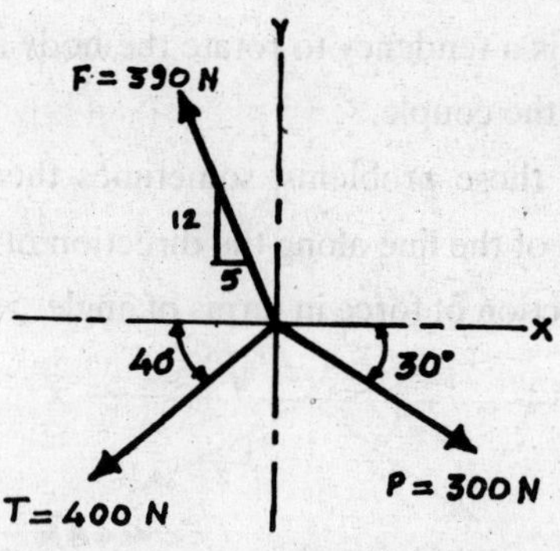

Solution : First find the inclination of force F with horizontal.

Let it be θ with negative direction of X–axis then

$$\theta = \tan^{-1}\left(\frac{12}{5}\right) = 67.4^{\circ}$$

Now refer the following figure for resolution.

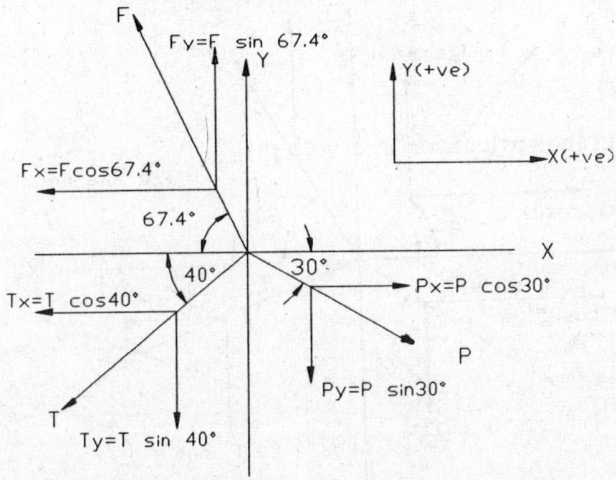

$$P_x = +P\cos 30 = +300\cos 30 = +259.8\text{ N}$$
$$P_y = -P\sin 30 = -300\sin 30 = -150\text{ N}$$
$$F_y = +F\sin 67.4 = +390\sin 67.4 = +360\text{ N}$$
$$F_x = -F\cos 67.4 = -390\cos 67.4 = -150\text{ N}$$
$$T_x = -T\cos 40 = -400\cos 40 = -306\text{ N}$$
$$T_y = -T\sin 40 = -400\sin 40 = -257.11\text{ N}$$

Problem 3. : Compute the X and Y components of each of the forces shown

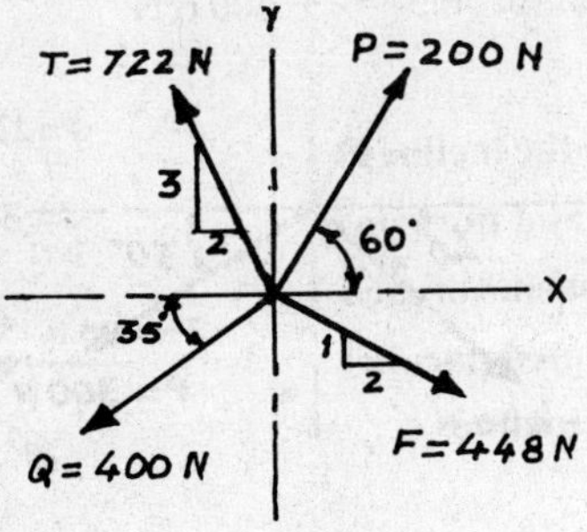

Solution : Let angle made by force T with negative X–axis be θ and angle made by F with positive X–axis be α.

then, $\theta = \tan^{-1}\left(\frac{3}{2}\right) = 56.3^{\circ}$

$\alpha = \tan^{-1}\left(\frac{1}{2}\right) = 26.56^{\circ}$

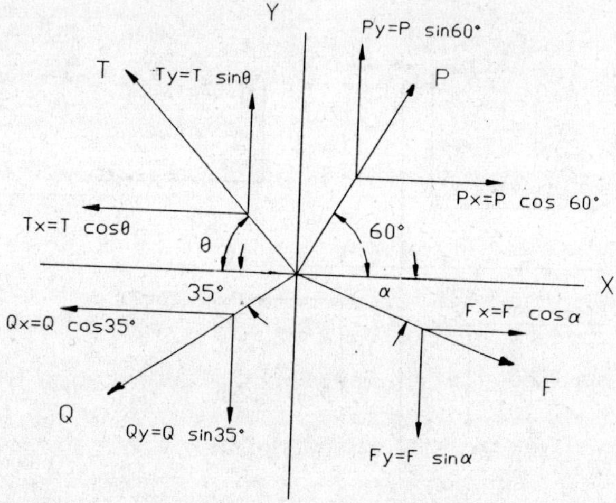

$P_x = +P\cos 60 = +200\cos 60 = +100$ N

$P_y = +P\sin 60 = +200\sin 60 = +173.2$ N

$F_x = +F\cos\alpha = +448\cos 26.56 = +400.72$ N

$F_y = -F\sin\alpha = -448\sin 26.56 = -200.3$ N

$Q_x = -Q\cos 35 = -400\cos 35 = -327.6$ N

$Q_y = -Q\sin 35 = -400\sin 35 = -229.43$ N

$T_x = -T\cos\theta = -722\cos 56.3 = -400.6$ N

$T_y = +T\sin\theta = +722\sin 56.3 = +600.6$ N

Problem 4. : The body on the incline is subjected to the vertical and horizontal forces. Find the components of each force along X–Y axis oriented parallel and perpendicular to the incline.

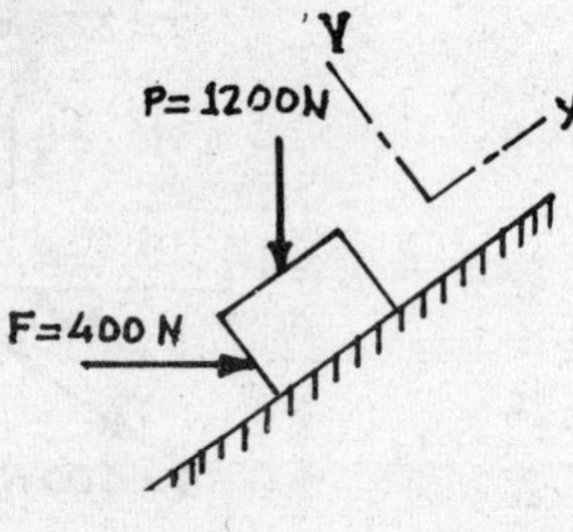

Solution : Let 'θ' be the angle of inclined plan with horizontal, then

$$\theta = \tan^{-1}\left(\frac{3}{4}\right) = 36.86^o$$

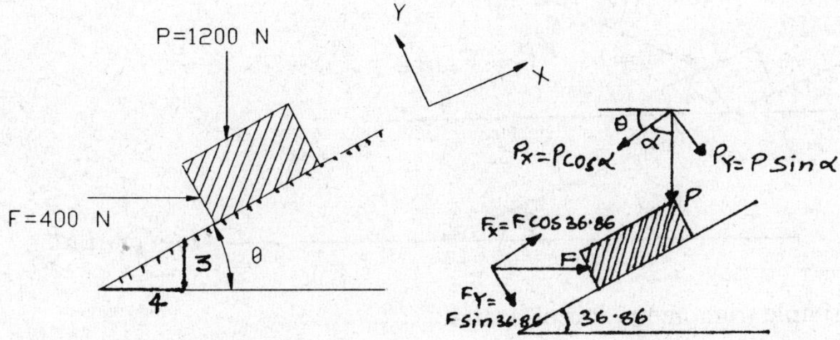

From Fig. we can write

$$\alpha = (90 - \theta) = (90 - 36.86) = 53.13^o$$

$$F_x = +F\cos 36.86 = +400\cos 36.86 = +320\text{ N}$$

$$F_y = -F\sin 36.86 = -400\sin 36.86 = -240\text{ N}$$

$$P_x = -P\cos\alpha = -1200\cos 53.13 = -720\text{ N}$$

$$P_y = -P\sin\alpha = -1200\sin 53.13 = -960\text{ N.}$$

Problem 5. : Determine the components of force P along the X–Y axis which are parallel and perpendicular to incline.

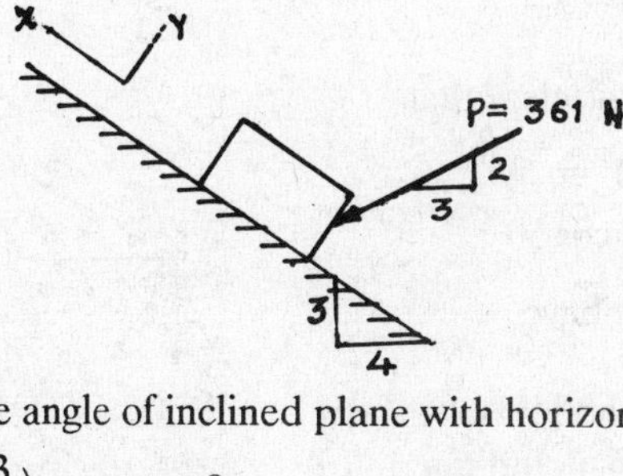

Solution : Let θ be angle of inclined plane with horizontal , then

$$\theta = \tan^{-1}\left(\frac{3}{4}\right) = 36.86^o$$

Let α be angle of P with horizontal, then

$$\alpha = \tan^{-1}\left(\frac{2}{3}\right) = 33.7^o$$

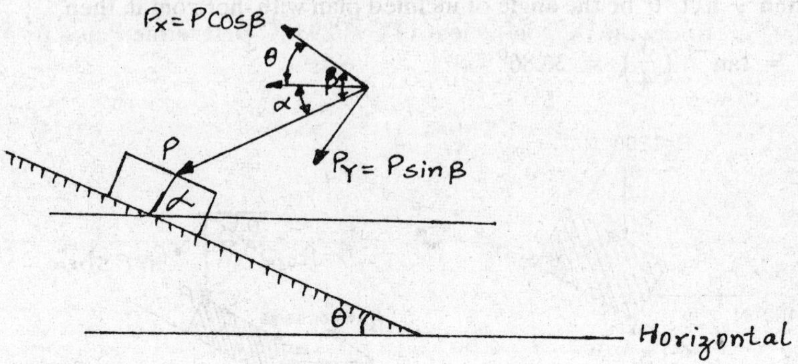

From simple trignometry it is clear that

$$\beta = (\alpha + \theta) = 33.7^o + 36.86^o = 70.56^o$$

$$P_x = P\cos\beta = 361 \cos 70.56 = 120.4 \text{ N}$$

$$P_y = -P\sin\beta = -361 \sin 70.56 = -340 \text{ N.}$$

Problem 6.: The body on 30° incline is acted upon by a force P inclined at 20° with horizontal. If P is resolved into components parallel and perpendicular to the incline and the value of parallel component is 300 N, compute the value of perpendicular component and of P.

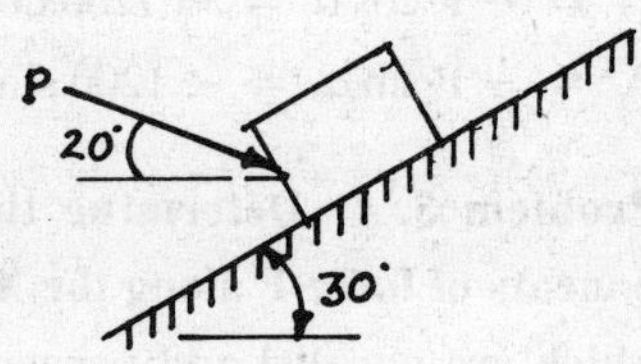

Solution : From fig. it is clear that

$$\theta = (20^o + 30^o) = 50^o$$

$$P_x = P\cos\theta = P\cos 50$$

$\because$ $P_x = 300$ N, we have

$$300 = P\cos 50,$$

$$\therefore \quad P = 466.7 \text{ N}$$

$$P_y = -P\sin 50,$$

$$\therefore \quad P_y = -466.7 \sin 50$$

$$= -357 \text{ N}$$

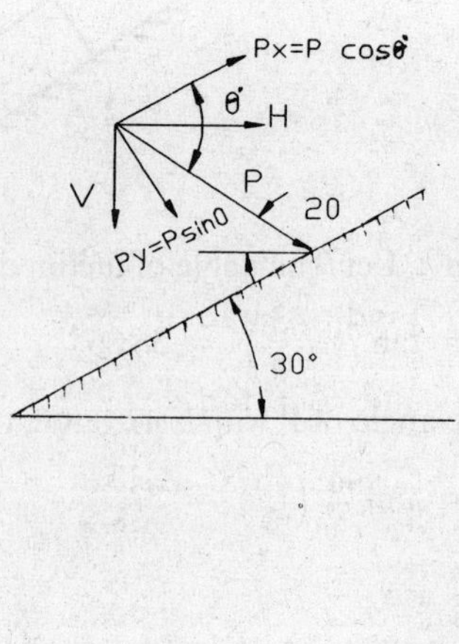

Problem 7. : In fig. the 'X' component of P is 893 N. Determine P and its Y component.

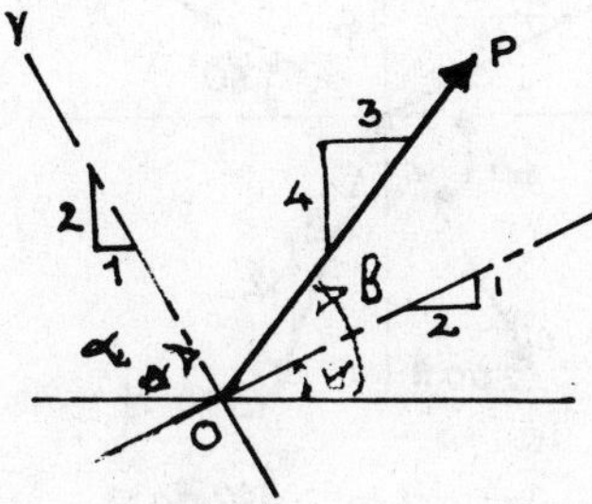

Solution : From fig. it is clear that

$$\theta = \tan^{-1}\left(\frac{1}{2}\right) = 26.56^{o}$$

$$\alpha = \tan^{-1}\left(\frac{2}{1}\right) = 63.43^{o}$$

$$\beta = \tan^{-1}\left(\frac{4}{3}\right) = 53.13^{o}$$

$$P_x = P\cos 26.57 = 893\text{ N}$$

$$\therefore\ P = 998.5\text{ N}$$

$$P_y = 998.5\ \sin 26.57 = 446.6\text{ N}$$

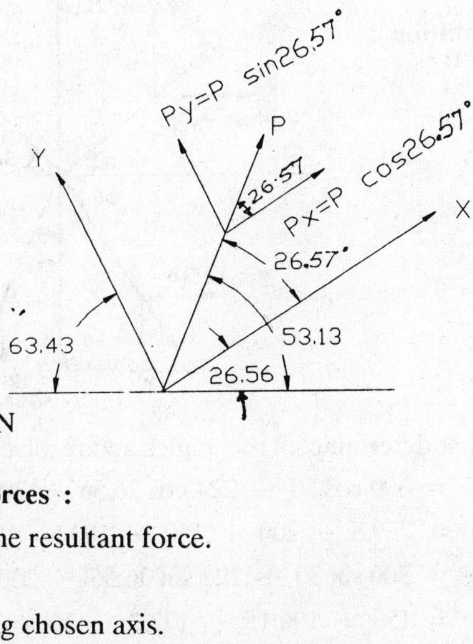

Resultant of Coplaner concurrent forces :

Steps to be followed in getting the resultant force.

Step (i) Choose reference axis.

Step (ii) Resolve all the forces along chosen axis.

Step (iii) Do summation of components of forces in X direction, i.e. $\Sigma X = R_x$ & similarly for y–direction, $\Sigma Y = R_y$ with standard direction convention

Step (iv) Find resultant as $R = \left[(\Sigma X)^2 + (\Sigma Y)^2\right]^{1/2}$

and $\theta = \tan^{-1}\left(\frac{\Sigma Y}{\Sigma X}\right)$; gives the direction of resultant with reference to X–axis.

Problem 8. : Determine the resultant of the four forces acting on the body shown.

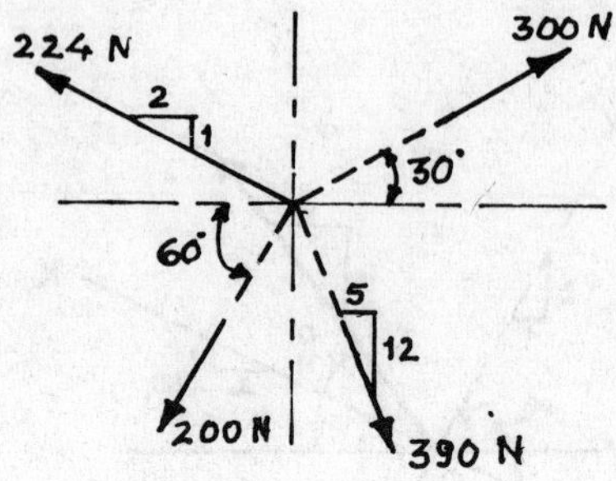

Solution :

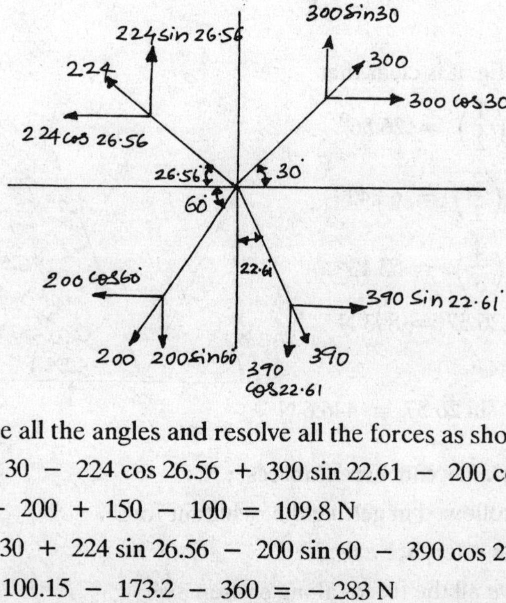

First determine all the angles and resolve all the forces as shown above.

$$\Sigma x = 300 \cos 30 - 224 \cos 26.56 + 390 \sin 22.61 - 200 \cos 60$$
$$= 259.8 - 200 + 150 - 100 = 109.8 \text{ N}$$

$$\Sigma y = 300 \sin 30 + 224 \sin 26.56 - 200 \sin 60 - 390 \cos 22.61$$
$$= 150 + 100.15 - 173.2 - 360 = -283 \text{ N}$$

$$R = \left[(\Sigma x)^2 + (\Sigma y)^2 \right]^{1/2}$$
$$= \left[(109.8)^2 + (-283)^2 \right]^{1/2} = 303.55 \text{ N}$$

$$\tan\theta = \frac{\Sigma y}{\Sigma x} = \frac{283}{109.8}, \quad \therefore \ \theta = 68.79$$

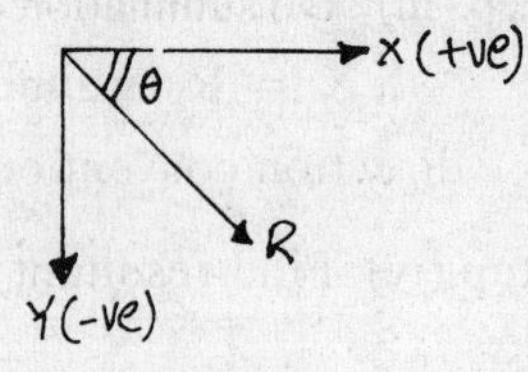

The resultant will be as shown

Problem 9. : The body shown is acted on by four forces. Determine the resultant.

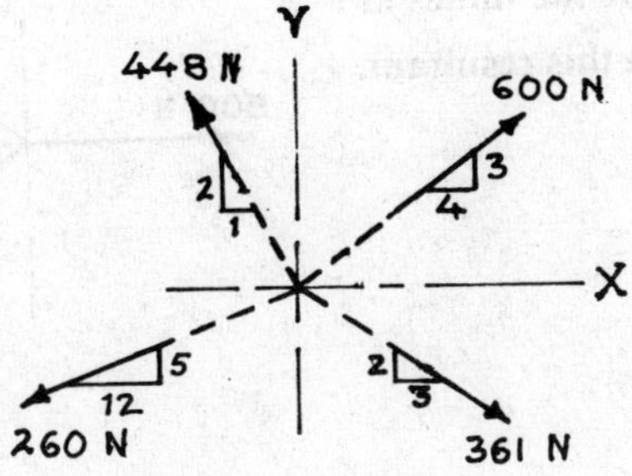

Solution : Determine all the angles and write the components of forces as shown below

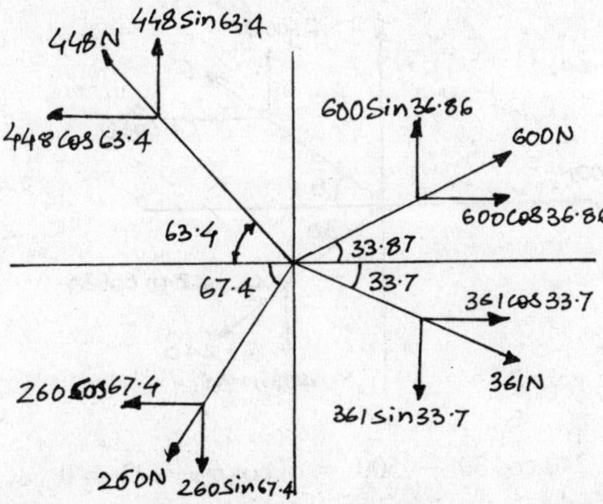

$$\Sigma x = 600 \cos 36.86 - 448 \cos 63.4 - 260 \cos 67.4 + 361 \cos 33.7$$

$$= 480 - 200.6 - 100 + 300.4 = 479.8 \text{ N}$$

$$\Sigma y = 600 \sin 36.86 + 448 \sin 63.4 - 260 \sin 67.4 - 361 \sin 33.7$$

$$= 360 + 400.6 - 240 - 200.3 = 320.3 \text{ N}$$

$$\text{Resultant } R = \left[(\Sigma x)^2 + (\Sigma y)^2\right]^{1/2}$$

$$= \left[(479.8)^2 + (320.3)^2\right]^{1/2} = 576.8 \text{ N}$$

$$\tan\theta = \frac{\Sigma Y}{\Sigma X} = \frac{320.3}{479.8};$$

$$\therefore \quad \theta = 33.72^{\circ}$$

Problem 10. : The force system shown has a resultant of 200 N pointing up along Y–axis. Compute the values of F and θ required to give this resultant.

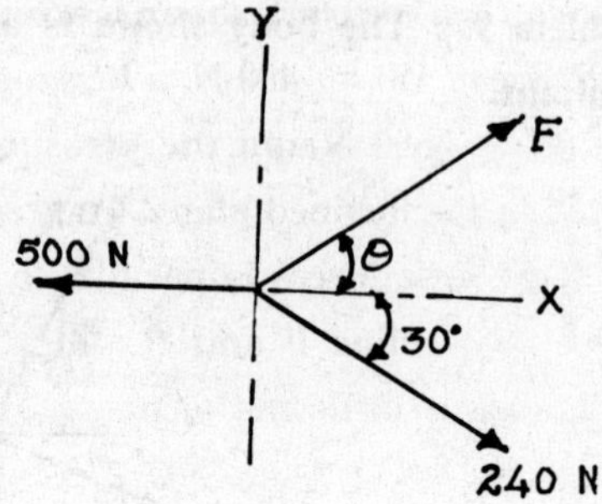

Solution : Resolving the given forces as shown below

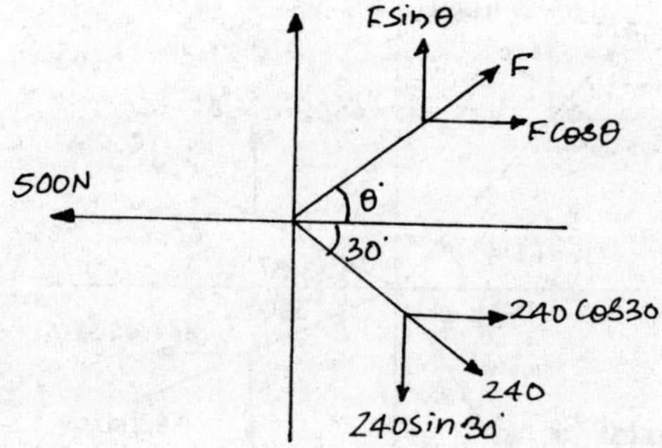

$\Sigma x = F\cos\theta + 240\cos 30 - 500 = F\cos\theta - 292 = 0$ . . . . (1)

Since resultant is acting vertically upwards

$\Sigma y = (F\sin\theta - 240\sin 30) = 200$ . . . . . (2)

From eq^n. (1) ⇒ $F\cos\theta = 292$ . . . . . (3)

From eq^n. (2) ⇒ $F\sin\theta = 320$ . . . . . (4)

Dividing equation (4) by (3), we get

$$\tan\theta = \frac{320}{292}, \quad \theta = 47.61^{\circ}$$

$$\therefore \quad F = \frac{292}{\cos 47.61} = 433\text{ N}.$$

Problem 11. : The block shown is acted on by its weight W = 400 N, a horizontal force F = 600 N and the pressure P exerted by the inclined plane. The resultant R of these forces is parallel to the incline. Determine P and R. Does the block move up or down the incline ?

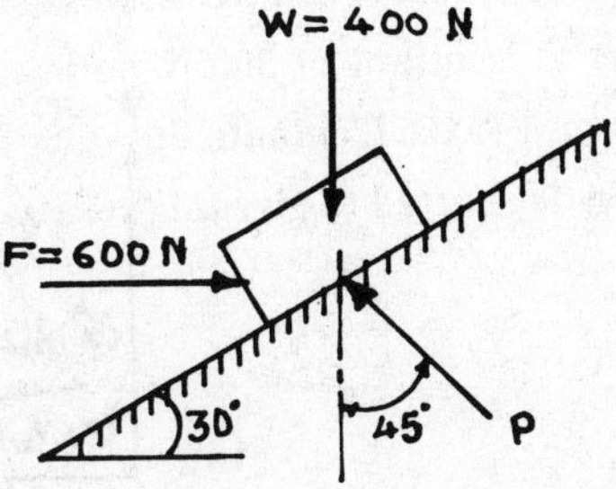

Solution :

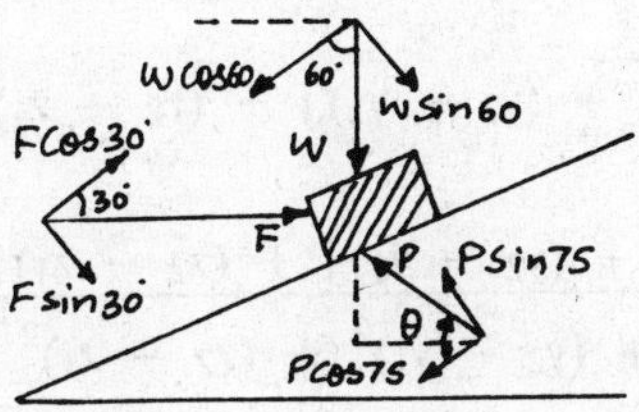

By geometry, $\theta = (45^\circ + 30^\circ) = 75^\circ$

For resolution refer fig shown. Let R be the resultant force acting along incline towards right, so that $\Sigma X = R$. Now as resultant lies along X direction summation of forces along Y direction should equal to zero.

$\therefore \quad \Sigma x = F\cos 30 - W\cos 60 - P\cos 75 = R \quad \ldots\ldots (1)$

$\Sigma y = P\sin 75 - W\sin 60 - F\sin 30 = 0 \quad \ldots\ldots (2)$

From eq^n. (1) $\Rightarrow$ $600\cos 30 - 400\cos 60 - P\cos 75 = R$

From eq^n. (2) $\Rightarrow$ $P\sin 75 = 400\sin 60 + 600\sin 30$

$$\therefore \quad P = \frac{(346.41 + 300)}{\sin 75} = 670\text{ N}$$

So from eq^n. (1) $\Rightarrow$ R = 146.2 N

As the value of R is positive, it means that the assumed direction was correct. So the block is moving upwards.

RESOLUTION AND RESULTANT OF FORCES IN SPACE :

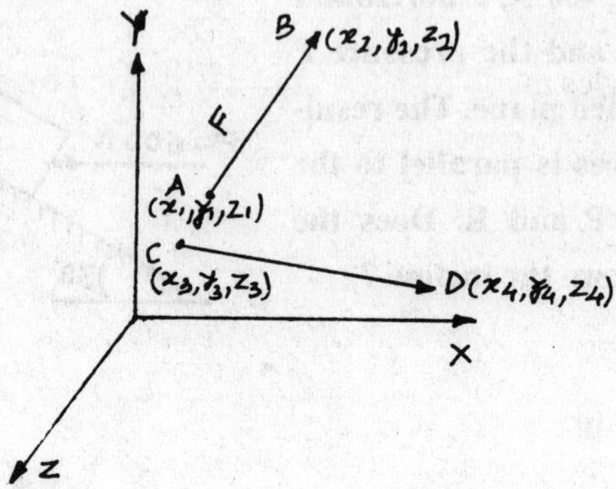

Vector AB is given by

$$\overline{AB} = (x_2 - x_1)\,i + (y_2 - y_1)\,j + (z_2 - z_1)\,k$$

unit vector along AB

$$\hat{n}_{AB} = \frac{(x_2 - x_1)\,i + (y_2 - y_1)\,j + (z_2 - z_1)\,k}{\left[(x_2 - x_1)^2 + (y_2 - y_1)^2 + (z_2 - z_1)^2\right]^{1/2}}$$

Force along AB is given by

$$\mathbf{F} = F\left[\frac{(x_2 - x_1)\,i + (y_2 - y_1)\,j + (z_2 - z_1)\,k}{\left[(x_2 - x_1)^2 + (y_2 - y_1)^2 + (z_2 - z_1)^2\right]^{1/2}}\right]$$

If $\mathbf{F}$ is given by $ai + bj + ck$

then $F = \left[a^2 + b^2 + c^2\right]^{1/2}$

Sign convention to be used : The standard direction convention used is that force $\mathbf{F} = ai + bj + ck$ is acting righward, upward and forward and is taken as in positive direction, whereas force $\mathbf{F} = -ai - bj - ck$ is acting lefward, downward and backward and is taken as in negative direction.

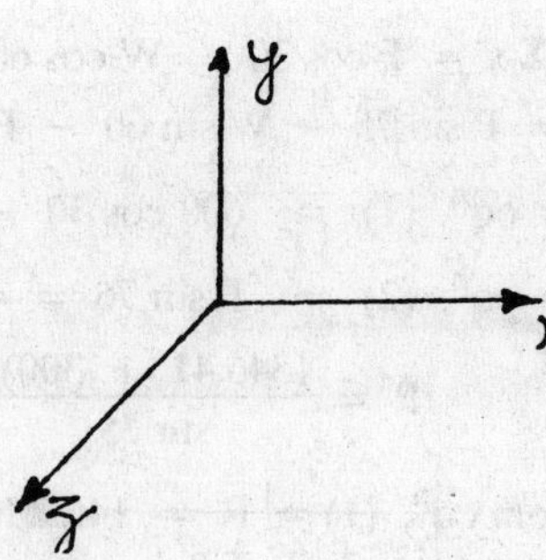

To find the angles made by force with axes x, y and z

$$F = \frac{F_x}{\cos\theta_x} = \frac{F_y}{\cos\theta_y} = \frac{F_z}{\cos\theta_z}$$

In terms of angles made with x, y and z, $\overline{F}$ is given by

$$\overline{F} = F(\cos\theta_x\, i + \cos\theta_y\, j + \cos\theta_z\, k)$$

To find component of force F along $\overline{AC}$

Let $\overline{F} = ai + bj + ck$ and $\overline{AC} = di + ej + fk$

$$F_{AC} = \overline{F} \cdot \hat{n}_{AC}$$

$$= (ai + bj + ck)\ \frac{(di + ej + fk)}{[d^2 + e^2 + f^2]^{1/2}}$$

$$= \frac{ad + be + cf}{[d^2 + e^2 + f^2]^{1/2}}$$

Same is used to find component of any vector like force, moment, couple along any direction.

To find moment of force F about any point A

Let $\overline{F} = ai + bj + ck$ and $\overline{r} = di + ej + fk$

Moment of $\overline{F}$ about A

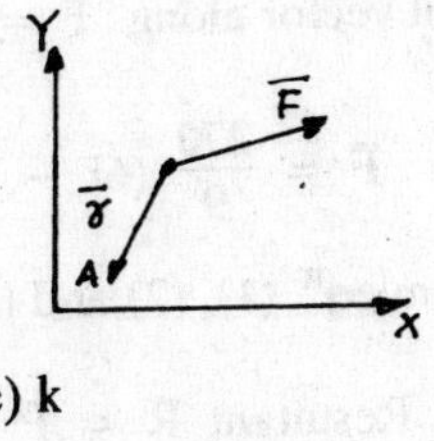

$$M_A^F = \overline{r} \times \overline{F} = \begin{vmatrix} i & j & k \\ d & e & f \\ a & b & c \end{vmatrix}$$

$$= (ec - bf)\, i - (dc - af)\, j + (db - ae)\, k$$

To find moment of force F about line AB

Let $\overline{AB} = li + mj + nk$ & $\overline{AC} = di + ej + fk$

$$\therefore \quad \hat{n}_{AB} = \frac{li + mj + nk}{[l^2 + m^2 + n^2]^{1/2}}$$

If $\overline{F} = ai + bj + ck$, then

Moment of force $\overline{F}$ about AB

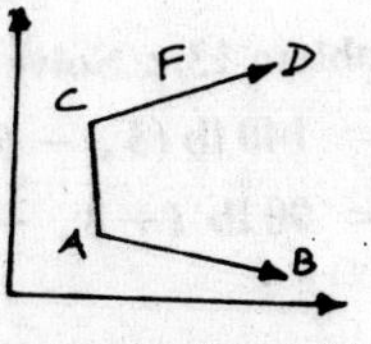

$$M_{AB}^F = \begin{vmatrix} \dfrac{l}{[l^2 + m^2 + n^2]^{1/2}} & \dfrac{m}{[l^2 + m^2 + n^2]^{1/2}} & \dfrac{n}{[l^2 + m^2 + n^2]^{1/2}} \\ d & e & f \\ a & b & c \end{vmatrix}$$

Problem 12. : Determine the resultant of the system of concurrent forces having the following magnitudes and passing through the origin and the indicated points : P = 280 lb (12, 6, – 4) ; T = 520 lb (–3, –4, 12) ; F = 270 lb (6, –3, –6).

Solution : Unit vector along $P = \dfrac{12i + 6j - 4k}{[\,12^2 + 6^2 + 4^2\,]^{1/2}}$

$$= \frac{1}{14}(12i + 6j - 4k)$$

$$\therefore \quad \overline{P} = \frac{280}{14}(12i + 6j - 4k) = 240i + 120j - 80k \qquad \ldots\ldots (1)$$

Unit vector along $T = \dfrac{-3i - 4j + 12k}{[\,3^2 + 4^2 + 12^2\,]^{1/2}}$

$$= \frac{1}{13}(-3i - 4j + 12k)$$

$$\therefore \quad \overline{T} = \frac{520}{13}(-3i - 4j + 12k)$$

$$= -120i - 160j + 480k \qquad \ldots\ldots (2)$$

Unit vector along $F = \dfrac{6i - 3j - 6k}{[\,6^2 + 3^2 + 6^2\,]^{1/2}} = \dfrac{1}{9}(6i - 3j - 6k)$

$$\overline{F} = \frac{270}{9}(6i - 3j - 6k) = 180i - 90j - 180k \qquad \ldots\ldots (3)$$

From eq^n. (1), (2) and (3), we get

$$\therefore \quad \text{Resultant } \overline{R} = \overline{P} + \overline{T} + \overline{F} = 300i - 130j + 220k$$

The resultant R is acting rightward, downward and forward.

Force $R = [\,300^2 + 130^2 + 220^2\,]^{1/2} = 394 \text{ lb}$

Problem 13. : Solve the above problem by using following data : P = 140 lb (3 , – 6 , 2) ; T = 260 lb (– 12 , + 4 , – 3) F = 90 lb (+ 3 , + 6 , – 6) . Also find the direction angle of the resultant.

Solution : Unit vector along $P = \dfrac{3i - 6j + 2k}{[\,3^2 + 6^2 + 2^2\,]^{1/2}}$

$$= \frac{1}{7}(3i - 6j + 2k)$$

$$\therefore \quad \bar{P} = \frac{140}{7}(3i - 6j + 2k) = 60i - 120j + 40k \qquad \ldots\ldots (1)$$

$$\text{Unit vector along } T = \frac{-12i + 4j - 3k}{[12^2 + 4^2 + 3^2]^{1/2}}$$

$$= \frac{1}{13}(-12i + 4j - 3k)$$

$$\bar{T} = \frac{260}{13}(-12i + 4j - 3k) = -240i + 80j - 60k \qquad \ldots\ldots (2)$$

$$\text{Unit vector along } F = \frac{3i + 6j - 6k}{[3^2 + 6^2 + 6^2]^{1/2}} = \frac{1}{9}(3i + 6j - 6k)$$

$$\bar{F} = \frac{90}{9}(3i + 6jk - 6k) = 30i + 60j - 60k \qquad \ldots\ldots (3)$$

From eq^n. (1), (2) & (3), we get

Resultant $\bar{R} = \bar{P} + \bar{T} + \bar{F} = -150i + 20j - 80k$

R is acting leftward , upward and backward.

$$\therefore \quad R = [150^2 + 20^2 + 80^2]^{1/2} = 171.1 \text{ lb}$$

$$\text{Now } R = \frac{|R_x|}{\cos\theta_x} = \frac{|R_y|}{\cos\theta_y} = \frac{|R_z|}{\cos\theta_z}$$

$$171.1 = \frac{150}{\cos\theta_x} = \frac{20}{\cos\theta_y} = \frac{80}{\cos\theta_z}$$

$$\therefore \quad \theta_x = 28.7^\circ \; ; \; \theta_y = 83.28^\circ \; ; \; \theta_z = 62.1^\circ$$

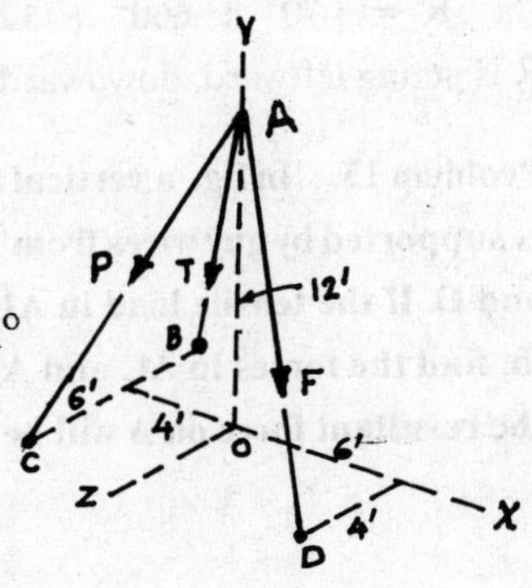

Problem 14. : Find the resultant of the force system shown in fig. in which P = 280 lb, T = 260 lb, and F = 210 lb.

Solution : Co–ordinates of points in figure are

A (0 , 12 , 0) ; B (– 4 , 0 , – 3) ;

C (– 4 , 0 , 6) ; D (6 , 0 , 4)

$$\overline{AC} = -4i - 12j + 6k$$

$$\therefore \quad \hat{n}_{AC} = \frac{-4i - 12j + 6k}{[16 + 144 + 36]^{1/2}} = \frac{1}{14}(-4i - 12j + 6k)$$

$$\overline{AB} = -4i - 12j - 3k$$

$$\therefore \quad \hat{n}_{AB} = \frac{1}{13}(-4i - 12j - 3k)$$

$$\overline{AD} = 6i - 12j + 4k$$

$$\therefore \quad \hat{n}_{AD} = \frac{1}{14}(6i - 12j + 4k)$$

$$\therefore \quad \overline{P} = \frac{280}{14}(-4i - 12j + 6k) = -80i - 240j + 120k$$

$$T = \frac{260}{13}(-4i - 12j - 3k) = -80i - 240j - 60k$$

$$F = \frac{210}{14}(6i - 12j + 4k) = 90i - 180j + 60k$$

Resultant $\overline{R} = \overline{P} + T + \overline{F} = -70i - 660j + 120k$

$$\therefore \quad R = [70^2 + 660^2 + 120^2]^{1/2} = 674.46 \text{ lb}$$

R is acting leftward, downward and forward

Problem 15.: In fig. a vertical boom AE is supported by guy wires from A to B, C, and D. If the tensile load in AD = 252 lb, find the forces in AC and AB so that the resultant force on A will be vertical.

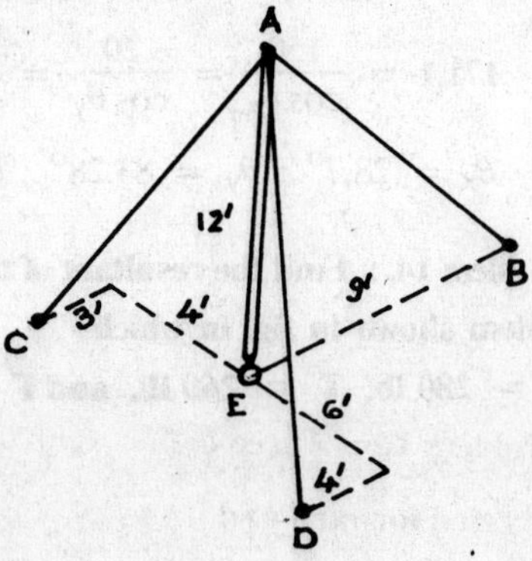

Solution : Co–ordinates of points in above figure

A (0 , 12 , 0) ; B (0 , 0 , – 9) ; C (– 4 , 0 ,3) ; D (6 , 0 , 4)

$\overline{AB} = -12j - 9k$; $\therefore \quad \hat{n}_{AB} = \frac{1}{15}(-12j - 9k)$

$\overline{AC} = -4i - 12j + 3k$;

$\therefore\ \hat{n}_{AC} = \frac{1}{13}(-4i - 12j + 3k)$

$\overline{AD} = 6i - 12j + 4k$;

$\therefore\ \hat{n}_{AD} = \frac{1}{14}(6i - 12j + 4k)$

$$F_{AB} = \frac{F_{AB}}{15}(-12j - 9k) = -0.8\,F_{AB}\,j - 0.6\,F_{AB}\,k \qquad \ldots\ldots(1)$$

$$\overline{F}_{AC} = \frac{F_{AC}}{13}(-4i - 12j + 3k) = (-0.307\,i - 0.923\,j + 0.23\,k)\,F_{AC} \qquad \ldots\ldots(2)$$

$$\overline{F}_{AD} = \frac{256}{14}(6i - 12j + 4k) = 108i - 216j + 72k \qquad \ldots\ldots(3)$$

As resultant being vertical, $\overline{R} = -R_y\,j$ and ΣX & ΣZ should be equal to zero.

From eq^n. (1), (2) & (3), we get

Resultant $\overline{R} = \Sigma F$

$$= (108 - 0.307\,F_{AC})\,i + (-216 - 0.8\,F_{AB} - 0.923\,F_{AC})\,j + (72 - 0.6\,F_{AB} + 0.23\,F_{AC})\,k$$

$$= -R_y\,j$$

$\therefore\ \Sigma X = 108 - 0.307\,F_{AC} = 0$; $F_{AC} = 351.8$ lb

& $\Sigma Z = 72 - 0.6\,F_{AB} + 0.23 \times 351.8 = 0$

$\therefore\ F_{AB} = 254.85$ lb

Problem 16 : Three concurrent forces P, T, and F have a resultant of 10 lb directed forward and up to the right at $\theta_x = 60^o$, $\theta_y = 60^o$, $\theta_z = 45^o$. P equals 21 lb and passes from the origin through point (3, 2, 6). The value of T is 18 lb and is directed from the origin toward point (–6, 6, –3). Determine the magnitude of the third force F and the angles it makes with the reference axes.

Solution : $R = \dfrac{|R_x|}{\cos\theta_x} = \dfrac{|R_y|}{\cos\theta_y} = \dfrac{|R_z|}{\cos\theta_z}$

$\therefore\ |R_x| = R\cos\theta_x = 10\cos 60 = 5\text{ lb}$

$|R_y| = R\cos\theta_y = 10\cos 60 = 5\text{ lb}$

$|R_z| = R\cos\theta_z = 10\cos 45 = 7.07\text{ lb}$

$\because$ R is pointing forward, up and right

$R_x = 5\text{ lb}\ ,\ R_y = 5\text{ lb}\ ,\ R_z = 7.07\text{ lb}$

$\therefore\ \bar{R} = 5i + 5j + 7.07\,k$ .....(1)

Unit Vector along $P = \dfrac{3i + 2j + 6k}{[3^2 + 2^2 + 6^2]^{1/2}} = \dfrac{1}{7}(3i + 2j + 6k)$

$\therefore\ \bar{P} = \dfrac{21}{7}(3i + 2j + 6k) = 9i + 6j + 18k$ .....(2)

Unit vector along $T = \dfrac{-6i + 6j - 3k}{[6^2 + 6^2 + 3^2]^{1/2}} = \dfrac{1}{9}(-6i + 6j - 3k)$

$\bar{T} = \dfrac{18}{9}(-6i + 6j - 3k) = -12i + 12j - 6k$ .....(3)

Let $\bar{F} = F_x i + F_y j + F_z k$ .....(4)

From eq^n. (1), (2) & (3), we get

Resultant $R = (-3 + F_x)\,i + (18 + F_y)\,j + (12 + F_z)\,k$

$= 5i + 5j + 7.07k$

$\therefore\ -3 + F_x = 5\ ;\ \Rightarrow\ F_x = 8$

$18 + F_y = 5\ ;\ \Rightarrow\ F_y = -13$

$12 + F_z = 7.07\ ;\ \Rightarrow\ F_z = -4.93$

$\therefore\ \bar{F} = 8i - 13j - 4.93k$

$\therefore\ F = [8^2 + 13^2 + 4.93^2]^{1/2} = 16.04\text{ lb}$

$F = \dfrac{|F_x|}{\cos\theta_x} = \dfrac{|F_y|}{\cos\theta_y} = \dfrac{|F_z|}{\cos\theta_z}$

$16.04 = \dfrac{8}{\cos\theta_x} = \dfrac{13}{\cos\theta_y} = \dfrac{4.93}{\cos\theta_z}$

$\therefore\ \theta_x = 60.1^\circ\ ;\ \theta_y = 35.85^\circ\ ,\ \theta_z = 72.1^\circ$

Problem 17 : In fig., a boom AC is supported by a ball–and–socket joint at C and by the cables BE and AD. If the force multiplier of a force F acting from B to E is F_m = 10 lb/ft and that of a force P acting from A to D is P_m = 20 lb/ft, find the component of each force along AC. Using the above data find the component of F and P along the direction DE.

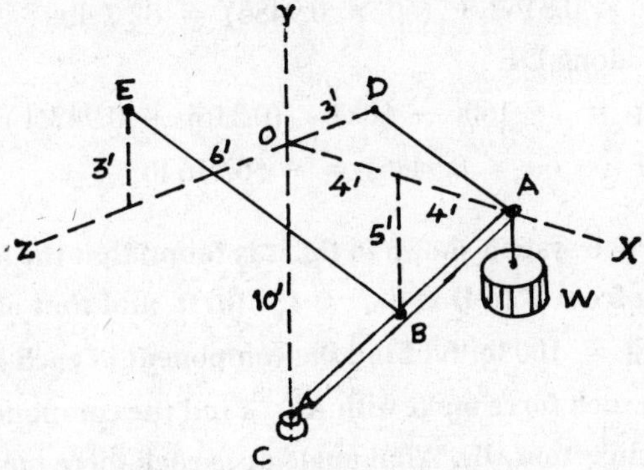

Solution : Co–ordinates of points in figure are A (8, 0, 0) ; B (4, −5, 0) ; C (0, − 10, 0) ; D (0, 0, − 3) ; E (0, 3, 6)

$\overline{BE} = -4i + 8j + 6k$; ∴ $\overline{F} = 10(-4i + 8j + 6k$

$= -40i + 80j + 60k$

$\overline{AD} = -8i - 3k$; ∴ $\overline{P} = 20(-8i - 3k) = -160i - 60k$

$\overline{AC} = -8i - 10j$; ∴ $\hat{n}_{AC} = \dfrac{-8i - 10j}{[8^2 + 10^2]^{1/2}}$

$= -0.6247i - 0.781j$

Component of F along AC $F_{AC} = \overline{F} \cdot \hat{n}_{AC}$

$= (-40i + 80j + 60k) \cdot (-0.6247i - 0.781j)$

$= -40 \times (-0.6247) + 80 \times (-0.781)$

$= -37.48$ lb

Component of P along AC $P_{AC} = \overline{P} \cdot \hat{n}_{AC}$

$= (-160i - 60k) \cdot (-0.6247i - 0.781j)$

$= -160 \times (-0.6247) + 0 + 0$

$= 99.95$ lb

Now $\overline{DE} = 3j + 9k \quad \therefore \quad \hat{n}_{DE} = \frac{1}{\sqrt{90}}(3j + 9k)$

$$= 0.316j + 0.9486k$$

Component of F along DE

$F_{DE} = \overline{F} \cdot \hat{n}_{DE} = (-40i + 80j + 60k) \cdot (0.316j + 0.9486k)$

$= 0 + (80 \times 0.316) + (60 \times 0.9486) = 82.2 \text{ lb}$

Component of P along DE

$P_{DE} = \overline{P} \cdot \hat{n}_{DE} = (-160i - 60k) \cdot (0.316j + 0.9486k)$

$= 0 + 0 + (-60 \times 0.9486) = -56.916 \text{ lb}$

Problem 18.: In the system shown in fig. it is found that the force multiplier of force D acting from B to D is $F_m = 150$ lb/ft and that of force P acting from A to E is $P_m = 100$ lb/ft. Find the component of each force along AC. What angle does each force make with AC ? Find the components of forces F and P along the direction DE. What angle does each force make with DE ?

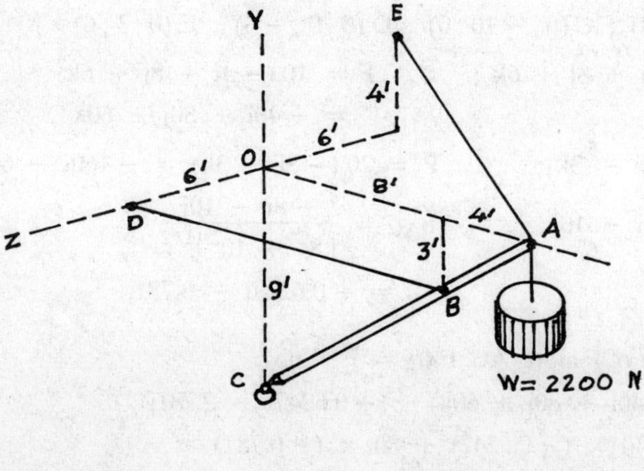

Solution : Co–ordinates of points in above figure are A (12, 0, 0) ; B (8, −3, 0) ; C (0, −9, 0) ; D (0, 0, 6) ; E (0, 4, − 6)

$\overline{BD} = -8i + 3j + 6k$

$\therefore \quad \overline{F} = 150(-8i + 3j + 6k) = -1200i + 450j + 900k$

$\overline{AE} = -12i + 4j - 6k$

$\therefore \quad \overline{P} = 100(-12i + 4j - 6k) = -1200i + 400j - 600k$

$\overline{AC} = -12i - 9j$

$\therefore \quad \hat{n}_{AC} = \frac{1}{15}(-12i - 9j) = -0.8i - 0.6j$

Component of F along AC

$F_{AC} = \overline{F} \cdot \hat{n}_{AC} = (-1200i + 450j + 900k) \cdot (-0.8i - 0.6j)$

$= -1200 \times (-0.8) + 450 \times (-0.6) = 690 \text{ lb}$

$\because \quad F_{AC} = F \cos\theta_F$ where θ_F is angle between $\overline{BD}$ & $\overline{AC}$

$\therefore \quad 690 = 150\,[8^2 + 3^2 + 6]^{1/2} \cos\theta_F$

$\therefore \quad \cos\theta_F = 0.4406 \qquad \therefore \quad \theta_F = 63.85^o$

Component of P along AC

$P_{AC} = \overline{P} \cdot \hat{n}_{AC} = (-1200i + 400j - 600k) \cdot (-0.8i - 0.6j)$

$= -1200 \times (-0.8) + 400 \times (-0.6) = 720 \text{ lb}$

$\because \quad P_{AC} = P \cos\theta_P$ where θ_P is angle between $\overline{AE}$ & $\overline{AC}$

$\therefore \quad 720 = 100\,[12^2 + 4^2 + 6^2]^{1/2} \cos\theta_P$

$\therefore \quad \cos\theta_P = 0.514 \qquad \therefore \quad \theta_P = 59.05^o$

Now $\overline{DE} = 4j - 12k$

$\therefore \quad \hat{n}_{DE} = \frac{1}{\sqrt{160}}(4j - 12k) = 0.316j - 0.948k$

$\therefore \quad F_{DE} = \overline{F} \cdot \hat{n}_{DE} = (-1200i + 450j + 900k) \cdot (0.316j - 0.948k)$

$= -711.61 \text{ lb}$

$\because \quad |F_{DE}| = F \cos\theta_F'$, where θ_F' is angle between $\overline{BD}$ & $\overline{DE}$

$\therefore \quad |-711.51| = 150\sqrt{109} \cos\theta_F'$

$\therefore \quad \cos\theta_F' = 0.4544$

$\therefore \quad \theta_F' = 62.98^o$

$P_{DE} = \overline{P} \cdot \hat{n}_{DE}$

$= (-1200i + 400j - 600k) \cdot (0.316j - 0.948k)$

$= 695.2 \text{ lb}$

$\because \quad P_{DE} = P \cos\theta_P'$, where θ_P' is angle between $\overline{AE}$ & $\overline{DE}$

$\therefore \quad 695.2 = 100 \times 14 \cos\theta_P'$

$\therefore \quad \cos\theta_{P'} = 0.496$

$\therefore \quad \theta_{P'} = 60.22^\circ$

Problem 19. : In the unsymmetrical cantilever framework shown in fig., points C and D are in the same vertical plane while B is 3 ft in front of this plane. Assume a force F to act from A to C and another force P to act from A to D. In terms of their force multipliers, find the components of these forces along the direction from A to B. What is the angle between F and P ?

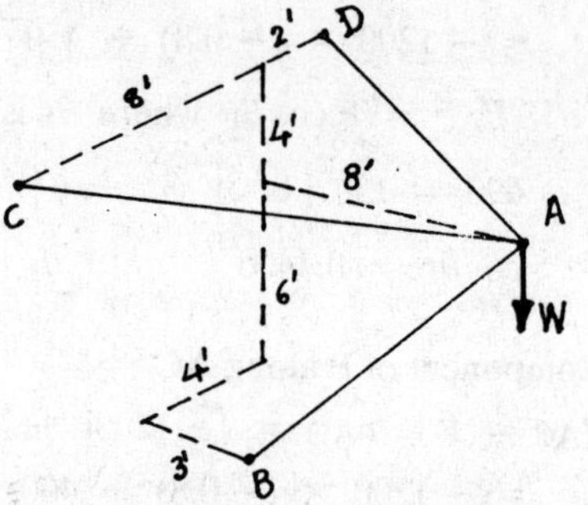

Solution : Co–ordinates of points in figure are A (8 , 0 , 0) ; B (3 , −6 , 4) ; C (0 , 4 , 8) ; D (0 , 4 , − 2)

$\overline{AC} = -8i + 4j + 8k$

$\therefore \quad \overline{F} = F_m(-8i + 4j + 8k)$

$\overline{AD} = -8i + 4j - 2k$

$\therefore \quad \overline{P} = P_m(-8i + 4j - 2k)$

$\overline{AB} = -5i - 6j + 4k$

$\therefore \quad \hat{n}_{AB} = \frac{1}{\sqrt{77}}(-5i - 6j + 4k)$

$= -0.57i - 0.683j + 0.455k$

Component of F along $\overline{AB}$

$F_{AB} = \overline{F} \cdot \hat{n}_{AB} = F_m(-8i + 4j + 8k) \cdot (-0.57i - 0.683j + 0.455k)$

$= 5.47\ F_m$

Component of P along AB

$P_{AB} = \overline{P} \cdot \hat{n}_{AB} = P_m(-8i + 4j - 2k) \cdot (-0.57i - 0.683j + 0.455k)$

$= 0.916\ P_m$

To find angle between $\overline{P}$ and F, we need to take component of P along F.

$\therefore \quad P_{on\ F} = \overline{P} \cdot \hat{n}_F$

$$= P_m(-8i + 4j - 2k) \cdot \frac{1}{\sqrt{144}}(-8i + 4j + 8k)$$

$$= 5.33\, P_m$$

$\because \quad P_{on\,F} = P\cos\theta$ where θ is angle between P & F

$\therefore \quad 5.33\, P_m = P_m\,[64 + 16 + 4]^{1/2}\cos\theta$

$\Rightarrow \quad \cos\theta = 0.5819$

or $\theta = 54.4^o$

Problem 20. : Refering to fig., find the magnitude of the force F acting from D towards A that will have a component of 100 lb along DB.

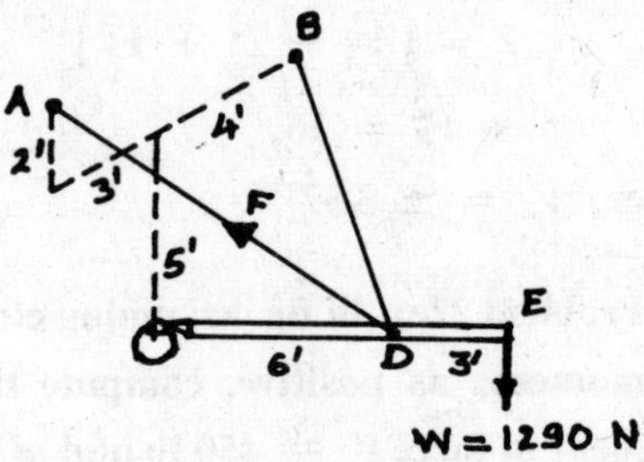

Solution : Taking C as origin co-ordinates of points are A (0 , 7 , 3) ; B (0 , 5 , − 4) ; C (0 , 0 , 0) ; D (6 , 0 , 0) ; E (9 , 0 , 0)

$\overline{AD} = 6i - 7j - 3k; \quad \therefore \quad F = F_m(6i - 7j - 3k)$

$\overline{DB} = -6i + 5j - 4k; \quad \therefore \quad \hat{n}_{DB} = \frac{1}{\sqrt{77}}(-6i + 5j - 4k)$

Component of F along DB

$$F_{DB} = 100 = F \cdot \hat{n}_{DB}$$

$$= F_m(6i - 7j - 3k) \cdot \frac{1}{\sqrt{77}}(-6i + 5j - 4k)$$

$$= 6.72\, F_m$$

$\therefore \quad F_m = 14.87$ lb/ft.

& $\quad F = 14.87\,[6^2 + 7^2 + 3^2]^{1/2}$

$= 144.2$ lb

Problem 21. : The force $F = \hat{i} + 2\hat{j} + F_z\hat{k}$ makes an angle of 60° with the line $L = 4\hat{i} + 3\hat{j}$. Find the value of F_z.

Solution : Component of $\bar{F}$ along $\bar{L}$

$$\bar{F}_{on\,L} = \bar{F} \cdot \hat{n}_L = (i + 2j + F_z k) \cdot \frac{(4i + 3j)}{[4^2 + 3^2]^{1/2}}$$

$$= \frac{4}{5} + \frac{6}{5} = 2$$

Again $F_{on\,L} = F \cos 60$

$$\therefore \quad 2 = [1^2 + 2^2 + F_z^2]^{1/2} \times 0.5$$

$$5 + F_z^2 = 16$$

or $F_z = \pm\ 3.32$

Problem 22. : In fig. assuming clockwise moments as positive, compute the moment of force F = 450 lb and of force P = 361 lb about points A, B, C, and D.

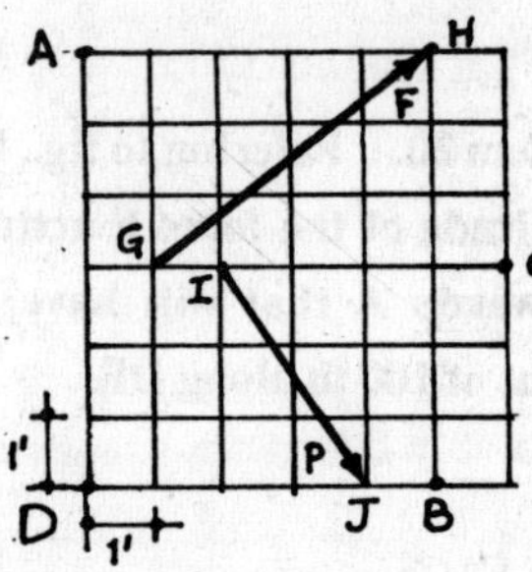

Solution : Taking D as origin, the coordinates of various points are
D (0, 0, 0) ; A (0, 6, 0) ; B (5, 0, 0) ; C (6, 3, 0)
G (1, 3, 0) ; H (5, 6, 0) ; I (2, 3, 0) ; J (4, 0, 0)
where in fig.

$\overline{GH} = 4i + 3j$

$$\therefore \quad \bar{F} = \frac{450}{[4^2 + 3^2]^{1/2}} (4i + 3j) = 360i + 270j$$

$\overline{IJ} = 2i - 3j$

$$\therefore \quad P = \frac{361}{[2^2 + 3^2]^{1/2}} (2i - 3j) = 200i - 300j$$

Now moment of F @ A

$$M_A^F = \overline{GA} \times \bar{F} = \begin{vmatrix} i & j & k \\ -1 & 3 & 0 \\ 360 & 270 & 0 \end{vmatrix} = -1350\,k\,.\ \text{lb} - \text{ft}$$

Moment of F about B

$$M_B^F = \overline{GB} \times \overline{F} = \begin{vmatrix} i & j & k \\ 4 & -3 & 0 \\ 360 & 270 & 0 \end{vmatrix} = +2160\,k\,.\ lb - ft$$

Moment of F about C

$$M_C^F = \overline{GC} \times \overline{F} = \begin{vmatrix} i & j & k \\ 5 & 0 & 0 \\ 360 & 270 & 0 \end{vmatrix} = +1350\,k.\ lb - ft$$

Moment of F about D

$$M_D^F = \overline{GD} \times \overline{F} = \begin{vmatrix} i & j & k \\ -1 & -3 & 0 \\ 360 & 270 & 0 \end{vmatrix} = +810\,k\,.\ lb - ft$$

Moment of P about A

$$M_A^P = \overline{IA} \times \overline{P} = \begin{vmatrix} i & j & k \\ -2 & 3 & 0 \\ 200 & -300 & 0 \end{vmatrix} = 0\ lb - ft.$$

Moment of P about B

$$M_B^P = \overline{IB} \times \overline{P} = \begin{vmatrix} i & j & k \\ 3 & -3 & 0 \\ 200 & -300 & 0 \end{vmatrix} = -300\,k\,.\ lb - ft.$$

Moment of P about C

$$M_C^P = \overline{IC} \times \overline{P} = \begin{vmatrix} i & j & k \\ 4 & 0 & 0 \\ 200 & -300 & 0 \end{vmatrix} = -1200\,k\,.\ lb - ft.$$

Moment of P about D

$$M_D^P = \overline{ID} \times \overline{P} = \begin{vmatrix} i & j & k \\ -2 & -3 & 0 \\ 200 & -300 & 0 \end{vmatrix} = 1200\,k\,.\ lb - ft$$

Problem 23. : In fig. if the moment of P is 480 ft / lb about D, find the moment of P about B. Also, if the moment of F about C is 750 ft / lb, find its moment about D.

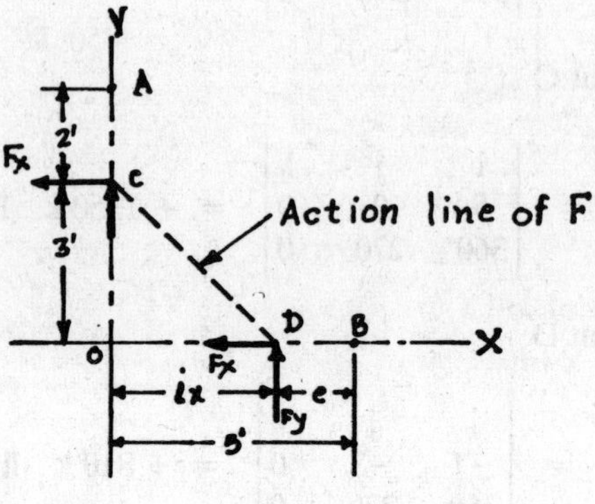

Solution : Taking 'D' as origin, co–ordinates of points are

D (0, 0, 0) ; A (0, 6, 0) ; B (5, 0, 0) ; C (6, 3, 0)

G (1, 3, 0) ; H (5, 6, 0) ; I (2, 3, 0) ; J (4, 0, 0)

$\overline{GH} = 4i + 3j \quad \therefore \quad F = F_m (4i + 3j)$

$\overline{IJ} = 2i - 3j \quad \therefore \quad \overline{P} = P_m (2i - 3j)$

Taking moment of P about D

$$M_D^P = \overline{ID} \times \overline{P} = \begin{vmatrix} i & j & k \\ -2 & -3 & 0 \\ 2 & -3 & 0 \end{vmatrix} \cdot P_m = 12\, P_m\, k = +480$$

$\therefore \quad P_m = 40$ lb/ft.

Moment of P about B

$$M_B^P = \overline{IB} \times \overline{P} = \begin{vmatrix} i & j & k \\ 3 & -3 & 0 \\ 2 & -3 & 0 \end{vmatrix} \cdot 40 = -120 \text{ lb} - \text{ft.}$$

Taking moment of F about C

$$M_C^F = \overline{GC} \times \overline{F} = \begin{vmatrix} i & j & k \\ 5 & 0 & 0 \\ 4 & 3 & 0 \end{vmatrix} \cdot F_m = 15\, F_m = 750$$

$\therefore \quad F_m = 50$ lb/ft.

Moment of F about D

$$M_D^F = \overline{GD} \times \overline{F} = \begin{vmatrix} i & j & k \\ -1 & -3 & 0 \\ 4 & 3 & 0 \end{vmatrix} \cdot 50 = 450 \text{ lb} - \text{ft.}$$

Problem 24. : In fig. find the y coordinate of point A so that the 361 lb force will have a clockwise moment of 400 ft-lb about 'O'. Also determine the Y intercept of the action line of the force.

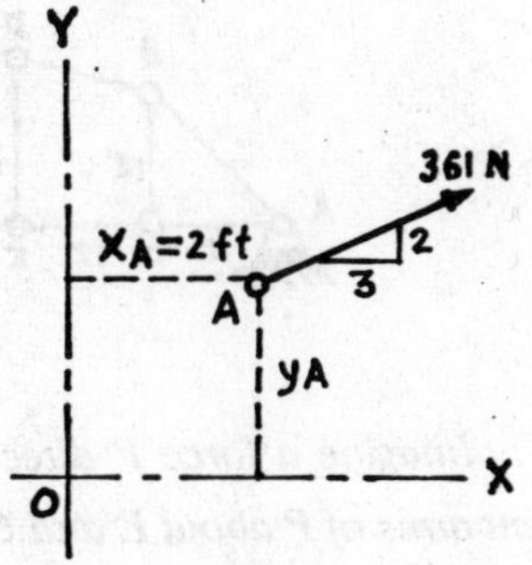

Solution : Co-ordinate of point A ≡ $(2, y_A, 0)$

Angle at which force applied, is given by

$$\tan\theta = \frac{2}{3} \quad \therefore \quad \theta = 33.69^o$$

Force through A = $\overline{F}$ = 361 (cos 33.69 i + sin 33.69 j)

= 300i + 200j

$\overline{AO} = -2i - y_A j$

Taking moment of F about O

$$M_O^F = 400 = \overline{AO} \times \overline{F} = \begin{vmatrix} i & j & k \\ -2 & -y_A & 0 \\ 300 & 200 & 0 \end{vmatrix}$$

i.e. $-400 + 300\, y_A$

$\therefore \quad 300\, y_A = 800$

$\therefore \quad y_A = 2.67$ ft

To find Y intercept of line of action of force F – Assume C is point of intersection of Y axis and force

$$M_O^F = 400 = \overline{CO} \times \overline{F} = \begin{vmatrix} i & j & k \\ 0 & -y' & 0 \\ 300 & 200 & 0 \end{vmatrix} = -300\, y'$$

$\therefore \quad y' = 1.33$ ft. (This is Y – intercept of force) .

Problem 25.: For the truss shown in fig. compute the perpendicular distance from E and from G to the line BD.

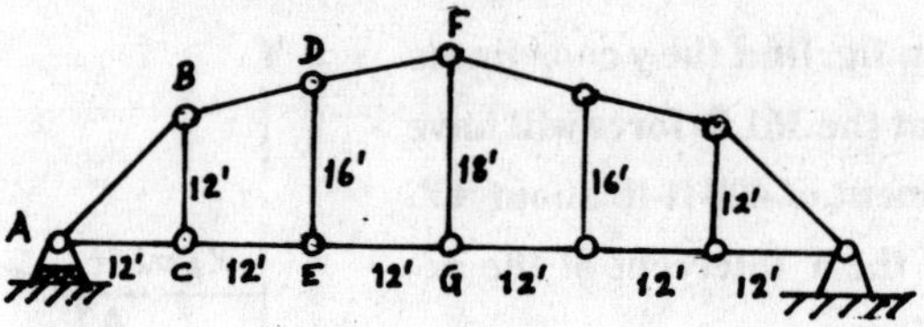

Hint : Imagine a force P directed along BD. The required distances are the moment arms of P about E and G.

Solution : Co–ordinates of points B, D, E and G are

B (12, 12, 0) ; D (24, 16, 0) ; E (24, 0, 0) ; G (36, 0, 0)

$\overline{BD} = 12i + 4j$

$\therefore$ force P along BD $= P_m\,(12i + 4j)$

Taking moment of P about E

$$M_E^P = \overline{BE} \times \overline{P} = \begin{vmatrix} i & j & k \\ 12 & -12 & 0 \\ 12 & 4 & 0 \end{vmatrix} P_m = 192\,P_m$$

$$= \frac{192\,P}{[\,12^2 + 4^2\,]^{1/2}} = 15.17\,P$$

Moment of P about E is also $= P \times r_e$, where r_e perpendicular distance of P from E.

$\therefore\ P \times r_e = 15.17\,P$

$\therefore\ r_e = 15.17$ ft

Taking moment of P about G

$$M_G^P = \overline{BG} \times \overline{P} = \begin{vmatrix} i & j & k \\ 24 & -12 & 0 \\ 12 & 4 & 0 \end{vmatrix} P_m = 240\,P_m$$

$$= \frac{240\ P}{[12^2 + 4^2]^{1/2}} = 18.97\ P$$

Moment of P about G is also $= P \times r_g$

where r_g is perpendicular distance of P from G

$\therefore\ P \times r_g = 18.97\ P$

$\therefore\ r_g = 18.97$ ft

Problem 26. : In the rocker arm shown in fig. the moment of F about O balances that of P about O. Find F.

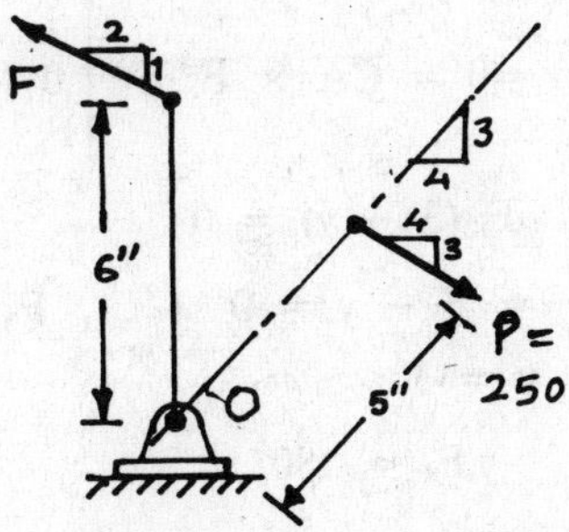

Solution :

$\tan\theta = \frac{1}{2} \quad \therefore\ \theta = 26.56$

$\tan\alpha = \frac{3}{4} \quad \therefore\ \alpha = 36.87$

$\therefore\ \varphi = 90 - 2 \times 36.87 = 16.26$

Equating the moment of forces F & P about O.

$F \cos 26.56 \times 6 = (250 \cos 16.26) \times 5$

$\therefore\ F = 223.6$ lb

Problem 27. : In Fig. the moment of a certain force F is 180 ft–lb clockwise about O and 90 ft–lb counterclockwise about B. If its moment about A is zero, determine the force.

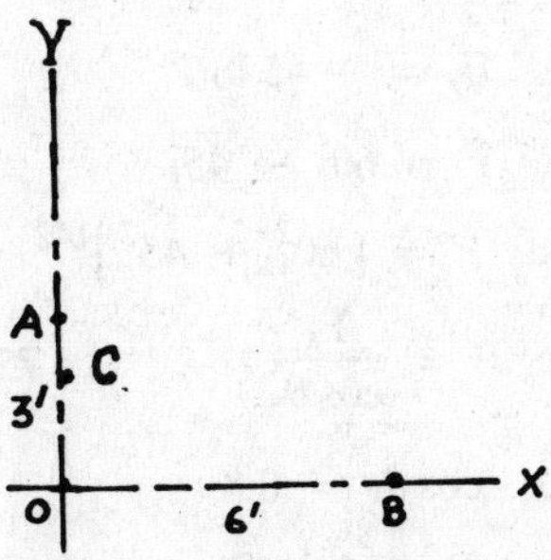

$\Sigma M_B = 90 \quad F_X \times (6 - x) = 90$

Solution : Let one point on y axis and force line be C (0, y, 0)

A (0, 3, 0) ; B (6, 0, 0)

Let the force be $\bar{F} = F_x i + F_y j$

Taking moment of F about O

$$M_O^F = 180 = \overline{CO} \times \bar{F} = \begin{vmatrix} i & j & k \\ 0 & -y & 0 \\ F_x & F_y & 0 \end{vmatrix} = y F_x$$

$\therefore \quad y F_x = 180$

Taking moment of F about A

$$M_O^F = 0 = \overline{CA} \times \bar{F} = \begin{vmatrix} i & j & k \\ 0 & 3-y & 0 \\ F_x & F_y & 0 \end{vmatrix} = F_x(3-y)$$

$\therefore \quad F_x(3 - y) = 0$

$\Rightarrow 3 - y = 0 \quad (\because \quad F_x \neq 0)$

or $\quad y = 3$

$\because \quad y F_x = 180$

$\therefore \quad 3F_x = 180$

$\therefore \quad F_x = 60$ lb

Taking moment of F about B

$$M_B^F = -90 = \begin{vmatrix} i & j & k \\ 6 & -y & 0 \\ F_x & F_y & 0 \end{vmatrix} = \begin{vmatrix} i & j & k \\ 6 & -3 & 0 \\ 60 & F_y & 0 \end{vmatrix}$$

$$= 6F_y + 180$$

$\therefore \quad F_y = -45$ lb

$\therefore \quad \bar{F} = 60i - 45j$

$\therefore \quad F = [60^2 + 45^2]^{1/2} = 75$ lb

$$F = \frac{F_x}{\cos\theta_x} \qquad \therefore \quad 75 = \frac{60}{\cos\theta_x}$$

$\therefore \quad \cos\theta_x = 0.8$

$\therefore \quad \theta_x = 36.85^{\circ}$

Problem 28. : In the fig. of above problem, a force P intersects the X axis at 4 ft to the right of O. If its moment about A is 170 ft–lb counterclockwise and its moment about B is 40 ft–lb clockwise, determine its y intercept.

Solution : C (4, 0, 0) is the point on line of force $\overline{P} = P_x i + P_y j$

co–ordination of points, A (0, 3, 0) ; B (6, 0, 0)

Taking moment of P about A

$$M_A^P = -170 = \overline{CA} \times \overline{P} = \begin{vmatrix} i & j & k \\ -4 & 3 & 0 \\ P_x & P_y & 0 \end{vmatrix} = -4P_y - 3P_x$$

$\therefore \; -4P_y - 3P_x = -170$

Taking moment of P about B

$$M_B^P = 40 = \overline{CB} \times \overline{P} = \begin{vmatrix} i & j & k \\ 2 & 0 & 0 \\ P_x & P_y & 0 \end{vmatrix} = 2P_y$$

$\therefore \; P_y = 20$ lb ; $P_x = 30$ lb

Let Point D be the point on line of force F and also on y axis and co–ordinates of D are $D \equiv (0, y, 0)$

Taking moment of P about B

$$M_B^P = 40 = \overline{DB} \times \overline{P} = \begin{vmatrix} i & j & k \\ 6 & -y & 0 \\ 30 & 20 & 0 \end{vmatrix} = 120 + 30y$$

$\therefore \; 120 + 30y = 40$

$\therefore \; y = -2.67$ ft

Problem 29. : In fig. a force F passing through C produces a clockwise moment of 600 ft - lb about A and a counterclock-wise moment of 300 ft - lb about B. Determine the moment of F about O.

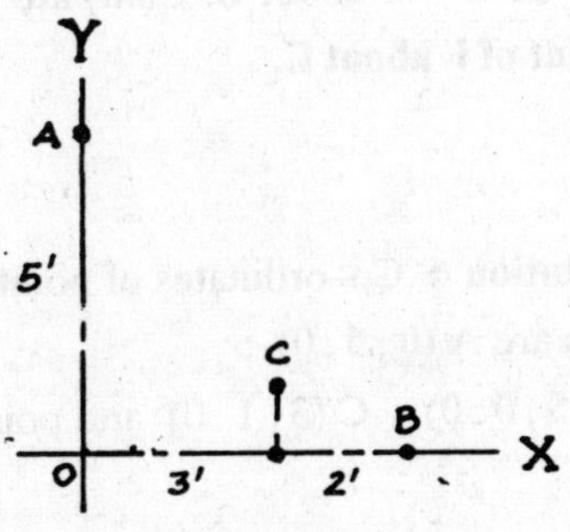

Solution : Co–ordinates of points in figure are A (0 , 5 , 0) ;
B (5 , 0 , 0) ; C (3 , 1 , 0)
Let force $\bar{F} = F_x i + F_y j$
Taking moment of F about A

$$M_A^F = 600 = \overline{CA} \times \bar{F} = \begin{vmatrix} i & j & k \\ -3 & 4 & 0 \\ F_x & F_y & 0 \end{vmatrix} = -3F_y - 4F_x$$

$$\therefore \quad -3F_y - 4F_x = 600 \qquad \ldots\ldots (1)$$

Taking moment of F about B

$$M_B^F = -300 = \overline{CB} \times \bar{F} = \begin{vmatrix} i & j & k \\ 2 & -1 & 0 \\ F_x & F_y & 0 \end{vmatrix} = 2F_y + F_x$$

$$\therefore \quad 2F_y + F_x = -300 \qquad \ldots\ldots (2)$$

Solving equations (1) and (2) we get
$F_x = -60$ lb & $F_y = -120$ lb
Now moment of $\bar{F}$ about O

$$M_O^F = \overline{CO} \times \bar{F} = \begin{vmatrix} i & j & k \\ -3 & -1 & 0 \\ -60 & -120 & 0 \end{vmatrix} = 300 \text{ lb} - \text{ft.}$$

Problem 30. : In Fig. a force F, having a slope of 2 vertical to 3 horizontal, produces a clockwise moment of 330 ft - lb about A and a counterclockwise moment of 420 ft - lb about B. Compute the moment of F about C.

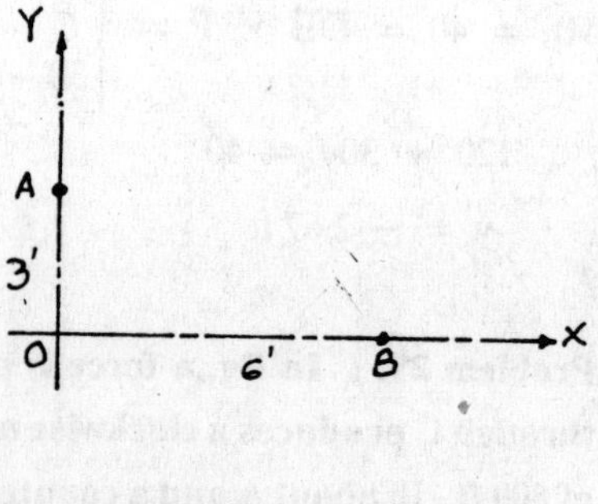

Solution : Co–ordinates of points in figure are A (0 , 5 , 0) ;
B (5 , 0 , 0) ; C (3 , 1 , 0) and point on line of force and x– axis be

G (x, 0, 0) force F from data is given by

$$F = \frac{F_m (3i + 2j)}{[3^2 + 2^2]^{1/2}} = F_m (0.832i + 0.5547j)$$

Taking moment of F about A

$$M_A^F = 330 = \overline{GA} \times \overline{F} = \begin{vmatrix} i & j & k \\ -x & 5 & 0 \\ 0.832 & 0.5547 & 0 \end{vmatrix} F_m = (-0.5547x - 4.16) F_m$$

$$\therefore \quad (0.5547x + 4.16) F_m = -330 \qquad \ldots\ldots (1)$$

Taking moment of F about B

$$M_B^F = -420 = \overline{GB} \times F = \begin{vmatrix} i & j & k \\ 5-x & 0 & 0 \\ 0.832 & 0.5547 & 0 \end{vmatrix} F_m$$

$$= (2.7735 - 0.5547\,x) F_m$$

$$\therefore \quad (2.7735 - 0.5547\,x)\ F_m = -420 \qquad \ldots\ldots (2)$$

Div iding eq^n. (1) by (2).

$0.5547\,x + 4.16 = 2.179 - 0.4358\,x$

$\therefore \quad x = -2$

Substituting $x = -2$ in equation (1) we get

$F_m = -108.17$ lb

$\therefore \quad \overline{F} = -90i - 60j$

Moment of F about C

$$M_C^F = \overline{GC} \times F = \begin{vmatrix} i & j & k \\ (3+2) & 1 & 0 \\ -90 & -60 & 0 \end{vmatrix} = -210 \text{ lb} - \text{ft}.$$

Problem 31.: In Fig., find the moment of force P about the diagonal AB. Use this result to show that the length of the common perpendicular between P and AB is 4 ft.

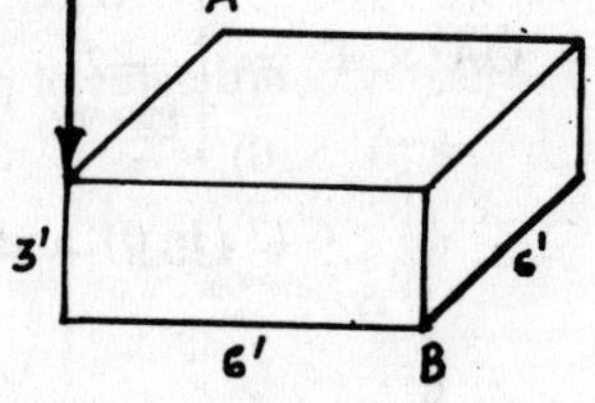

Solution : Co–ordinates of points in figure are A (0, 3, −6) ; B (6, 0, 0)

$$AB = 6i - 3j - 6k \quad \therefore \quad \hat{n}_{AB} = \frac{1}{9}(6i - 3j - 6k)$$

Moment of P about $\overline{AB}$

$$M_{AB}^{P} = \begin{matrix} \overline{AB} \Rightarrow \\ \overline{r_A} \Rightarrow \\ P \Rightarrow \end{matrix} \begin{vmatrix} 6 & -3 & -6 \\ 0 & 0 & -6 \\ 0 & -P & 0 \end{vmatrix} \times \frac{1}{9}$$

$$= \frac{1}{9}(-36P) = -4P$$

Also moment of P about AB

$= P \times r_{AB}$ where r_{AB} is perpendicular distance between P and AB

$\therefore \quad P \times r_{AB} = 4P$

$\therefore \quad r_{AB} = 4$ ft .

Problem 32. : F is 400 lb force directed from A (2, 1, –4) to B (4, 4, 1). (a) Find the moment of F about the origin and the magnitude of this moment. (b) Find the moment of F about the point C (3, 4, 2) and its moment arm from C. (c) Find the moment arm of F about the axis passing through the origin and point C.

Solution : Given coordinates of points are A (2 , 1 , −4) ; B (4 , 4 , 1) ; C (3 , 4 , 2)

$$\overline{AB} = 2i + 3j + 5k \quad \therefore \quad \hat{n}_{AB} = \frac{1}{\sqrt{38}}(2i + 3j + 5k)$$

$$\therefore \quad \overline{F} = \frac{400}{\sqrt{38}}(2i + 3j + 5k)$$

$$= 129.78i + 194.67j + 324.44k$$

Moment of F about O

$$M_O^F = \overline{OA} \times F = \begin{vmatrix} i & j & k \\ -2 & -1 & 4 \\ 129.78 & 194.67 & 324.44 \end{vmatrix}$$

$$= 1103i - 1168j + 259.56k$$

$$\therefore\ M_O^F = [(1103)^2 + (-1168)^2 + (259.56)^2]^{1/2}$$
$$= 1627.33 \text{ lb} - \text{ft}.$$

Moment of F about C

$$M_C^F = \overline{CA} \times F = \begin{vmatrix} i & j & k \\ -1 & -3 & -6 \\ 129.78 & 194.67 & 324.44 \end{vmatrix}$$

$$= 194.7i - 454.24j + 194.67k$$

$$\therefore\ M_C^F = [(194.7)^2 + (454.24)^2 + (194.67)^2]^{1/2} = 531.16 \text{ lb} - \text{ft}.$$

Also moment of F about C = F × moment arm of F to C

M_C^F = 400 × perpendicular distance of C from F

= 531.16

$\therefore$ Perpendicular distance of C from F $= \dfrac{531.16}{400} = 1.328$ ft

Unit vector along OC $= \dfrac{1}{\sqrt{29}}(3i + 4j + 2k)$

Taking moment of F about OC

$$M_{OC}^F = \begin{vmatrix} 3 & 4 & 2 \\ -1 & -3 & -6 \\ 129.78 & 194.67 & 324.44 \end{vmatrix} \frac{1}{\sqrt{29}}$$

$$= (584.1 - 1816.96 + 389.34)\frac{1}{\sqrt{29}} = -156.63 \text{ lb} - \text{ft}.$$

Also, M_{OC}^F = 400 × moment arm of F about OC

= − 156.63

$\therefore$ Moment arm of F about OC = 0.392 ft.

Problem 33. : In fig. , a boom AC is supported by a ball–and–socket joint at C and by the cables BE and AD. If the force multiplier of force F acting from B to E is F_m = 10 lb/ft , find (a) the moment of F about point C, (b) the moment of F about point D, (c) the moment of F about a line directed from C to D.

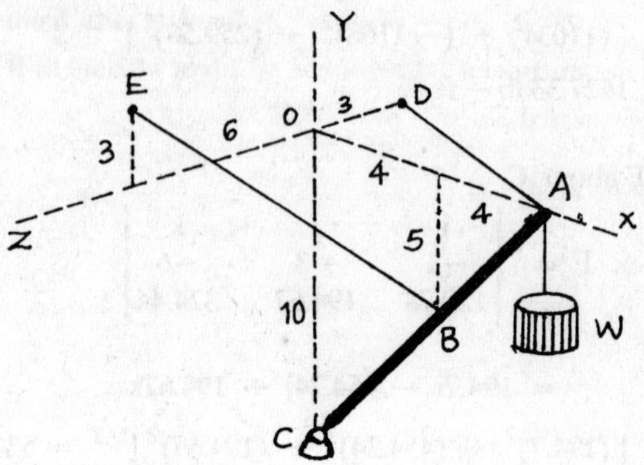

Solution : Co–ordinates of points in figure are A(8 , 0 , 0) ; B(4 , −5 , 0) ; C(0, −10 , 0) ; D(0 , 0 , −3) ; E(0 , 3 , 6)

$\overline{BE} = -4i + 8j + 6k$

$\therefore \quad \overline{F} = 10(-4i + 8j + 6k)$

Moment of F about C

$$M_C^F = \overline{CB} \times F = \begin{vmatrix} i & j & k \\ 4 & 5 & 0 \\ -40 & 80 & 60 \end{vmatrix} = 300i - 240j + 520k$$

Moment of F about D

$$M_D^F = \overline{DB} \times \overline{F} = \begin{vmatrix} i & j & k \\ 4 & -5 & 3 \\ -40 & 80 & 60 \end{vmatrix} = -540i - 360j + 120k$$

Now, $\overline{CD} = 10j - 3i \quad \therefore \quad \hat{n}_{CD} = \frac{1}{\sqrt{109}}(10j - 3i)$

Moment of F about $\overline{CD}$

$$M_{CD}^F = \begin{vmatrix} 0 & 10 & -3 \\ 4 & 5 & 0 \\ -40 & 80 & 60 \end{vmatrix} \frac{1}{\sqrt{109}}$$

$$= \frac{-2400 - 1560}{\sqrt{109}} = -379.3 \text{ lb} - \text{ft.}$$

Problem 34. : In fig. of above problem, a force P acts from A towards D. Determine the magnitude of P to cause a moment of 2000 lb-ft about the line directed from B toward E.

Solution : Co–ordinates of points are

A(8, 0, 0) ; B(4, –5, 0) ; C(0, –10, 0) ; D(0, 0, –3) ; E(0, 3, 6)

$\overline{AD} = -8i - 3k \quad \therefore \quad \overline{P} = P_m(-8i - 3k)$

$\overline{BE} = -4i + 8j + 6k \quad \therefore \quad \hat{n}_{BE} = \dfrac{1}{\sqrt{116}}(-4i + 8j + 6k)$

Moment of P about BE

$$M_{BE}^{P} = \begin{vmatrix} -4 & 8 & 6 \\ 4 & 5 & 0 \\ -8 & 0 & -3 \end{vmatrix} \times \frac{P_m}{\sqrt{116}} = 2000 \text{ lb} - \text{ft.}$$

$$\frac{P_m}{\sqrt{116}}(60 + 96 + 240) = 2000$$

$\therefore \quad P_m = 54.4$ lb

$\therefore \quad \overline{P} = 54.4(-8i - 3k)$

$\therefore \quad P = 54.4\,[8^2 + 3^2]^{1/2} = 464.75$ lb

Problem 35. : For the system shown in fig., the force multiplier of P acting from A to E is P_m = 100 lb – ft. Find the moment of P about C and about D. What is the moment of P about an axis directed from C to D ? Compute the magnitude of the force F acting from B to D that will cause a moment of 960 lb-ft about an axis directed from A to E.

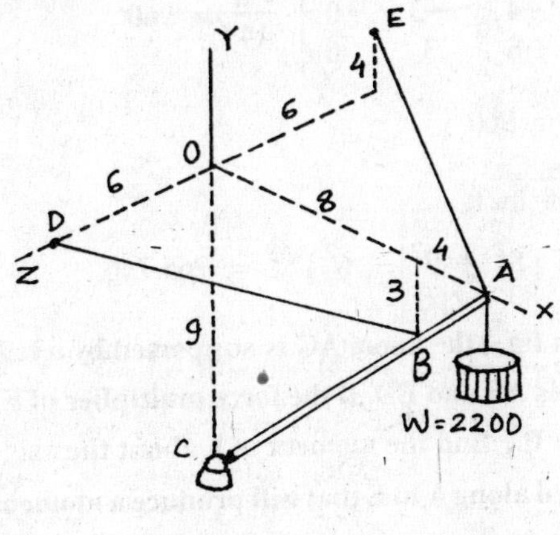

Solution : Co–ordinates of points in figure are $A(12, 0, 0)$; $B(8, -3, 0)$; $C(0, -9, 0)$; $D(0, 0, 6)$; $E(0, 4, -6)$

$\overline{AE} = -12i + 4j - 6k \quad \therefore \quad \overline{P} = 100(-12i + 4j - 6k)$

Taking moment of P about D

$$M_D^P = \overline{DA} \times \overline{P} = \begin{vmatrix} i & j & k \\ 12 & 0 & -6 \\ -12 & 4 & -6 \end{vmatrix} 100 = (24i - 144j + 48k)\,100$$

Moment of P about C

$$M_C^P = \overline{CA} \times \overline{P} = \begin{vmatrix} i & j & k \\ 12 & 9 & 0 \\ -12 & 4 & -6 \end{vmatrix} 100 = (-54i + 72j + 156k)\,100$$

Moment of P about CD

$$M_{CD}^P = \begin{vmatrix} 0 & 9 & 6 \\ 12 & 9 & 0 \\ -12 & 4 & -6 \end{vmatrix} \frac{100}{\sqrt{117}} = \frac{100}{\sqrt{117}}(-648 + 936)$$

$$= 2662 \text{ lb} - \text{ft.}$$

Now, taking moment of F about AE

$\overline{AE} = -12i + 4j - 6k \quad \therefore \quad \hat{n}_{AE} = \frac{1}{14}(-12i + 4j - 6k)$

$\overline{BD} = -8i + 3j + 6k \quad \therefore \quad F = F_m(-8i + 3j + 6k)$

$$\therefore \quad M_{AE}^F = \begin{vmatrix} -12 & 4 & -6 \\ -4 & -3 & 0 \\ -8 & 3 & 6 \end{vmatrix} \frac{F_m}{14} = 960$$

$$\therefore \quad 528 \cdot \frac{F_m}{14} = 960$$

$$\therefore \quad F_m = 25.45 \text{ lb/ft}$$

$$\therefore \quad F = 25.45\,[8^2 + 3^2 + 6^2]^{1/2} = 265.7 \text{ lb}$$

Problem 36. : In fig., the boom AC is supported by a ball–and–socket joint at C and by cables AE and BD. If the force multiplier of F acting from B to D is F_m = 10 lb/ft, find the moment of F about the axis AE. Determine the value of P directed along A to E that will produce a moment of 280 lb - ft about the axis DB.

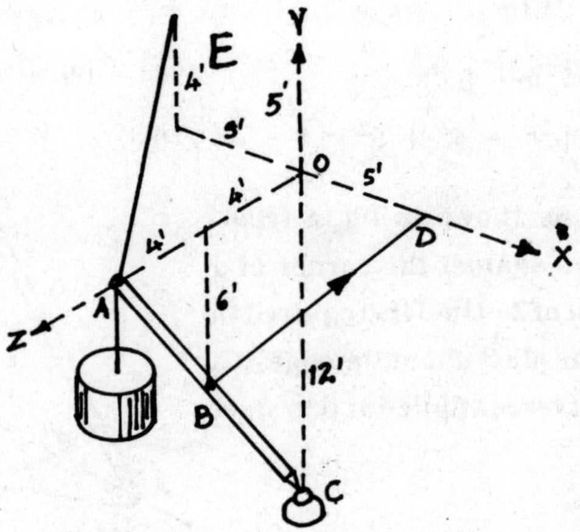

Solution : Co–ordinates of points in figure are A (0 , 0 , 8) ;
B (0 , −6 , 4) ; C (0 , −12 , 0) ; D (5 , 0 , 0) ; E (−5 , 4 , 0)

$\overline{BD} = 5i + 6j - 4k \qquad \therefore \quad \overline{F} = 10\,(5i + 6j - 4k)$

$\overline{AE} = -5i + 4j - 8k \qquad \therefore \quad \hat{n}_{AE} = \frac{1}{\sqrt{105}}\,(-5i + 4j - 8k)$

$\overline{AB} = -6j - 4k$

Moment of F about AE.

$$M_{AE}^{F} = \begin{vmatrix} -5 & 4 & -8 \\ 0 & -6 & -4 \\ 50 & 60 & -40 \end{vmatrix} \frac{1}{\sqrt{105}} = -546.5 \text{ lb} - \text{ft.}$$

Now, taking moment of P about DB

$\overline{P} = P_m\,(-5i + 4j - 8k)$

$\overline{DB} = -5i - 6j + 4k$

$\therefore \quad \hat{n}_{DB} = \frac{1}{\sqrt{77}}\,(-5i - 6j + 4k)$

$\overline{DA} = -5i + 8k$

$$M_{DB}^{P} = \begin{vmatrix} -5 & -6 & 4 \\ -5 & 0 & 8 \\ -5 & 4 & -8 \end{vmatrix} \frac{P_m}{\sqrt{77}} = 280$$

$$\therefore \quad \frac{560\ P_m}{\sqrt{77}} = 280$$

$$\therefore \quad P_m = 4.387\ \text{lb} - \text{ft.}$$

$$\therefore \quad P = 4.387\ [\,5^2 + 4^2 + 8^2\,]^{1/2} = 44.95\ \text{lb}$$

Problem 37. : As shown in fig. a triangular plate rests against the corner of a room. A moment of 260 lb–ft is required to start rotating the plate about the edge AC. What is the least force applied at B to start this rotation ?

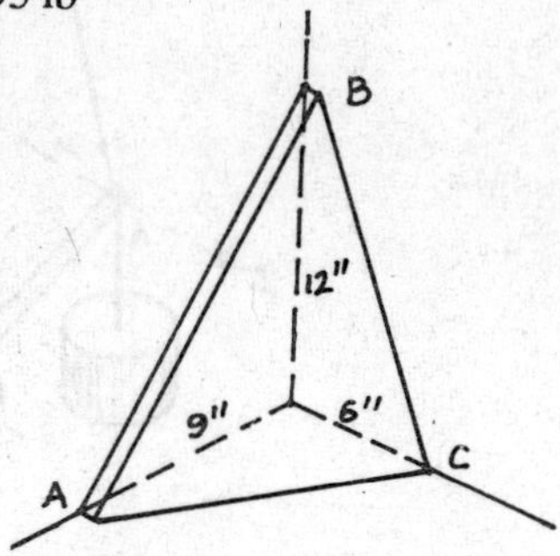

Solution : Co–ordinates of points are A (0 , 0 , 9) ; B (0 , 12 , 0) ; C (6 , 0 , 0)

$\overline{AC} = 6i - 9k \qquad \therefore \quad \hat{n}_{AC} = \frac{1}{\sqrt{117}}\,(6i - 9k)$

$\overline{AB} = 12j - 9k$

Force F will be minimum if it acts perpendicular to place at point B i.e. if should act along $\overline{M}$ which is cross product of $\overline{BC}$ & $\overline{BA}$

$$\therefore \quad \overline{M} = \overline{BC} \times \overline{BA} = \begin{vmatrix} i & j & k \\ 6 & -12 & 0 \\ 0 & -12 & 9 \end{vmatrix}$$

$$= -108i - 54j - 72k$$

$$\therefore \quad F = F_m\,(-108i - 54j - 72k)$$

Taking moment of F about AC

$$M_{AC}^{F} = \begin{vmatrix} 6 & 0 & -9 \\ 0 & 12 & -9 \\ -108 & -54 & -72 \end{vmatrix} \frac{F_m}{\sqrt{117}} = 260$$

$$\Rightarrow -\frac{19764}{\sqrt{117}}\,F_m = 260$$

$$\therefore \quad F_m = -0.1423\ \text{lb/ft}$$

$$\therefore \quad F = 0.1423\ [\,108^2 + 54^2 + 72^2\,]^{1/2} = 20\ \text{lb}$$

Problem 38. : Replace the system of forces acting on the frame in fig. by a resultant force R through A and a couple acting horizontally through B and C.

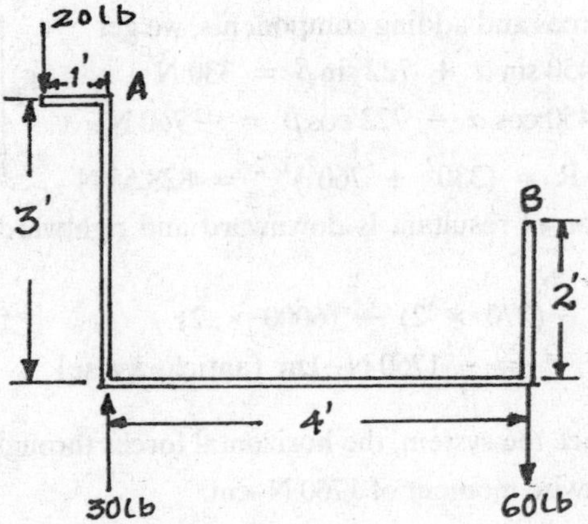

Soluiton : The system has the resultant force R = 20 – 30 + 60 = 50 N acting downward.

Taking moment about A.

$$\Sigma M_A = -20 + (60 \times 4) = 220 \text{ N} - \text{m (clockwise)}$$

In order to replace the system, the horizontal forces through B and C should give a moment of 220 N–m in clockwise direction.

$$\therefore \text{ Horizontal force at B or C} = \frac{220}{d\,(BC)} = \frac{220}{2} = 110 \text{ N}.$$

The direction of horizontal force at B is rightward and that at C is leftward to produce clockwise moment.

Problem 39. : Fig. shows a plate subjected to loads. Replace the loads by an equivalent force through C and a couple acting through A and B. Solve if the forces of the couple are (a) horizontal and (b) vertical.

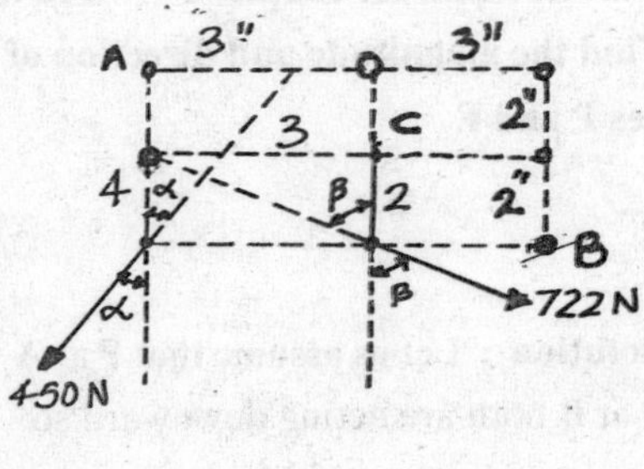

Solution : From fig.

$$\alpha = \tan^{-1}\frac{3}{4} = 36.869$$

$$\beta = \tan^{-1}\frac{3}{2} = 56.309$$

Resolving the forces and adding components, we get

$$\Sigma F_x = -450 \sin\alpha + 722 \sin\beta = 330 \text{ N}$$

$$\Sigma F_y = -450 \cos\alpha - 722 \cos\beta = -760 \text{ N}.$$

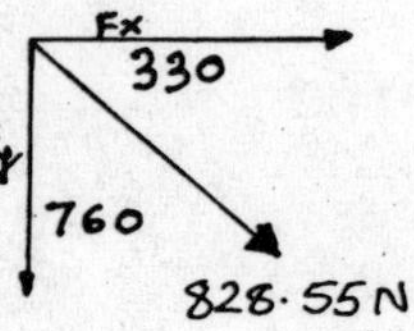

$\therefore$ Resultant $R = (330^2 + 760^2)^{1/2} = 828.55$ N

The direction of resultant is downward and rightward. Taking moment about C

$$\Sigma M_C = -(270 \times 2) - (6000 \times 2)$$

$$= -1760 \text{ N–cm (anticlockwise)}$$

In order to replace the system, the horizontal forces through A and B should give an anticlockwise moment of 1760 N–cm.

$\therefore$ Horizontal forces thorugh A & B $= \dfrac{1740}{4} = 435$ N.

The direction of horizontal force at A is leftward and that at B is righward. To replace the system by vertical forces through A and B vertical forces thorugh A & B $= \dfrac{1740}{6} = 290$ N

The direction of vertical force at A is donward and at B it is upward.

Problem 40. : A vertical force P at A and another vertical force F at B which act on the bar shown in fig. produce a resultant force of 150 lb down at D and a counterclockwise couple C = 300 lb–ft. Find the magnitude and direction of forces P and F.

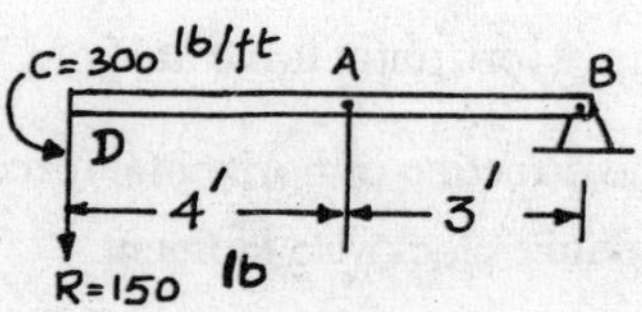

Solution : Let us assume that P at A and F at B both are acting downward so

$P + F = 150$ N ..... (1)

Taking moment about D

$\Sigma M_D = 4P + 7F = -300$ ($\because$ Counterclockwise) .... (2)

Solving equation (1) and (2) we get.

$F = -300$ N & $P = 450$ N.

Thus F at B is acting upward.

Problem 41.: Determine the amount and position of the resultant of the loads acting on the Fink truss shown in fig.

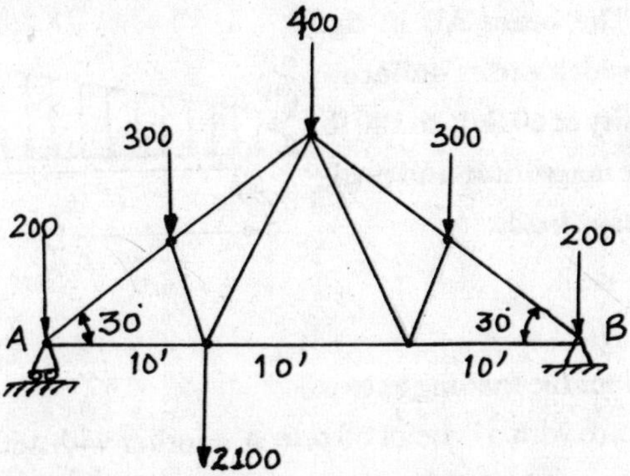

Solution : Let the resultant force R be acting downward at a distance x from joint A.

$$R = 200 + 300 + 2100 + 400 + 300 + 200 = 3500 \text{ N}$$

Taking moment of all forces about A and equating it to $3500 \times x$, we get

$$(300 \times 7.5) + (400 \times 15) + (300 \times 22.5) + (200 \times 30) + (2100 \times 10) = 3500 \times x$$

$$\text{or} \quad x = \frac{42000}{3500} = 12 \text{ m}$$

Problem 42.: Find the values of P and F so that the four forces shown in fig. produce an upward resultant of 300 lb acting at 4 ft from the left end of the bar.

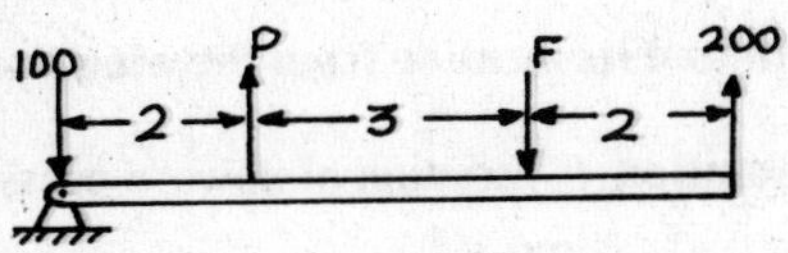

Solution : The sum of moments about the point 4m from leftward (let it be A)

must addup to zero, since R passes through it.

Taking moment about A

$$\Sigma M_A = -(100 \times 4) + 2P + F - (200 \times 3) = 0$$

i.e. $2P + F = 1000$ .....(1)

Also $P + 200 - F - 100 = 300$

$\therefore$ $P - F = 200$ .....(2)

Solving equation (1) and (2) we get,

$P = 400$ N , $F = 200$N.

Problem 43. : The beam AB in fig. supports a load which varies uniformly from an intensity of 60 lb/ft to 180 lb / ft. Calculate the magnitude and position to the resultant load.

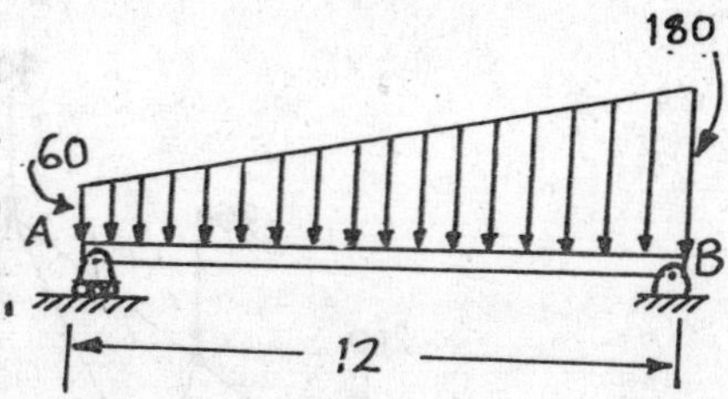

Solution : Consider the loading as two triangular loads, are with a base of 60 n/m & another with 180 N/m.

$$\therefore \text{ Total load } R = \frac{1}{2} \times 60 \times 12 + \frac{1}{2} \times 180 \times 12$$

$$= 360 + 1080 = 1440 \text{ N}.$$

These two loads acts at the centroids of respective triangles, i.e. 4m from either supports. If we assume resultant acts at a distance x from A then

$$x = \frac{(360 \times 4) + (1080 \times 8)}{1440} = 7 \text{ metre}.$$

Problem 44. : The 16–ft wing of a small airplane is subjected to a lift which varies from zero at the tip of 360 lb/ft at the fuselage according to $\omega = 90\,x^{1/2}$ where x is measured in feet from the tip. Compute the resultant lift and its location from the wing tip.

Solution : Variation of upward pressure intensity is given by

$$\omega = 90\,x^{1/2}$$

Total lift R = Area of load distribution diagram.

$$= 90 \int_0^{16} x^{1/2}\, dx = 90 \times \frac{2}{3} (16)^{3/2} = 3840 \text{ N}$$

Let the total lift 'R' be acting at a distance 'x' from the wing H.P. Now taking moment about tip

$$M_{tip}^{R} = R \cdot x = \int_0^{16} \omega \cdot x\, dx$$

$$\therefore \quad x = \text{Distance of R from Tip} = \frac{\int_0^{16} 90\, (x)^{3/2}\, dx}{3840}$$

$$= \frac{2}{5} \times (16)^{5/2} \times \frac{90}{3840} = 9.6 \text{ m}.$$

Problem 45. : Determine the resultant of the three forces acting on the dam shown in fig. and locate its intersection with the base AB. For good design, this intersection should occur within the middle third of the base. Does it ?

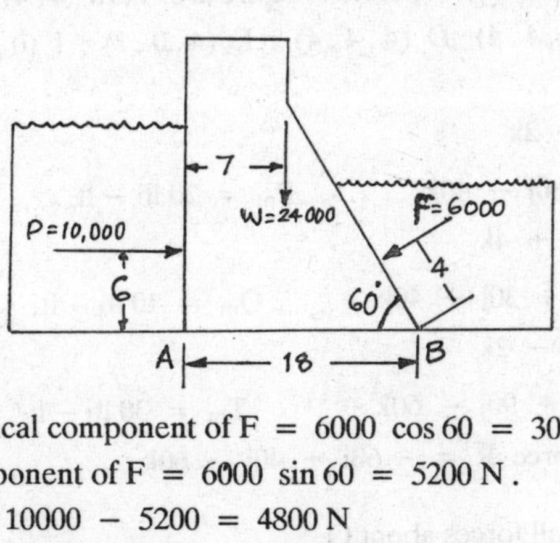

Solution : Vertical component of F $= 6000 \cos 60 = 3000$ N.

Horizontal component of F $= 6000 \sin 60 = 5200$ N.

Now $\quad \Sigma F_x = 10000 - 5200 = 4800$ N

$\quad \Sigma F_y = 24000 + 3000 = 27000$ N,.

$\therefore$ Resultant $R = [(4800)^2 + (27000)^2]^{1/2} = 27423.34$ N.

Taking moments about B

$$\Sigma M_B = (10000 \times 6) - (24000 \times 11) - (6000 \times 4)$$

$$= 228000 \text{ N} - \text{m}$$

If resultant intersects base at a distance x from B then x

$$= \frac{228000}{27423.34} = 8.31 \text{ m}$$

(distance from mid point of base) = 0.69 m. It lies within the middle third of base.

Problem 46. : The rectangular block of fig. is acted upon by a set of forces whose force multipliers and directions are as follows : P_m = 20 lb/ft directed from A to E, Q_m = 10 lb/ft directed from B to C, T_m = 30 lb/ft directed from G to F. Reduce these forces to a single resultant force acting at O and a couple.

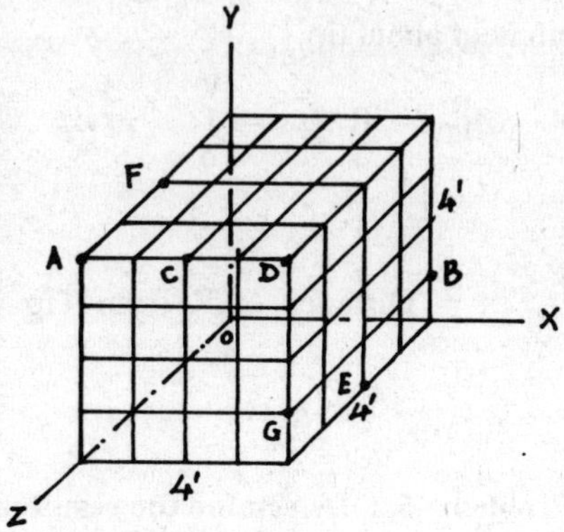

Solution : Co–ordinates of points in figure are A (0 , 4 , 4) ; B (4 , 1 , 0) ; C (2 , 4 , 4) ;D (4 , 4 , 4) ; E (4 , 0 , 2) ; F (0 , 4 , 2) ; G (4 , 1 , 4)

$\overline{AE} = 4i - 4j - 2k$

$\therefore \ \overline{P} = 80i - 80j - 40k \quad \because \ P_m = 20 \text{ lb} - \text{ft}.$

$\overline{BC} = -2i + 3j + 4k$

$\therefore \ \overline{Q} = -20i + 30j + 40k \quad \because \ Q_m = 10 \text{ lb} - \text{ft}.$

$\overline{GF} = -4i + 3j - 2k$

$\therefore \ T = -120i + 90j - 60k \quad \because \ T_m = 30 \text{ lb} - \text{ft}.$

$\therefore$ Resultant of force $\overline{R} = -60i + 40j - 60k$

Taking moment of all forces about O

$$\Sigma M_o = (\overline{OA} \times \overline{P}) + (\overline{OB} \times \overline{Q}) + (\overline{OG} \times \overline{T})$$

$$= \begin{vmatrix} i & j & k \\ 0 & 4 & 4 \\ 80 & -80 & -40 \end{vmatrix} + \begin{vmatrix} i & j & k \\ 4 & 1 & 0 \\ -20 & 30 & 40 \end{vmatrix} + \begin{vmatrix} i & j & k \\ 4 & 1 & 4 \\ -120 & 90 & -60 \end{vmatrix}$$

$$\Sigma M_O = -220i - 80j + 300k \text{ lb} - \text{ft}.$$

Thus $\overline{R} = -60i + 40j - 60k$

and couple $= -220i - 80j + 330k$ at point O.

Problem 47. : In Fig., a force T acts along BE and a force P acts along AD. Assuming their force multipliers to be

T_m = 20 lb/ft and P_m = 10 lb/ft, find the force F to be applied at C to reduce their resultant to a couple. What is this resultant couple ?

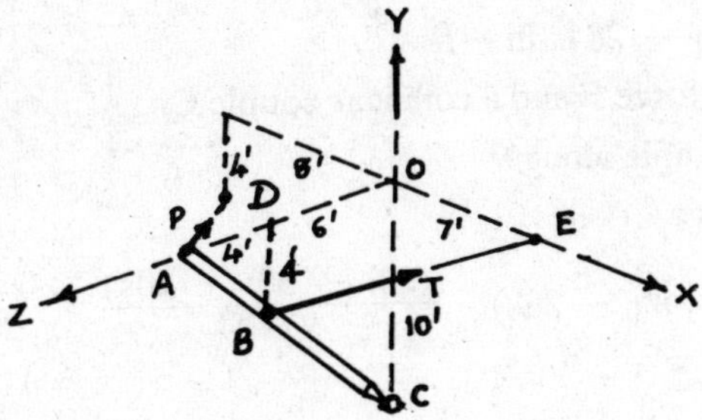

Solution : Co–ordinates of points in figure are A (0 , 0 , 10) ;
B (0 , −4 , 6) ; C (0 , −10 , 0) ; D (−8 , −4 , 0) ; E (7 , 0 , 0)

$\overline{BE} = 7i + 4j - 6k$

$\therefore\ T = 140i + 80j - 120k \qquad \because\ T_m = 20$ lb/ft

$\overline{AD} = -8i - 4j - 10k$

$\therefore\ P = -80i - 40j - 100k \qquad \because\ P_m = 10$ lb/ft

Resultant of these forces $\overline{R} = 60i + 40j - 220k$

To reduce resultant of T , $\overline{P}$ & F to couple, F must be $= -\overline{R}$

$= -60i - 40j + 220k$ at C

To find the couple, we will take moment of all forces about O.

$\Sigma M_O = (\overline{OB} \times T) + (\overline{OA} \times \overline{P}) + (\overline{OC} \times F)$

$$= \begin{vmatrix} i & j & k \\ 0 & -4 & 6 \\ 140 & 80 & -120 \end{vmatrix} + \begin{vmatrix} i & j & k \\ 0 & 0 & 10 \\ -80 & -40 & -100 \end{vmatrix} + \begin{vmatrix} i & j & k \\ 0 & -10 & 0 \\ -60 & -40 & 220 \end{vmatrix}$$

$= -1800i + 40j - 40k$

Problem 48. : The resultant of a certain force system is $F = 20\hat{i} - 30\hat{j} + 10\hat{k}$ lb at the origin of coordinate axes and a couple $C = 56\hat{i} + 42\hat{j} - 28\hat{k}$ lb/ft . Replace this force and couple by an equivalent system (known as a wrench) consisting of the force F and a collinear couple C_F.

Solution : $\bar{F} = 20i - 30j + 10k$ lb and
$\bar{C} = 56i + 42j - 28k$, lb – ft.
Wrench means a force F and a collinear couple C_F.
Component of couple along F

$$C_{on\,F} = \bar{C} \cdot \hat{n}_F$$

$$= (56i + 42j - 28k) \cdot \frac{(20i - 30j + 10k)}{\sqrt{1400}}$$

$$= -11.225$$

$$\therefore \bar{C}_{onF} = -11.225 \cdot \frac{(20i - 30j + 10k)}{\sqrt{1400}}$$

$$\bar{C}_F = -6i + 9j - 3k$$

$$\bar{F} = 20i - 30j + 10k$$

Let the point of intersection of F with XY plane be (x, y, 0)
Taking moment of force about

$$C - C_F = \overline{OA} \times \bar{F} = \begin{vmatrix} i & j & k \\ x & y & 0 \\ 20 & -30 & 10 \end{vmatrix}$$

$$= (56i + 42j - 28k) - (-6i + 9j - 3k)$$

$$\Rightarrow 62i + 33j - 25k = 10y\,i - 10x\,j + (-30x - 20y)\,k$$

$$\therefore -10x = 33 \quad \Rightarrow x = -3.3$$

$$10y = 62 \quad \Rightarrow = 6.2$$

Thus F is passing through point (– 3.3 , 6.2 , 0).

CHAPTER TWO

EQUILIBRIUM OF FORCE SYSTEMS

Equilibrium is the term used to designate the condition where the resultant of a force system is zero. Perhaps the most significant difference between the previous chapter and this one is that the force system to be analysed was specified explicitly whereas here the student must orginate the force system.

The physical meaning of equilibrium, as applied to a body, is that the body is either at rest or is moving along a straight line path with constant velocity.

Equation of Equilibrium : (For general or spatial force system).
Basic Vector equation of equilibrium is

$$\Sigma F = 0 \text{ and } \Sigma M = 0$$

Which is equivalent to the following six scalar equations

$$\Sigma X = 0 \text{ and } \Sigma M_x = 0$$

$$\Sigma Y = 0 \text{ and } \Sigma M_y = 0$$

$$\Sigma Z = 0 \text{ and } \Sigma M_z = 0$$

Thus general or spatial force system in equilibrium can be solved for six unknown quantities.

Free Body Diagram :

An isolated view of a body which shows only the external forces acting on the body is called a Free Body diagram (abbreviated as FBD).

Type of body removed	Sketch of reacting bodies	Action of body removed on the free body
Earth		
Flexible cord rope Cable weight neglected	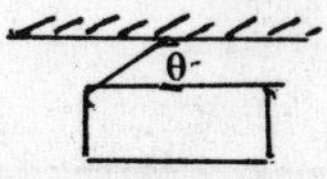	
Smooth surface		
Roller		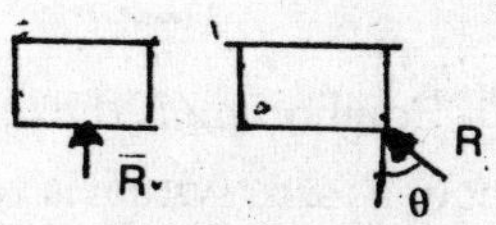
Smooth pin or hinge	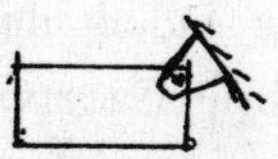	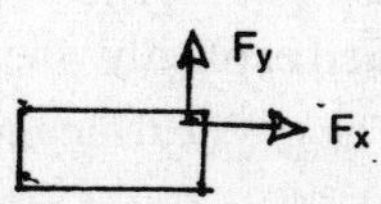
Ball and socket joint	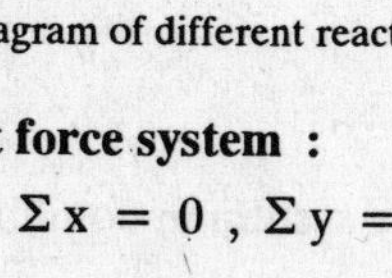	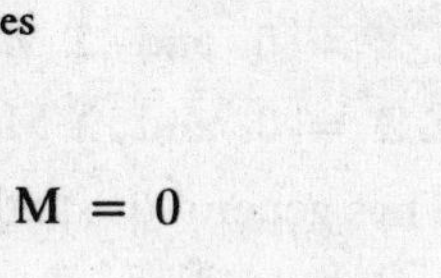

Diagram of different reacting bodies

Coplaner non–concurrent force system :

The equations will be $\Sigma x = 0$, $\Sigma y = 0$, $\Sigma M = 0$

Coplaner con–current force system :

The equations are $\Sigma X = 0$ and $\Sigma Y = 0$

$\Sigma X = 0$ means the summation of components of all forces in X–direction is zero.

Lami's Theorem :

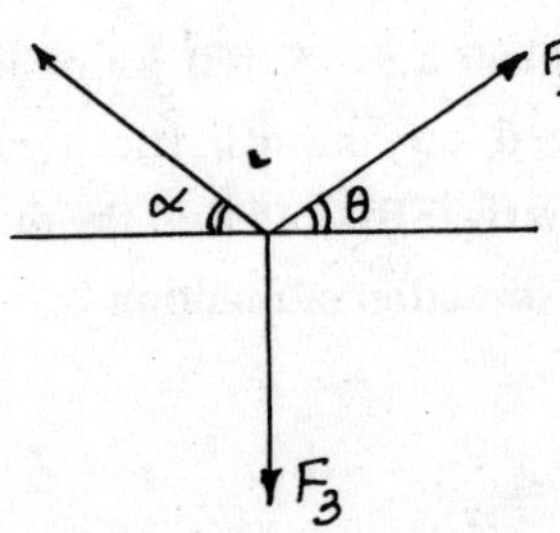

If a body is in equilibrium under the action of only three con–current forces F_1, F_2 and F_3 as shown in fig.

then according to Lami's theorem the ratio of

$$\frac{\text{force}}{\text{sine of angle bet. remaining two forces in a point diag.}} \text{ is constant}$$

$$\text{i.e } \frac{F_1}{\sin(90 + \alpha)} = \frac{F_2}{\sin(90 + \theta)} = \frac{F_3}{\sin(180 - \alpha - \theta)} = \text{constant}$$

Sine Rule :

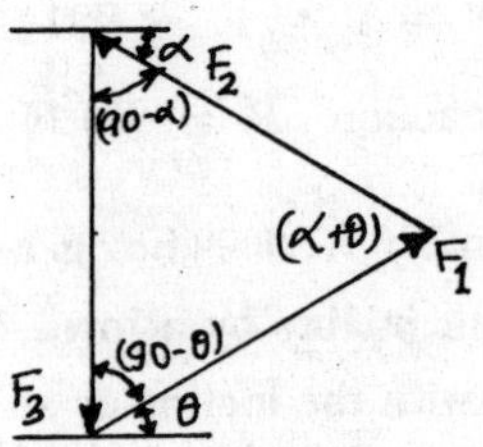

The above system of forces if added graphically in a tip to tail fashion forms a closed triangle as shown

Now according to sine rule the ratio

$$\frac{\text{force}}{\text{sine of angle (opposite) in vector diagram}}$$

is constant .

$$\text{or } \frac{F_1}{\sin(90 - \alpha)} = \frac{F_2}{\sin(90 - \theta)} = \frac{F_3}{\sin(\alpha + \theta)} = \text{constant}$$

Procedure to solve the Equilibrium Problems :

(i) First Draw the Free Body Diagram of the given system. Show all the known and unknown forces in the diagram. The direction of unknown forces is arbitrarily chosen. Later on if it is found that the numerical value of the unknown force works out to be negative then the conclusion is that arbitrarily assumed direction was exactly opposite. If the value is positive then assumed direction is correct.

(ii) Write the equilibrium equations such that the number of equations must be equal to number of unknowns.

(iii) Solve the formulated system of equations.

Problem 1.: A 100 kg of weight is placed on smooth inclined plane. Draw the FBD and find the magnitude and direction of reaction ?

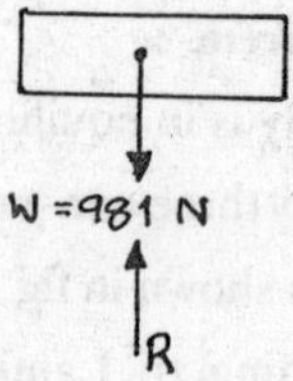

FBD Will be as shown

Solution :

100 kg mass will have a weight force of

$$F = 100 \times 9.81 \text{ Newtons}$$

$$= 981 \text{ N}$$

$$\Sigma Y = 0 = R - 981$$

$\therefore$ Reaction $R = 981$ N (upwards)

Problem 2. : A 300 N box is held at rest on a smooth incline by a force P making an angle θ with the incline as shown in fig. If $\theta = 45^o$, determine the value of P

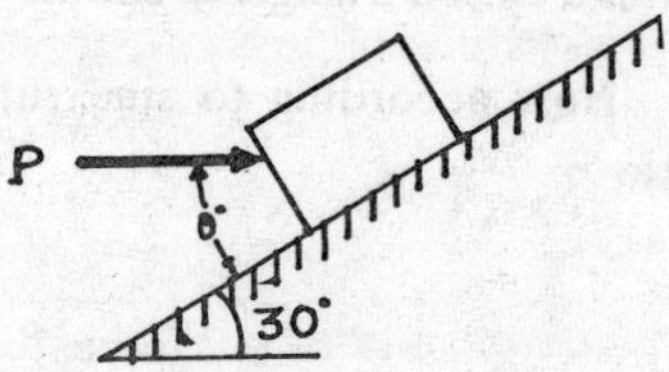

Solution : Draw FBD of the given system.

Forces involved will be P, weight force 'W' and a normal reaction 'R' from inclined plane.

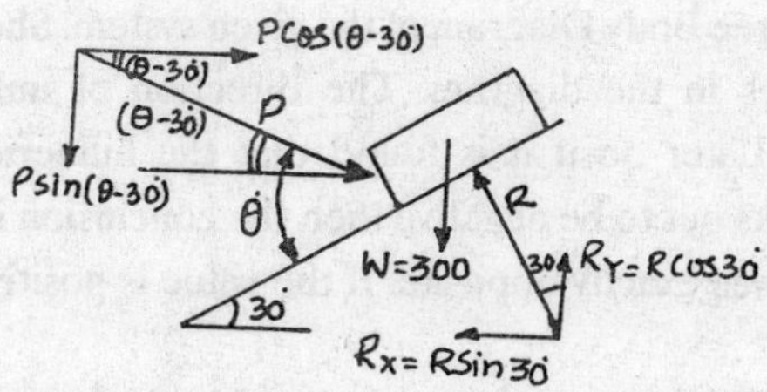

$$\Sigma X = P\cos(\theta - 30) - R\sin 30 = 0 \quad \ldots\ldots (1)$$

$$\Sigma Y = R\cos 30 - 300 - P\sin(\theta - 30) = 0 \quad \ldots\ldots (2)$$

Given that $\theta = 45^o$

from eq^n. (1) $\Rightarrow$ $P \cos 15 - R \sin 30 = 0$ ..... (3)

from eq^n. (2) $\Rightarrow$ $R \cos 30 - 300 - P \sin 15 = 0$ ..... (4)

from eq^n. (3) $\Rightarrow$ $R = 1.932\,P$

putting this value in equation (4)

$$(1.932\,P) \cos 30 - 300 - P \sin 15 = 0$$

$\therefore$ $P = 212.2$ N

Problem 3. : If in the above problem the value of P = 180 N, determine the angle θ it must make with smooth incline to hold the 300 N box at rest.

Solution : Rewriting the same equations (1) and (2)

$$P \cos (\theta - 30) - R \sin 30 = 0 \quad \text{..... (1)}$$

$$R \cos 30 - 300 - P \sin (\theta - 30) = 0 \quad \text{..... (2)}$$

from eq^n. (1) $\Rightarrow$ $R = \dfrac{P \cos (\theta - 30)}{\sin 30}$

Put this value of R in equation (2)

$$\therefore \quad \frac{P \cos (\theta - 30)}{\sin 30} \cos 30 - 300 - P \sin (\theta - 30) = 0$$

$$180 \times \cot 30 \cdot \cos (\theta - 30) - 300 - 180 \sin (\theta - 30) = 0$$

$$311.7 \cos (\theta - 30) - 180 \sin (\theta - 30) = 300$$

Use Trial and Error method to solve above equation.

Solving above equation by trial and error method, we get $\theta = 33.5^\circ$

Problem 4. : The 300 N sphere is supported by pull P = 160 N and a weight of 200 N passing over frictionless pulley. Determine the values of α & θ that maintain the equilibrium.

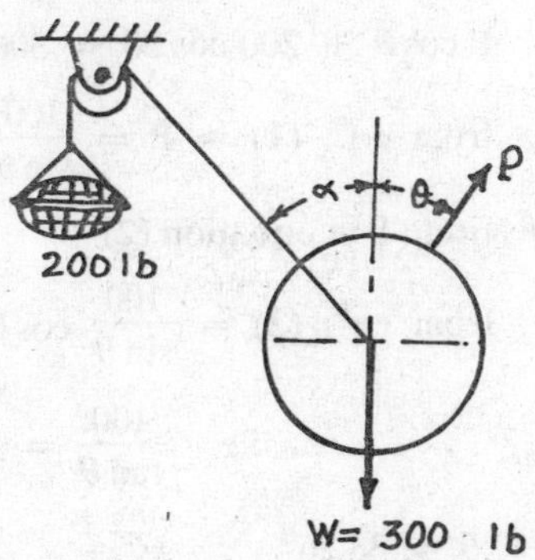

Solution : Draw the FBD showing all the forces as shown in fig.

Resolve the forces as shown, with the assumption $\uparrow Y \rightarrow X$ is positive.

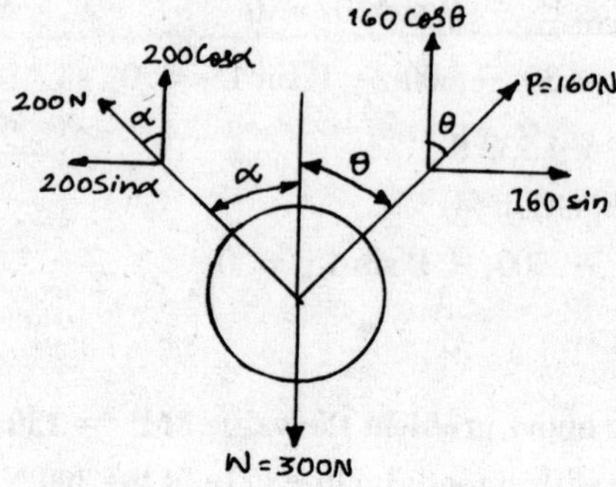

$\Sigma X = 160 \sin\theta - 200 \sin\alpha = 0$ ..... (1)

$\Sigma Y = 160 \cos\theta + 200 \cos\alpha - 300 = 0$ ..... (2)

from eq^n. (1) $\Rightarrow \sin\theta = 1.25 \sin\alpha$ ..... (3)

from eq^n. (2) $\Rightarrow \cos\theta = 1.875 - 1.25 \cos\alpha$ ..... (4)

Squaring (3) and (4) and adding we get

$\theta = 38.4^o$ & $\alpha = 29.54^o$

Problem 5. : In above problem if P is not known and $\alpha = 30^o$, compute P and θ.

Solution : Equations of equilibrium will be

$P \sin\theta - 200 \sin 30 = 0$ ......(1)

$P \cos\theta + 200 \cos 30 = 300$ ...... (2)

from eq^n. (1) $\Rightarrow P = \dfrac{100}{\sin\theta}$

Substitute P in equation (2)

from eq^n. (2) $\Rightarrow \dfrac{100}{\sin\theta} \cos\theta + 200 \cos 30 = 300$

i.e. $\dfrac{100}{\tan\theta} = 126.8$

$\therefore \theta = 38.26^o$

and from eq^n. (1) $\Rightarrow P = 161.5$ N .

Problem 6. : Cords are looped around a small spacer separating two cylinders each weighing 400 N and pass, as shown, over a frictionless pulley to weights 200 N and 600 N. Determine the angle θ and the normal reaction N between the cylinder and smooth horizontal surface.

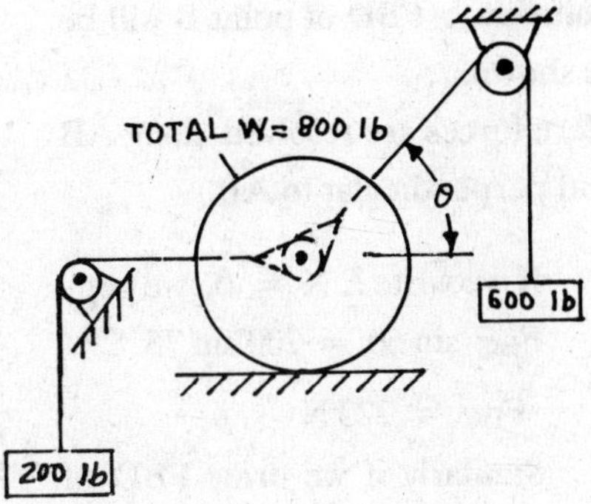

Solution : Draw the FBD

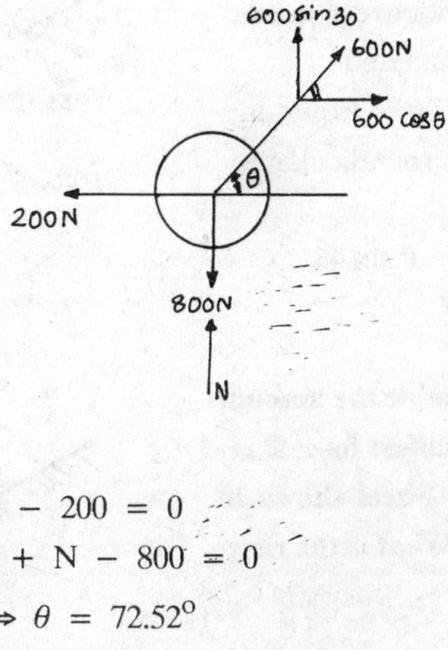

$$\Sigma X = 600 \cos\theta - 200 = 0 \quad \ldots\ldots (1)$$

$$\Sigma Y = 600 \sin\theta + N - 800 = 0 \quad \ldots\ldots (2)$$

from eq^n. (1) $\Rightarrow \theta = 72.52^o$

$\therefore$ from eq^n. (2) $\Rightarrow N = 234.32$ N

Problem 7. : Three bars, pinned together at B and C and supported by hinges at A and D as shown in fig. form a four–link mechanism. Determine the value of P that will prevent motion.

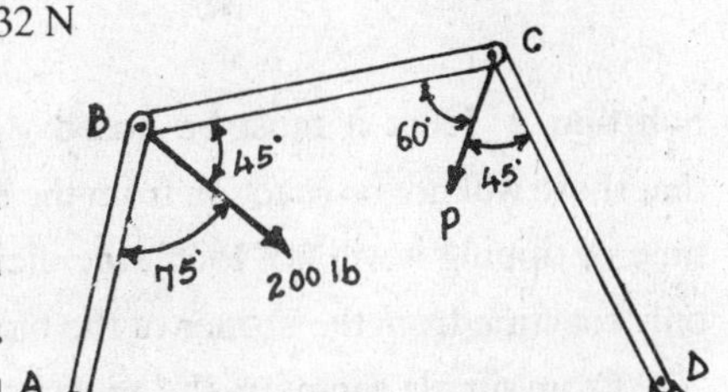

Solution : FBD of point B will be as shown.

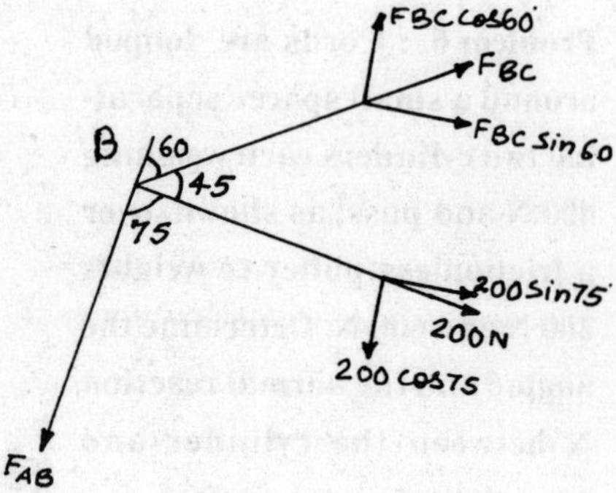

Here forces are resolved along AB and perpendicular to AB.

If we write $\Sigma X = 0$, we get

$F_{BC} \sin 60 = 200 \sin 75$

$\therefore \quad F_{BC} = 223$ N

Similarly if we draw FBD of point C.

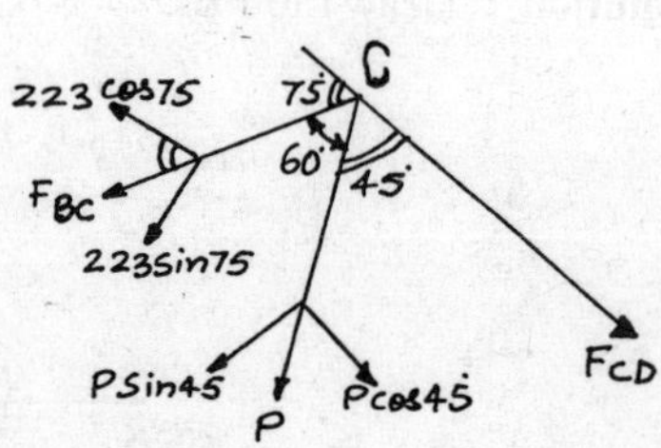

Here forces are resolved along link CD, for the obvious reason that one unknown is eliminated.

Again $\Sigma X = 0$ (Equating forces which are perpendicular to CD) we get,

$223 \sin 75 = - P \sin 45$

$\therefore \quad P = 304.62$ N

Problem 8.: Determine the amount and direction of smallest force P required to start the wheel shown in fig. over the block. What is the reaction at the block ?

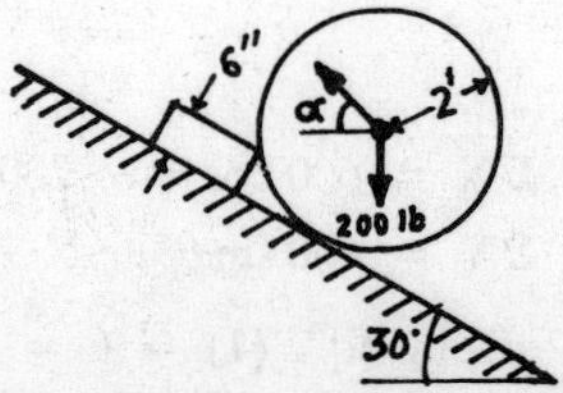

Solution : Here it must be noted that there will not be reaction from the inclined plane on the wheel since at the time of tipping it will not touch the inclined plane. While tipping it will have only reaction from the corner of the block.

From simple geometry the angle between reaction and the vertical, will be $(\theta + 30)$

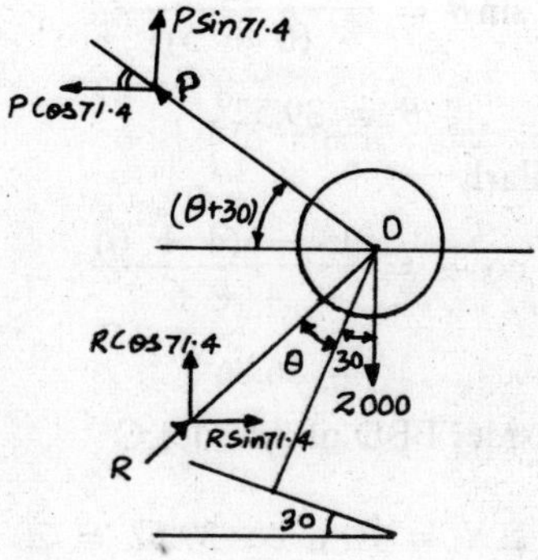

we have $\cos\theta = \left(\frac{2.0 - 0.5}{2}\right)$

$\theta = 41.4^o$

Also in the problem it is asked to find P_{min} for the wheel to overcome the step block. Now P will be minimum when it will be at right angle to reaction. Thus angle between P and horizontal will be$(41.4 + 30) = 71.4^o$.

$\Sigma X = R \sin 71.4 - P \cos 71.4 = 0$

$\therefore \quad R = 0.336 P$ .... (1)

$\Sigma Y = P \sin 71.4 + R \cos 71.4 = 2000$ ..... (2)

from eq^n. (1) & (2) ⇒ P = 1895.5 N & R = 636.9 N

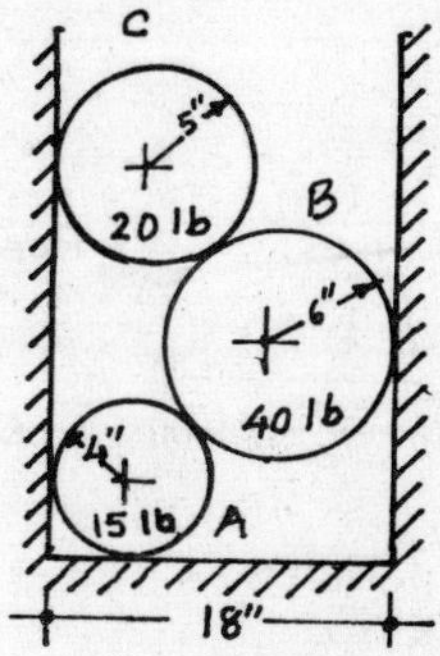

Problem 9.: Three cylinders are piled in a rectangular ditch as shown in fig. Neglecting friction determine the reaction between cylinder A and the vertical wall.

W_A = 15 N ;

R_A = 4 cm ;

W_B = 40 N ;

R_B = 6 cm ;

W_C = 20 N ;

R_C = 5 cm .

Solution : Considering equilibrium of all three cylinders in a ditch we have

$\Sigma X = R_C + R_{AH} = R_B$ ..... (1)

$\Sigma Y = W_A + W_B + W_C = R_{AV}$

$\therefore \quad R_{AV} = 75 \text{ N}$

Angle θ will be

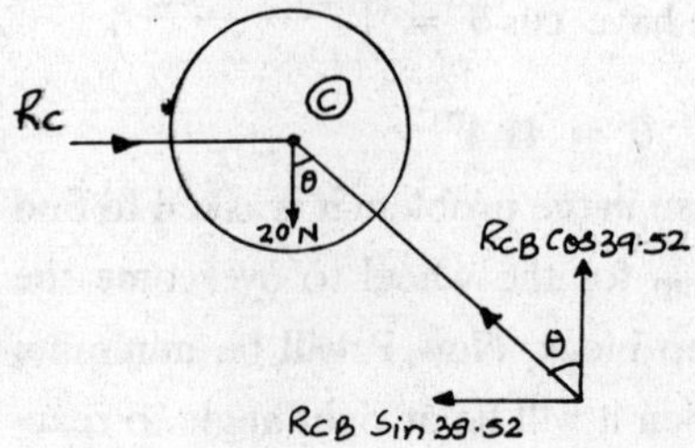

$$\sin\theta = \frac{18 - (6 + 5)}{(6 + 5)}$$

$$\therefore \quad \theta = 39.52^o$$

Similarly

$$\cos\alpha = \frac{18 - (4 + 6)}{4 + 6}$$

$$\therefore \quad \alpha = 36.86^o$$

Consider FBD of cylinder C

$$\Sigma Y = R_{CB} \cos 39.52 = 20$$

$$\Rightarrow R_{CB} = 25.92 \text{ N}$$

$$\Sigma X = R_C - R_{CB} \sin 39.52 = 0$$

$$\Rightarrow R_C = 16.5 \text{ N}$$

Consider FBD of cylinder B

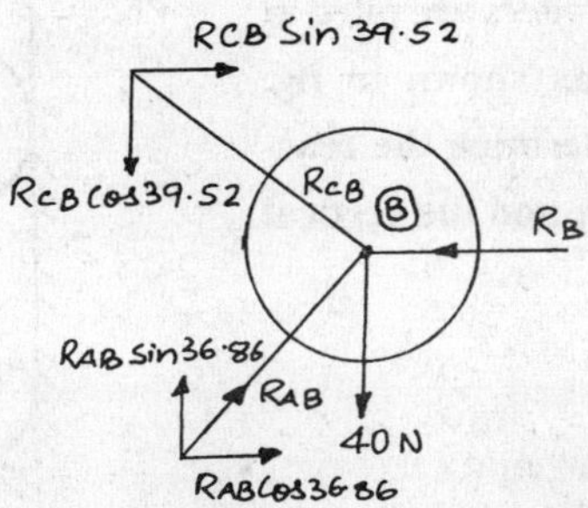

$$\Sigma Y = R_{AB} \sin 36.86 - R_{CB} \cos 39.52 - 40 = 0$$

$$R_{AB} \sin 36.86 - 25.92 \cos 39.52 = 40$$

$$\Rightarrow R_{AB} = 100 \text{ N}$$

$$\Sigma X = R_{AB} \cos 36.86 + R_{CB} \sin 39.52 - R_B = 0$$

$$100 \cos 36.86 + 25.92 \sin 39.52 - R_B = 0$$

from equation $R_B = 96.5$ N

from eq^n. (1) $\Rightarrow R_{AH} = R_B - R_C = 96.5 - 16.5 = 80$ N

Problem 10. : Determine the value of T if the force system in fig. is in equilibrium.

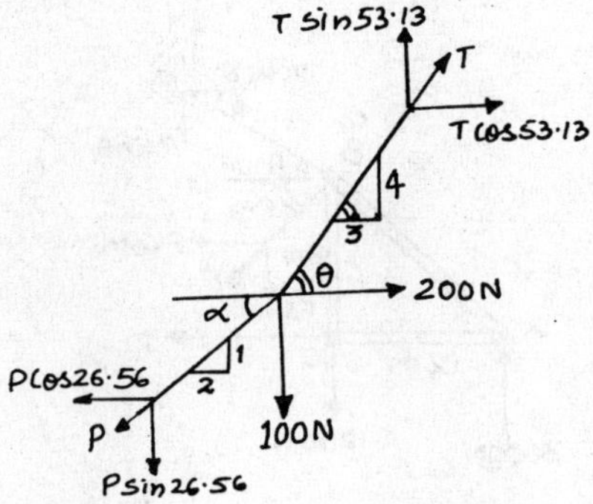

Solution : From fig., $\tan\theta = \frac{4}{3}$ $\therefore$ $\theta = 53.13^{\circ}$

Also $\tan\alpha = \frac{1}{2}$, $\therefore$ $\alpha = 26.56^{\circ}$

$\Sigma X = T\cos 53.13 + 200 - P\cos 26.56 = 0$ ..... (1)

$\Sigma Y = T\sin 53.13 - 100 - P\sin 26.56 = 0$ ..... (2)

Solving eq^n. (1) and (2), we get

T = 400 N and P = 492 N

Problem 11. : Determine the forces P, F and T required to keep the triangular frame ABC shown in fig. in equilibrium.

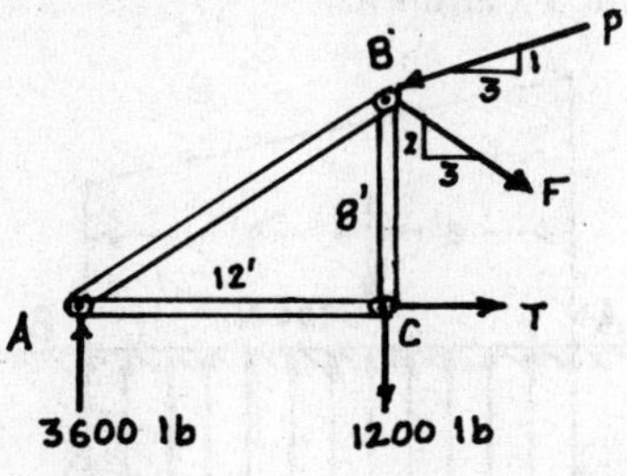

Solution : From fig., it is found $\theta = 18.43^{\circ}$ & $\alpha = 33.7^{\circ}$

Taking moment about B (since two unknowns are involved at this point).

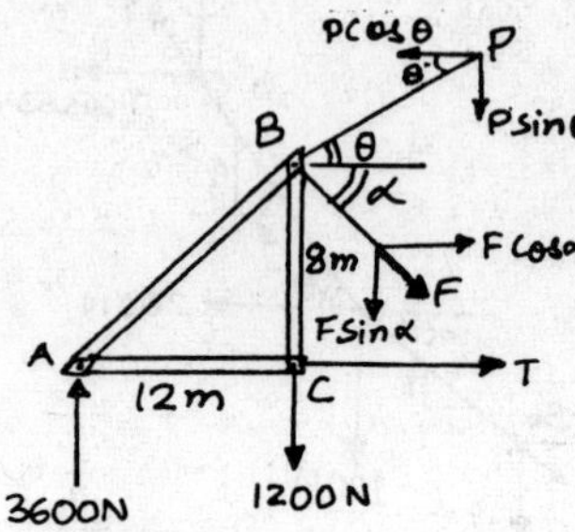

$$\Sigma M_B = T \times 8 = 3600 \times 12$$

$$\Rightarrow T = 5400 \text{ N}$$

$$\Sigma X = T + F\cos\alpha - P\cos\theta = 0$$

$$5400 + F\cos 33.7 - P\cos 18.43 = 0 \quad \ldots (1)$$

$$\Sigma Y = 3600 - 1200 - F\sin\alpha - P\sin\theta = 0$$

$$F\sin 33.7 + P\sin 18.43 = 2400 \quad \ldots (2)$$

Solving for F and P equations (1) and (2), we get

P = 6330 N and F = 722 N

Problem 12. : The weight of the trapezoidal block is 7200 N acting were shown in fig. The ground reaction varies uniformly from an intensity of P_A N/m at A to P_B N/m at B. Determine P_A and P_B.

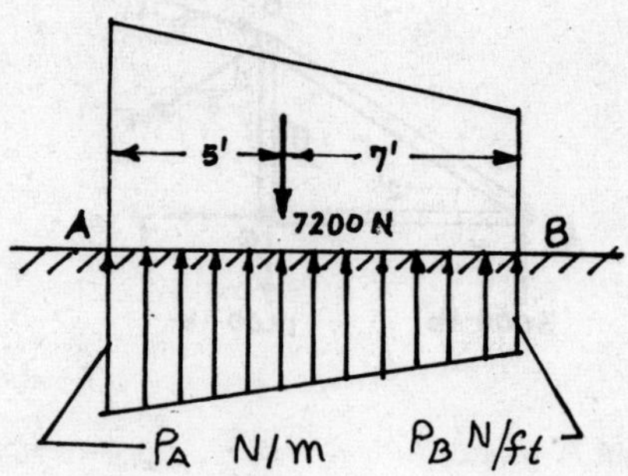

Solution : Dividing the pressure diagram into two triangles with bases P_A and P_B respectively.

$$\text{Total upward force} = \left(\frac{1}{2} \times P_A \times 12\right) + \left(\frac{1}{2} \times P_B \times 12\right)$$

$$= 6\,(P_A + P_B) = 7200$$

i.e. $P_A + P_B = 1200$ .....(1)

As the centroid of the triangle is at a distance of (1 / 3 altitude) from the base, the forces $6P_A$ and $6P_B$ will act at their respective centroids.

Taking moments about A :

$$\Sigma M_A = (6\,P_A \times 4) + (6\,P_B \times 8) = 7200 \times 5 = 36000$$

i.e. $P_A + 2P_B = 1500$ .....(2)

Solving equation (1) and (2), we get

$P_A = 900$ N/m

$P_B = 300$ N/m

Problem 13.: As shown in fig. the intensity of loading on a simply supported beam 10 m long is given by $y = 10\,x^3$ where y is in N / m and x is in meter, measured from A. Find the reactions at A and B.

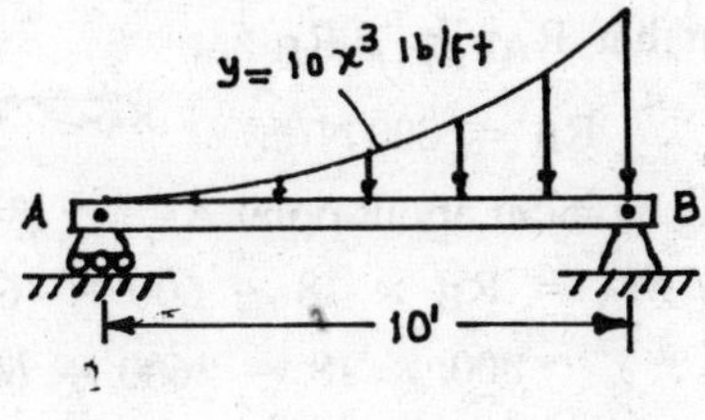

Solution : Total load $= \int_0^{10} 10\,x^3\,dx = 25000$ N

$\therefore \quad R_A + R_B = 25000$ ..... (1)

Centre of gravity of the load is at a distance X given by

$$X = \frac{\int_0^{10} 10\,x^4\,dx}{\int_0^{10} 10\,x^3\,dx} = \left(\frac{4 \times 10}{5}\right) = 8\text{ m}.$$

Taking moment about A we get

$$\Sigma M_A = 25000 \times 8 = R_B \times 10$$

i.e. $R_B = 20000$ N

$\therefore$ $R_A = 5000$ N .

Problem 14. : The wheel loads on a small tractor crossing on 18 m span are given in fig. Determine the distance x at which the support reaction at A is twice that of B.

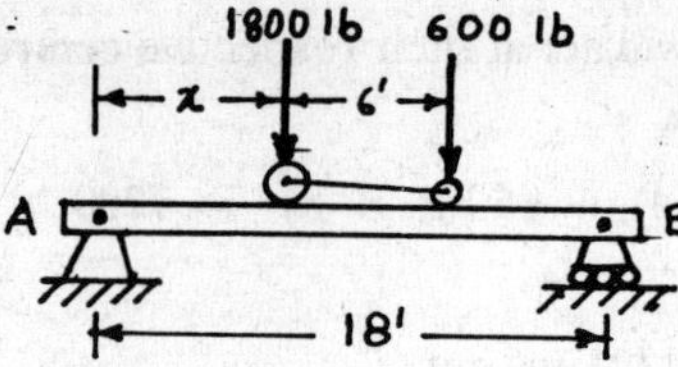

Solution : The FBD is shown below,

R_{AH} will be zero, since no other horizontal force is present.

$\Sigma Y = R_{AV} + R_B = 1800 + 600$ ..... (1)

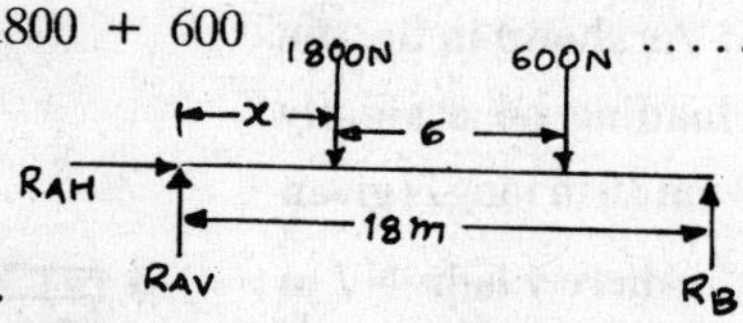

Given that $R_{AV} = 2 R_B$

$\therefore$ $R_B = 800$ N

Taking moment about point A.

$\Sigma M_A = R_B \times 18 - 600 \times (6 + x) - 1800 (x) = 0$

$800 \times 18 - 3600 - 600 x - 1800 x = 0$

$\therefore$ $x = 4.5$ m

Problem 15. : The differential chain hoist in fig. consists of two concentric pulleys rigidly fastened together. The pulleys form two sprockets for an endless chain looped over them in two loops. In one loop is mounted a movable pulley supporting a load W. Neglecting friction, determine the maximum load W that can just be raised by a pull P applied as shown.

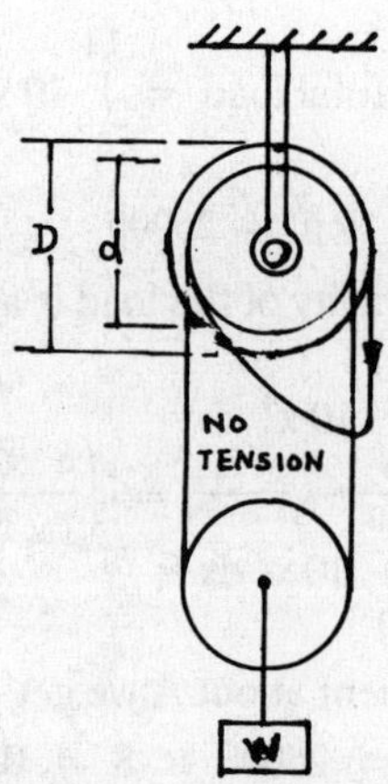

Solution : Pull in each segment of the chain supporting the movable pulley is

$$T = \frac{W}{2}$$

Consider the FBD of upper pulley.

Take moment about 'O'.

$$P\frac{D}{2} = \frac{W}{3}\left(\frac{D}{2} - \frac{d}{2}\right)$$

From which $W = \dfrac{2PD}{D - d}$

Problem 16.: If each pulley shown in fig. weighs 36 N and W = 720 N, find P to maintain equilibrium.

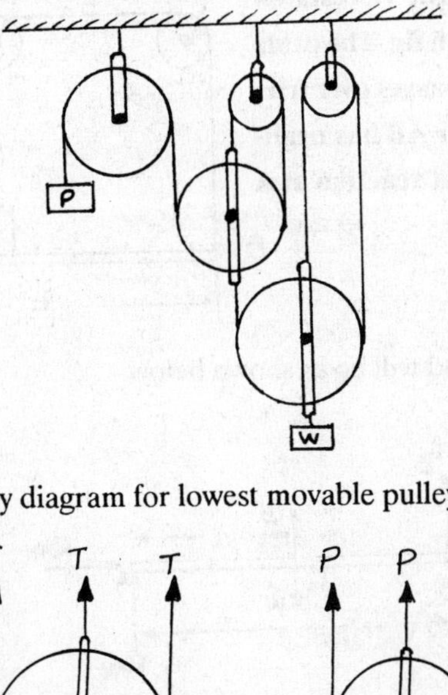

Solution : Free body diagram for lowest movable pulley is shown in fig.

$$3T = W + 36$$

i.e. $T = \frac{W}{3} + 12$ ......(1)

FBD for the next movable pulley is drawn adjacent in fig.

$$3P = T + 36$$

i.e. $P = \frac{T}{3} + 12$ ..... (2)

Substituting for T from (1)

$$P = \frac{W}{9} + 4 + 12 = \frac{W}{9} + 16$$

Problem 17. : A weight W rests on the bar AB as shown in fig. The cable connecting W and B passes over frictionless pulleys. If bar AB has negligible weight, show that reaction at A is $W\frac{(L - a)}{(L + a)}$.

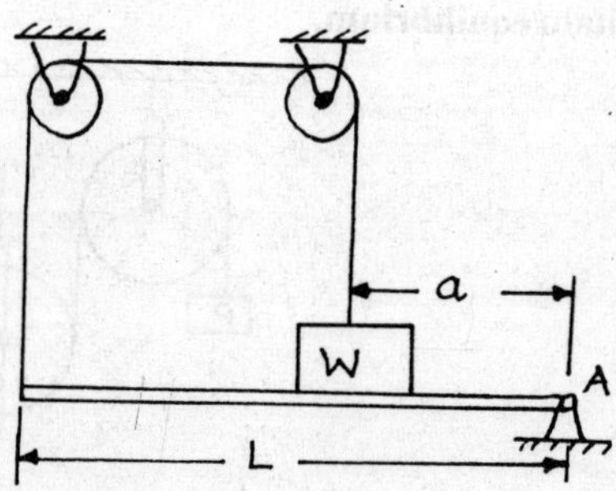

Solution : FBD of rod will be as shown below

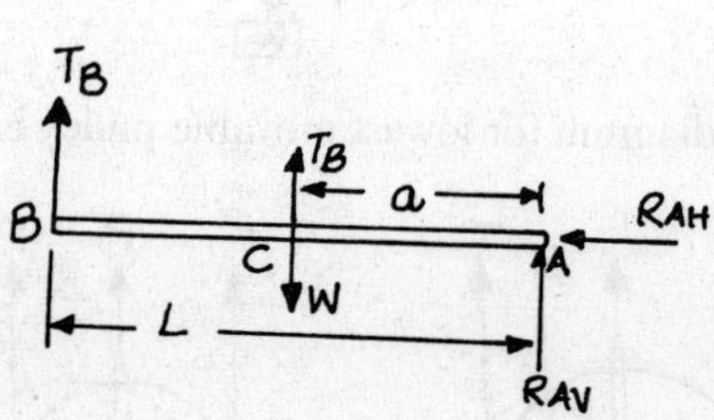

$$\Sigma M_A = 0 = -T_B \times L - (T_B - W) \times a = 0$$

$$\therefore \quad T_B = \frac{Wa}{(L + a)}$$

$$\Sigma M_B = 0 = (T_B - W) \times (L - a) + R_{AV} \times L = 0$$

$$\left(\frac{Wa}{L + a} - W\right) \times (L - a) + R_{AV} \times L = 0$$

$$R_{AV} \times L = -\left(\frac{Wa - WL - Wa}{(L + a)}\right) \times (L - a)$$

$$\therefore \quad R_{AV} = \frac{WL\ (L - a)}{(L + a) \times L} = W\frac{(L - a)}{(L + a)}.$$

Problem 18. : A boom AB is supported in a horizontal positon by a hing 'A' and a cable which runs from C over a small pulley at D as shown in fig. Compute the tension T in the cable and the horizontal and vertical components of the reaction at A. Neglect the weight of the boom and the size of the pulley at D.

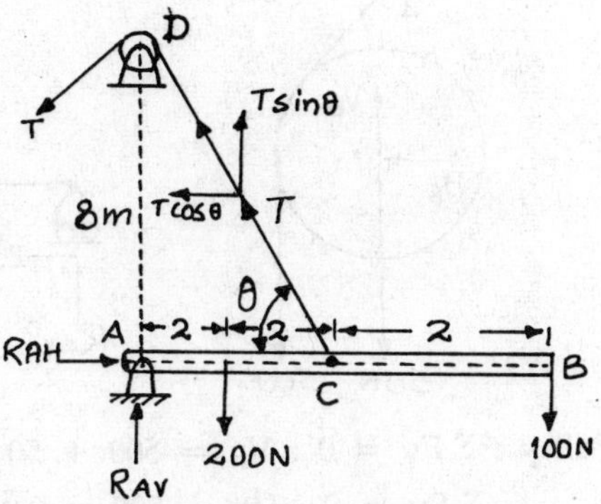

Solution : From FBD , the angle θ will be

$$\tan\theta = \frac{8}{4}, \quad \therefore \quad \theta = 63.43^{o}$$

$$\Sigma X = R_{AH} - T\cos\theta = 0 \qquad \ldots\ldots (1)$$

$$\Sigma Y = R_{AV} - 200 - 100 + T\sin\theta = 0 \qquad \ldots\ldots (2)$$

$$\Sigma M_A = T\sin\theta \times 4 - 200 \times 2 - 100 \times 6 = 0 \qquad \ldots\ldots (3)$$

$$\therefore \quad T = 279.5\ N$$

from eq^n. (2) $\Rightarrow$ $R_{AV} - 300 + 279.5 \times \sin 63.43 = 0$

$$\therefore \quad R_{AV} = 50\ N$$

from eq^n. (2) $\Rightarrow$ $R_{AH} = 125.14\ N$

Problem 19. : A pulley of 1ft. radius, supporting a load of 500 N is mounted at B on a horizontal beam as shown in fig. The beam weighs 200 N and the pulley weighs 50 N, find the hinge force at C.

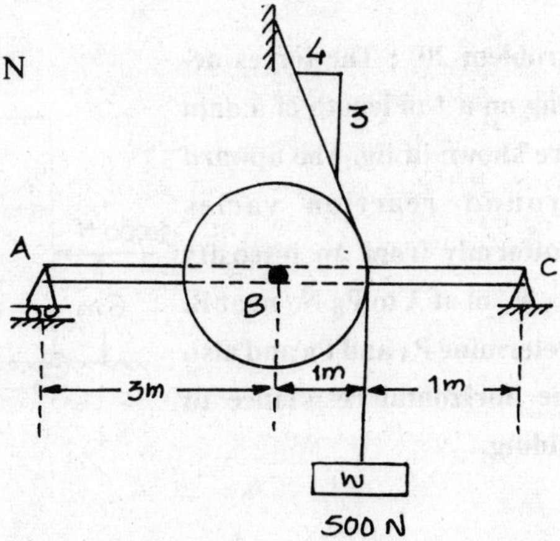

Solution : Free body diagram for pulley and for the beam is shown in fig. (a) & (b).

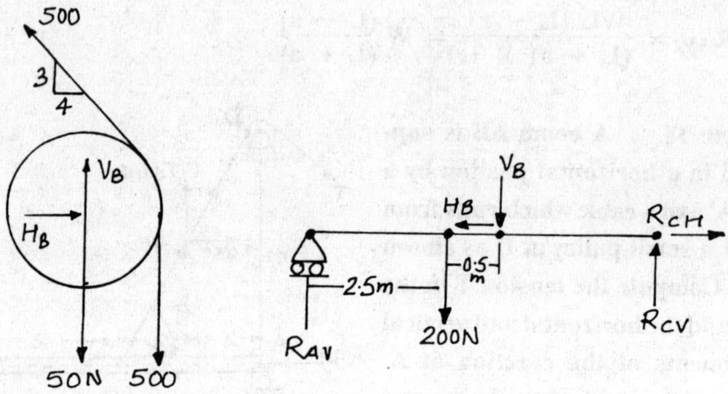

Pulley : $\Sigma F_V = 0 : V_B = 500 + 50 - (0.6 \times 500) = 250$ N

$\Sigma F_H = 0 : H_B = 500 \times 0.8 = 400$ N

Beam : $\Sigma F_H = 0$

$R_{CH} = 400$

$R_{CV} \times 5 = (250 \times 3) + (200 \times 2.5)$

i.e. $R_{CV} = 250$ N .

$R_{AV} = 250 + 200 - 250 = 200$ N .

$R_C = (400^2 + 250^2)^{1/2} = 471.69$ N

R_C inclined with horizontal at $\theta = \tan^{-1}\left(\frac{250}{400}\right) = 32^o$.

Problem 20. : The forces acting on a 1 m length of a dam are shown in fig. The upward ground reaction varies uniformly from an intensity P_A N / m at A to P_B N / m at B. Determine P_A and P_B and also the horizontal resistance to sliding.

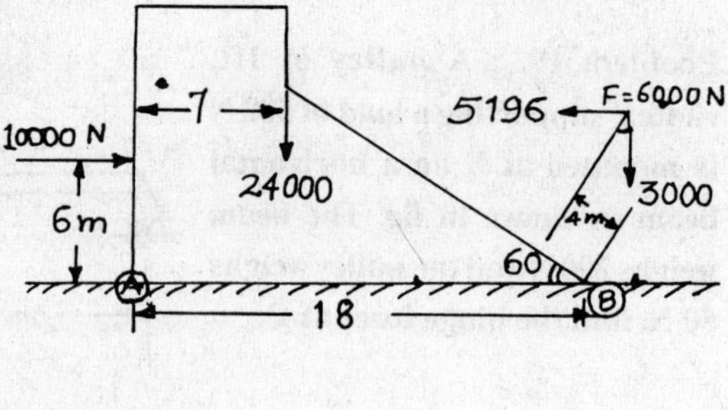

Solution : The FBD of diagram is shown below

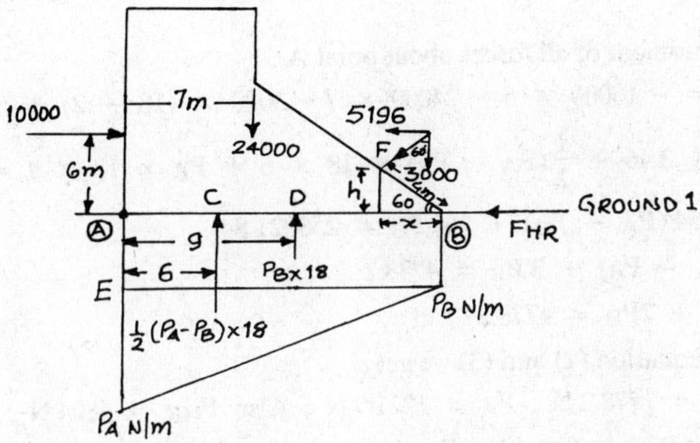

Force F = 6000 N is resolved into its two components. The distance x and h shown in fig. is calculated as

$$\sin 60 = \frac{h}{4} \Rightarrow h = 3.46 \text{ m}$$

$$\& \quad \cos 60 = \frac{x}{4} \Rightarrow x = 2 \text{ m}$$

Now P_A and P_B are the reactions exerted by the ground per unit length. Now we can divide the area AB P_B P_A EA into two parts,

(i) Rectangle AB P_B E : Area of this rectangle will represent a force of P_B (N/m) × 18 (m) and will act at the centroid of rectangle i.e. 9 m away from point A. It is shown in fig. by point D.

(ii) Triangle E P_B P_A. : Area of this triangle will represent a force of $\frac{1}{2} \times (P_A - P_B) \times 18$ and this will act at one third of the length from point A. i.e. 6 m away from point A. It is shown in fig. by point C.

Applying equilibrium equations,

$$\Sigma Y = \frac{1}{2}(P_A - P_B) \times 18 + P_B \times 18 - 24000 - 3000 = 0$$

$$9 P_A + 9 P_B = 27000$$

$$P_A + P_B = 3000 \qquad \ldots\ldots(1)$$

$$\Sigma X = 10000 - 5196 - F_{HR} = 0$$

Where F_{HR} is Horizontal resistance force

$$F_{HR} = 4804 \text{ N} \quad \text{.....(2)}$$

Taking moment of all forces about point A.

$$\Sigma M_A = -10000 \times 6 - 24000 \times 7 - 3000 \times (18 - 2) + 5196 \times 3.46 + \frac{1}{2}(P_A - P_B) \times 18 \times 6 + P_B \times 18 \times 9 = 0$$

$$\Rightarrow 54(P_A - P_B) + 162 P_B = 258021.84$$

$$(P_A - P_B) + 3P_B = 4778.2$$

$$P_A + 2P_B = 4778.2 \quad \text{..... (3)}$$

Solving equation (1) and (3) we get

$P_B = 1778.2$ N ; $P_A = 1221.9$ N ; Also $F_{HR} = 4804$ N

Note that P_B is greater than P_A, hence assumed triangle must be reversed.

Problem 21. : The Fink truss in fig. is supported by a hinge at A and by rollers on a 1 to 2 slope at B. The bracing bars are normal to the inclined members at their midpoints. The wind loads are perpendicular to AC. Determine the horizontal and vertical components of the reactions at A and B.

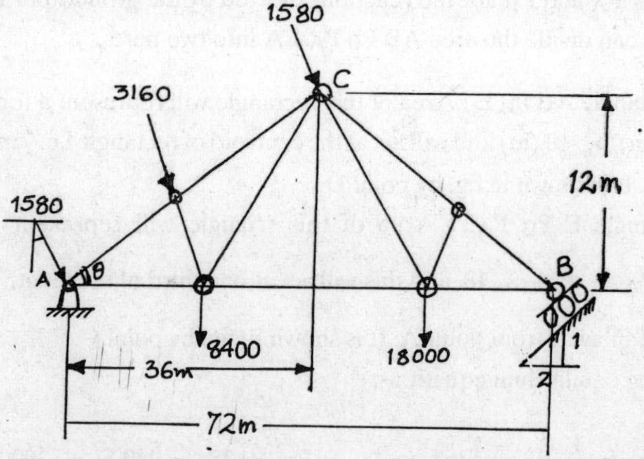

Solution : Angle θ & α will be given be

$$\theta = \tan^{-1} \frac{12}{36} = 18.43^\circ$$

and $\alpha = 26.56^{\circ}$

Applying equilibrium equations ,

$$\Sigma Y = R_{AV} - 1499 - 2998 - 1499 - 8400 - 18000 + R_B \cos 26.56 = 0 \quad \ldots\ldots (1)$$

$$\Sigma X = R_{AH} + 499 + 999 + 499 - R_B \sin 26.56 = 0 \quad \ldots\ldots (2)$$

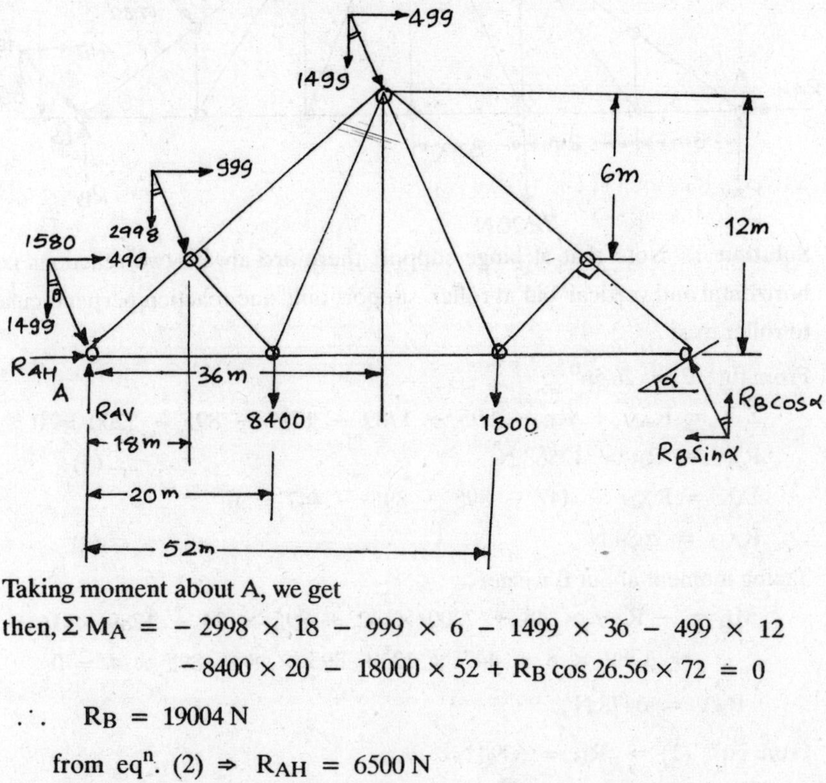

Taking moment about A, we get

$$\text{then, } \Sigma M_A = -2998 \times 18 - 999 \times 6 - 1499 \times 36 - 499 \times 12 - 8400 \times 20 - 18000 \times 52 + R_B \cos 26.56 \times 72 = 0$$

$\therefore \quad R_B = 19004$ N

from eq^n. (2) $\Rightarrow$ $R_{AH} = 6500$ N

from eq^n. (1) $\Rightarrow$ $R_{AV} = 15397$ N

Vertical reaction at B = 19004 cos 26.56 = 17000 N

Horizontal reaction at A = R_{AH} = 6500 N

Problem 22. : The roof truss show in fig. is supported by a hinge at A and a horizontal roller support at B. The wind loads are perpendicular to BC. Compute the total reactions at A and B.

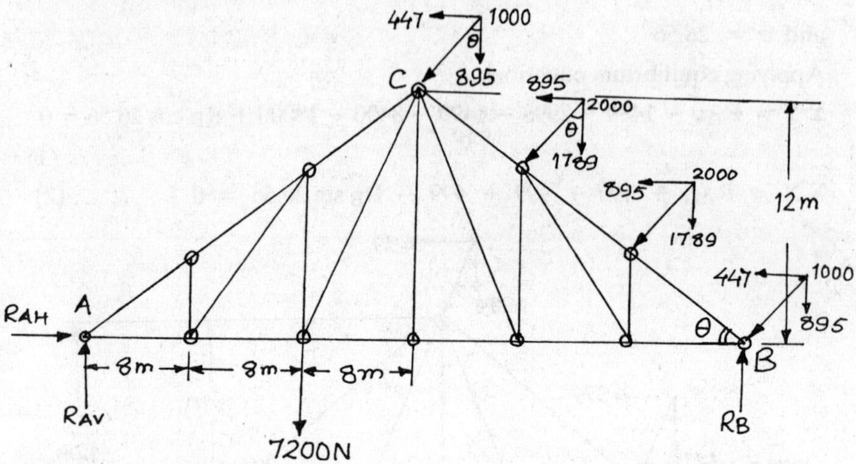

Solution : Note that at hinge support there are always two reactions i.e. horizontal and vertical but at roller support only one reaction perpendicular to roller exist .

From fig., $\theta = 26.56^\circ$.

$$\Sigma Y = R_{AV} + R_B - 895 - 1789 - 1789 - 895 - 7200 = 0$$

$$R_{AV} + R_B = 12568 \text{ N} \qquad \ldots\ldots (1)$$

$$\Sigma X = R_{AH} - 447 - 895 - 895 - 447 = 0$$

$$R_{AH} = 2684 \text{ N} \qquad \ldots\ldots (2)$$

Taking moment about B we get ,

$$\Sigma M_B = - R_{AV} \times 48 + 7200 \times 32 + 895 \times 24 + 1789 \times 16$$
$$+ 1789 \times 8 + 447 \times 12 + 895 \times 8 + 895 \times 4 = 0$$

$$\therefore \quad R_{AV} = 6478 \text{ N},$$

From eq^n. (1) $\Rightarrow$ $R_B = 6090$ N

Problem 23. : A 12 m bar of negligible weight rests in a horizontal position on the smooth incline as shown in fig. Compute the distance x at which the load T = 100 N should be placed from point B to keep the bar horizontal.

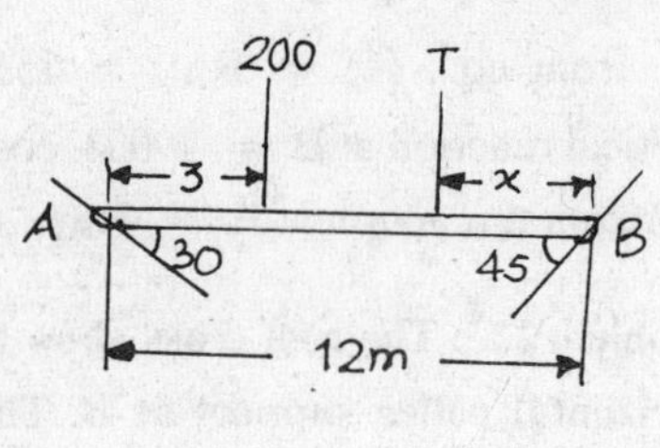

Solution : FBD of bar will be

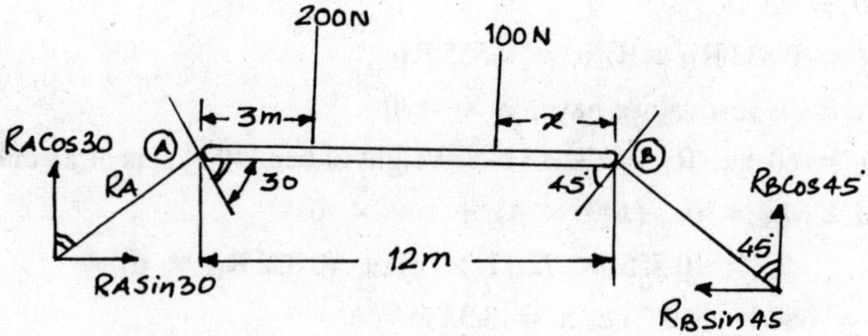

$$\Sigma X = R_A \sin 30 - R_B \sin 45 = 0 \quad \ldots\ldots (1)$$

$$\Sigma Y = R_A \cos 30 - 200 - 100 + R_B \cos 45 = 0$$

i.e. $R_A \cos 30 + R_B \cos 45 = 300 \quad \ldots\ldots (2)$

Taking moment about A we get (↻ positive)

$$\Sigma M_A = -200 \times 3 - 100(12 - x) + R_B \cos 45 \times 12 = 0$$

i.e. $R_B \cos 45 \times 12 + 100 \times x = 1800 \quad \ldots\ldots (3)$

Solve (1) and (2) to get

$R_A = 219.6$ N , $R_B = 155.3$ N

Using these values in eq^n. (3) , we get

$x = 4.82$ m .

Problem 24. : Find the distance x (measured along AB) at which a horizontal force of 60 N should be applied to hold the uniform bar AB in the position shown in fig. Bar AB is 10 m long and weights 140 N. The incline and the floor are smooth.

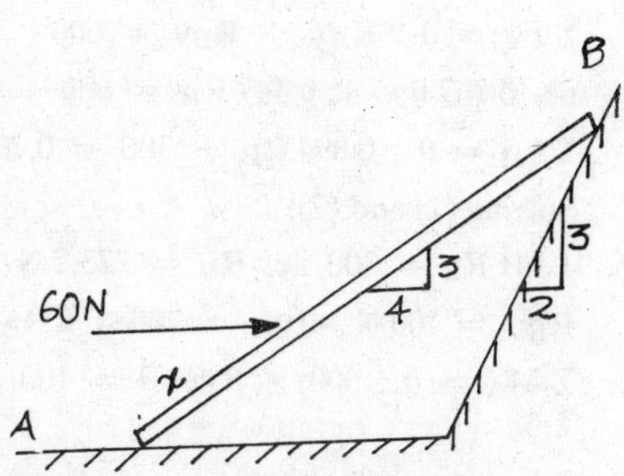

Solution : R_A and R_B are normal to respective planes.

R_B makes an angle of $\tan^{-1}\left(\frac{3}{2}\right)$ with vertical

i.e. $\theta = 56.31^o$

$R_{BH} = 0{:}832\ R_B$; $R_{BV} = 0.555\ R_B$;

Since R_A is vertical we have, $\Sigma X = 0$

$R_{BH} = 60$ i.e. $R_B = 72.11$ N . Weight of bar 140 N acts at its mid point.

Take $\Sigma M_A = 0$; $(140 \times 4) + (60 \times 6x)$

$$= (0.555 \times 72.11 \times 8) + (0.832\ R_B \times 6)$$

$36x = 680 - 560$ i.e. $x = 3.33$ m

Problem 25. : Bar AB of negligible weight is subjected to a vertical force of 600N and a horizontal force of 300 N applied as shown in fig. Find the angle θ at which equilibrium exits. Assume smooth inclined surfaces.

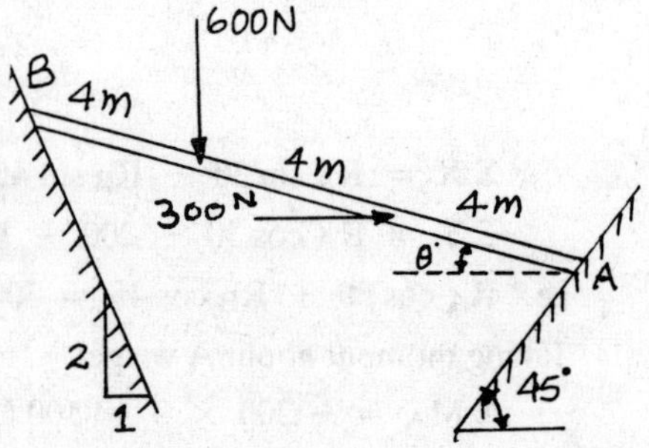

Solution : $R_{BV} = \frac{1}{\sqrt{5}} R_B = 0.447\ R_B$

$R_{BH} = 0.894\ R_B$; $R_{AH} = R_{AV} = 0.707\ R_A$.

$\Sigma F_V = 0 : R_{AV} + R_{BV} = 600$

i.e. $0.707\ R_A + 0.447\ R_B = 600$ .....(1)

$\Sigma F_H = 0$; $0.894\ R_B + 300 = 0.707\ R_A$ .....(2)

Solving (1) and (2),

$1.341\ R_B = 300$ i.e. $R_B = 223.7$ N

$R_{BV} = 100$ N , $R_{BH} = 200$ N ; $R_{AV} = 500\text{ N} = R_{AH}$

$\Sigma M_A = 0 : 600 \times 8 \cos\theta = 100 \times 12 \cos\theta + 200 \times 12 \sin\theta + 300 \times 4 \sin\theta$

which gives $\sin\theta = \cos\theta$ i.e. $\theta = 45^o$.

Problem 26. : A concurrent spatial system consists of three forces $P = P_m(3\hat{i} - 2\hat{j} + 4\hat{k})$, $Q = Q_m(2\hat{i} + \hat{j} + 3\hat{k})$ $T = T_m(4\hat{i} + \hat{j} - 2\hat{k})$ held in equilibrium by a fourth force $R = -400\hat{j}$ lb. Find values of P_m, Q_m & T_m.

Solution : Given $P = 3P_m i - 2P_m j + 4P_m k$

$$Q = 2Q_m i + Q_m j + 3Q_m k$$
$$T = 4T_m i + T_m j - 2T_m k$$
$$R = -400 j$$

For equilibrium, $\Sigma F = 0$

i.e. $\Sigma F_x = 3P_m + 2Q_m + 4T_m = 0$ ..... (1)

$\Sigma F_y = -2P_m + Q_m + T_m - 400 = 0$ ..... (2)

$\Sigma F_z = 4P_m + 3Q_m - 2T_m = 0$ ..... (3)

Solving equation (1), (2) and (3) we get

$P_m = -116.3636$, $Q_m = 160$, $T_m = 7.2727$

Problem 27. : The shear–leg derrick shown in fig. supports a vertical load of 3000 N applied at A. Points B, C, and D are in the same horizontal plane and A, O, and D are in the XY plane. Determine the force in each member of the derrick.

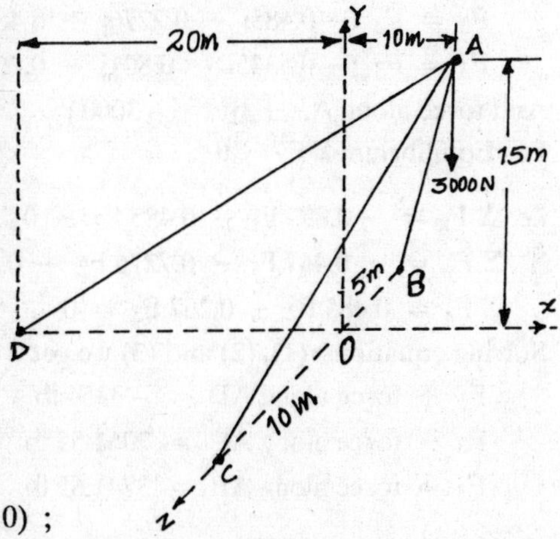

Solution : Co–ordinates of points in figure are A (10, 15, 0) ;
B (0, 0, −5) ; C (0, 0, 10) ; D (− 20, 0, 0) ; E (10, 0, 0)

Let F_1 be the force acting along AD

F_2 be the force acting along AC

F_3 be the force acting along AB

Let us first find unit vectors along AD, AC and AB

$$\hat{n}_{AD} = \frac{(-20 - 10)\,i + (0 - 15)\,j}{[(-30)^2 + (-15)^2]^{1/2}}$$
$$= -0.894\,i - 0.447\,j$$

$$\hat{n}_{AC} = \frac{(0 - 10)\,i + (0 - 15)\,j + (10 - 0)\,k}{[(-10)^2 + (-15)^2 + (10)^2]^{1/2}}$$
$$= -0.485i - 0.7276j + 0.485k$$

$$\hat{n}_{AB} = \frac{(0 - 10)i + (0 - 15)\,j + (-5 + 0)\,k}{[(-10)^2 + (-15)^2 + (-5)^2]^{1/2}}$$
$$= -0.5345i - 0.801j - 0.267k$$

$$\hat{n}_{AE} = \frac{(10 - 10)\,i + (0 - 15)j}{[15^2]^{1/2}} = -j$$

$\therefore$ Forces F_1, F_2 & F_3 are given as

$\overline{F}_1 = F_1\,(-0.897i - 0.447j)$

$\overline{F}_2 = F_2\,(-0.485i - 0.7276j + 0.485k)$

$\overline{F}_3 = F_3\,(-0.5345i - 0.801j - 0.267k)$

And force along A, $F_{AE} = -3000\,j$

For Equilibrium $\Sigma F = 0$

i.e. $\Sigma F_x = -0.897\,F_1 - 0.485\,F_2 - 0.5345\,F_3 = 0$ .... (1)

$\Sigma F_y = -0.447\,F_1 - 0.7276\,F_2 - 0.801\,F_3 - 3000 = 0$ .... (2)

$\Sigma F_z = 0.485\,F_2 - 0.267\,F_3 = 0$ ..... (3)

Solving equations (1), (2) and (3) we get

$F_1 \rightarrow$ force along AD $= -3354$ lb

$F_2 \rightarrow$ force along AC $= 2061.55$ lb

$F_3 \rightarrow$ force along AB $= 3741.65$ lb

Problem 28. : A vertical load of 1100 lb is supported by the three bars shown in fig. Find the force in each bar. Points C, O, and D are in the XZ plane while B is 5 ft above this plane.

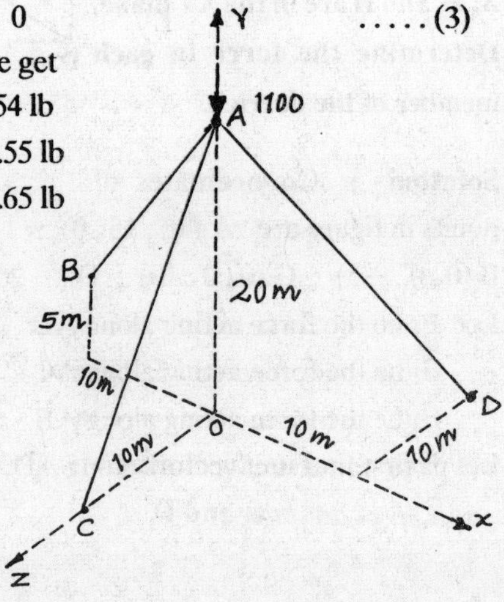

Solution : Co–ordinates of points in figure are

A (0 , 20 , 0) ; B (– 10 , 5 , 0) ; C (0 , 0 , 10) ; D (10 , 0 , – 10)

$$\overline{AB} = -10i - 15j$$

$$\therefore \hat{n}_{AB} = \frac{-10i - 15j}{[10^2 + 15^2]^{1/2}} = -0.5547i - 0.832j$$

$$\overline{AC} = -20j + 10k$$

$$\hat{n}_{AC} = \frac{-20j + 10k}{[20^2 + 10^2]^{1/2}} = -0.8944j + 0.4472k$$

$$\overline{AD} = 10i - 20j - 10k$$

$$\hat{n}_{AD} = \frac{10i - 20j - 10k}{[10^2 + 20^2 + 10^2]^{1/2}} = 0.408i - 0.8165j - 0.408k$$

$$\therefore \overline{F}_{AB} = F_{AB}(-0.5547i - 0.832j)$$

$$\overline{F}_{AC} = F_{AC}(-0.8944j + 0.4472k)$$

$$\overline{F}_{AD} = F_{AD}(0.408i - 0.8165j - 0.408k)$$

$$\overline{P} = +1100j$$

For Equilibrium $\Sigma F = 0$

$$\therefore -0.5547 F_{AB} + 0.408 F_{AD} = 0 \quad \ldots\ldots (1)$$

$$-0.832 F_{AB} - 0.8944 F_{AC} - 0.8165 F_{AD} + 1100 = 0 \quad \ldots\ldots (2)$$

$$0.4472 F_{AC} - 0.408 F_{AD} = 0 \quad \ldots\ldots (3)$$

Solving equations (1), (2) and (3) we get

F_{AB} = 361 lb , F_{AC} = 448 lb , F_{AD} = 490 lb .

Problem 29. : A vertical load P = 800 lb applied to the tripod shown in fig. causes a compressive force of 256 lb in leg AB and a compressive force of 283 lb in leg AC. Determine the force in leg AD and the coordinates x_D and z_D of its lower end D.

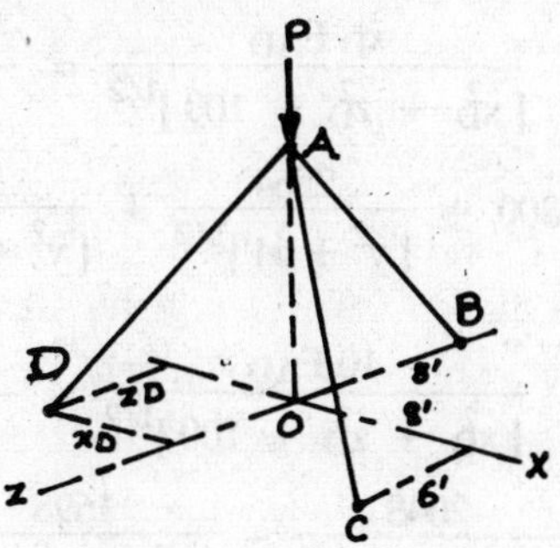

Solution : Co–ordinates of points in figure are

A (0 , y , 0) ; B (0 , 0 −8) ; C (8 , 0 , 6) ; D (− x_D , 0 , z_D)

$$\overline{AB} = -yj - 8k \; ; \quad \therefore \; \hat{n}_{AB} = \frac{-yj - 8k}{[y^2 + 64]^{1/2}}$$

$$\overline{AC} = 8i - yj + 6k ; \quad \therefore \; \hat{n}_{AC} = \frac{8i - yj + 6k}{[y^2 + 100]^{1/2}}$$

$$\overline{AD} = -x_D - yj + z_D k \; ; \therefore \; \hat{n}_{AD} = \frac{-x_D i - yj + z_D k}{[x_D^2 + y^2 + z_D^2]^{1/2}}$$

$$\overline{AO} = -yj$$

$$\therefore \; F_{AB} = \frac{-256(-yj - 8k)}{[y^2 + 64]^{1/2}}$$

$$F_{AC} = \frac{-283(8i - yj + 6k)}{[y^2 + 100]^{1/2}}$$

$$F_{AD} = \frac{F_{AD}(-x_D i - yj + z_D k)}{[x_D^2 + y^2 + z_D^2]^{1/2}}$$

$$F = -800j$$

For Equilibrium $\Sigma F = 0$

$$\therefore \; \frac{-x_D F_{AD}}{[x_D^2 + y^2 + z_D^2]^{1/2}} - \frac{2264}{[y^2 + 100]^{1/2}} = 0$$

$$\Rightarrow \; \frac{x_D F_{AD}}{[x_D^2 + y^2 + z_D^2]^{1/2}} = \frac{-2264}{[y^2 + 100]^{1/2}}$$

$$\Rightarrow \; \frac{x_D F_{AD}}{[x_D^2 + z_D^2 + 100]^{1/2}} = -160 \quad \ldots\ldots(1)$$

$$-800 + \frac{2564}{[y^2 + 64]^{1/2}} + \frac{2834}{[y^2 + 100]^{1/2}} - \frac{y F_{AD}}{[x_D^2 + y^2 + z_D^2]^{1/2}} = 0$$

$$\Rightarrow \; \frac{10 F_{AD}}{[x_D^2 + z_D^2 + 100]^{1/2}} = -400 \quad \ldots\ldots(2)$$

$$\frac{2048}{[y^2 + 64]^{1/2}} - \frac{1698}{[y^2 + 100]^{1/2}} + \frac{Z_D F_{AD}}{[x_{sup}2 + y^2 + Z_D^2]^{1/2}} = 0$$

$$\Rightarrow \frac{Z_D F_{AD}}{[x_D^2 + Z_D^2 + 100]^{1/2}} = -40 \qquad \ldots\ldots (3)$$

From equation (1), (2) and (3)

$x_D = 4$ ft. , $z_D = 1$ ft. & $F_{AD} = -432.67$ lb

Problem 30. : Determine the maximum safe vertical load W that can be supported by the tripod shown in fig. without exceeding a compressive load of 2400 lb in any member.

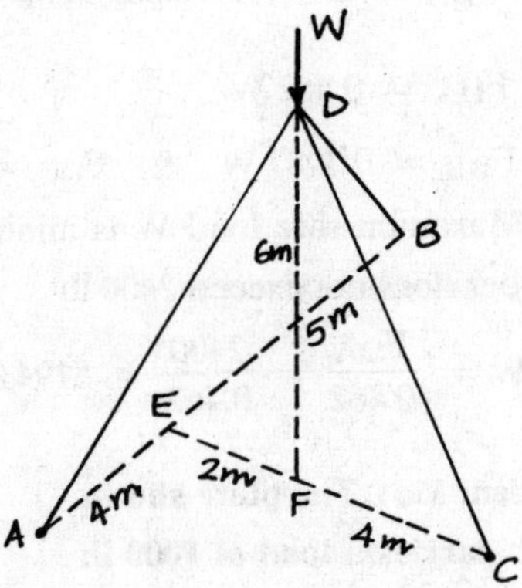

Solution : Taking F as origin, co-ordinates of points in fig. are

A (−2, 0, 4) ; B (−2, 0, −5) ;

C (4, 0, 0) ; D (0, 6, 0)

$\therefore \; \overline{DA} = -2i - 6j + 4k$

$\overline{DB} = -2i - 6j - 5k$

$\overline{DC} = 4i - 6j$

$\therefore \; \hat{n}_{DA} = -0.2672i - 0.8j + 0.534k$

$\hat{n}_{DB} = -0.248i - 0.744j - 0.620k$

$\hat{n}_{DC} = 0.554i - 0.832j$

$\therefore \; \bar{F}_{DA} = -F_{DA}(-0.2672i - 0.8j + 0.5345k)$

$\bar{F}_{DB} = -F_{DB}(-0.248i - 0.744j - 0.620k)$

$\bar{F}_{DC} = -F_{DC}(0.554i - 0.832j)$

$F_{DF} = -Wj$

For Equilibrium $\Sigma F = 0$

$\therefore \; 0.2672 F_{DA} + 0.248 F_{DB} - 0.554 F_{DC} = 0 \qquad \ldots (1)$

$0.8 F_{DA} + 0.744 F_{DB} + 0.832 F_{DC} - W = 0 \qquad \ldots\ldots (2)$

$-0.5345 F_{DA} + 0.620 F_{DB} = 0 \qquad \ldots\ldots (3)$

$\therefore \; F_{DA} = 1.159 F_{DB}$ or $F_{DB} = 0.863 F_{DA} \qquad \ldots\ldots (4)$

Substituting above relations in equation (1)

$- 0.31 F_{DB} - 0.248 F_{DB} + 0.554 F_{DC} = 0$

$\therefore \quad F_{DC} = 1.006 F_{DB}$

$= 1.006 \times 0.863 F_{DA}$

$F_{DC} = 0.868 F_{DA}$ .... (5)

From equation (2), (4) and (5)

$- 0.8 F_{DA} - 0.744 \times 0.863 F_{DA} - 0.832 \times 0.868 F_{DA} + W = 0$

$\Rightarrow F_{DA} = 0.462 W$

$\therefore \quad F_{DB} = 0.3987 W \quad \& \quad F_{DC} = 0.401 W$

Maximum safe load W is minimum value of W for which force in any member does not exceed 2400 lb

$\therefore \quad W = \frac{F_{DA}}{0.462} = \frac{2400}{0.462} = 5194.8 \text{ lb.}$

Problem 31. : The plate shown in fig. carries a load of 1000 lb applied at E and is supported in a horizontal position by three vertical cables attached at A, B and C. Compute the tension in each cable.

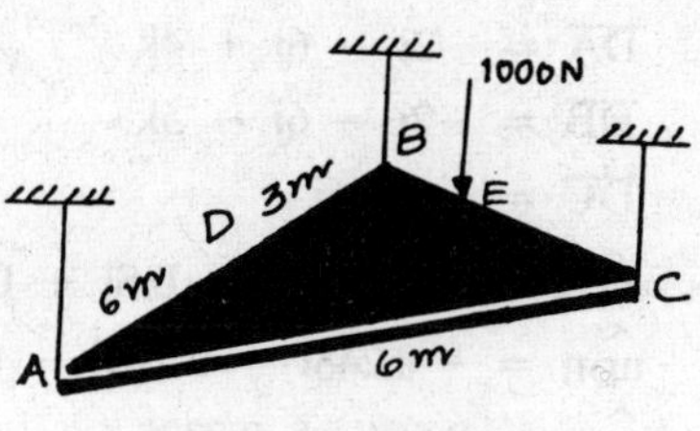

Solution : Let T_1, T_2, T_3 be the tensions in cables A, B, C, respectively.

From figure $T_1 + T_2 + T_3 = 1000$ ..... (1)

Now taking moment of all forces about BA

$\Sigma M_{BA} = T_3 \times 10 - 1000 \times 4 = 0$

$\therefore \quad T_3 = 400 \text{ lb}$

Now for taking moment of all forces about BC we need to know perpendicular distance of point A and E from BC.

$\angle BCD = \tan^{-1}\left(\frac{3}{10}\right) = 16.7^\circ$

$\therefore \quad d(EF) = 6 \sin(16.7) = 1.724$

$\angle ABC = 90 - 16.7 = 73.3^\circ$

$d(AG) = 9 \sin(73.3) = 8.62$

Taking moment of all forces about BC

$\Sigma M_{BC} = T_1 \times 8.62 - 1000 \times 1.724 = 0$

$\therefore \; T_1 = 200$ lb

Now substituting values of T_1 & T_3 in equation (1,) we get

$T_2 = 400$ lb

Problem 32. : A vertical mast AE is supported in a ball–and–socket joint at A and by cables CE and BD as shown in fig. Find the components of the reaction at A.

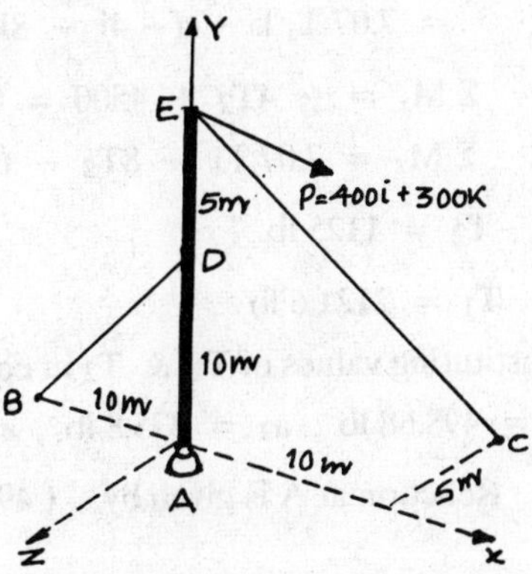

Solution : Taking A as origin co–ordinates of points in fig. are

A (0 , 0 , 0) ; B (– 10 , 0 , 0) ;

C (10 , 0 , –5) ; D (0 , 10 , 0) ; E (0 , 15 , 0)

Let T_1 and T_2 be tensions in cables BD and EC respectively.

$\overline{DB} = -10i - 10j \; ; \quad \therefore \; \hat{n}_{DB} = 0.707i - 0.707j$

$\overline{EC} = 10i - 15j - 5k \; ; \quad \therefore \; \hat{n}_{EC} = 0.5345i - 0.8j - 0.267k$

$\therefore \; T_1 = T_1(-0.707i - 0.707j)$

$T_2 = T_2(0.5345i - 0.8j - 0.267k)$

Let reaction at $A = a_1 i + a_2 j + a_3 k$

For Equilibrium

$\therefore \; -0.707\,T_1 + 0.5345\,T_2 + 400 + a_1 = 0 \quad \ldots\ldots (1)$

$-0.707\,T_1 - 0.8\,T_2 + a_2 = 0 \quad \ldots\ldots (2)$

$-0.267\,T_2 + 300 + a_3 = 0 \quad \ldots\ldots (3)$

Now taking moment of all forces about A

$$\Sigma M_A = T_1 \begin{vmatrix} i & j & k \\ 0 & 10 & 0 \\ -0.707 & -0.707 & 0 \end{vmatrix}$$

$$+ T_2 \begin{vmatrix} i & j & k \\ 0 & 15 & 0 \\ 0.5345 & -0.8 & -0.267 \end{vmatrix} + \begin{vmatrix} i & j & k \\ 0 & 15 & 0 \\ 400 & 0 & 300 \end{vmatrix}$$

$$= 7.07\,T_1\,k + (-4i - 8k)\,T_2 + 4500i - 6000k = 0$$

$\therefore \quad \Sigma M_x = -4T_2 + 4500 = 0$

$\Sigma M_z = 7.07\,T_1 - 8T_2 - 6000 = 0$

$\therefore \quad T_2 = 1125$ lb

$\therefore \quad T_1 = 2121.6$ lb

Substituting values of T_1 & T_2 in equations (1), (2) and (3) we get

$a_1 = 498.68$ lb , $a_2 = 375.8$ lb , $a_3 = 0$ lb

$\therefore$ Reaction at A is given by $(498.68\,i + 375.8\,j)$

Problem 33. : A mast AD is supported in a ball–and–socket joint at A and by two guy wires BD and CD as shown in fig. Find the magnitude of the force at A if F = 600 $\hat{i}$ lb & P = 400 $\hat{k}$ lb.

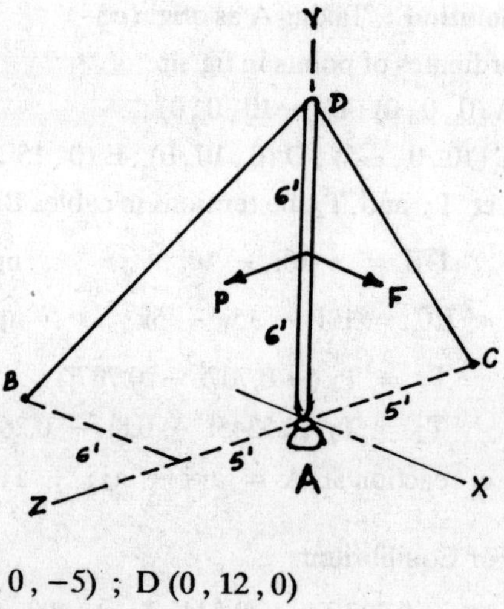

Solution : Taking A as origin, co–ordinates of points in fig. are

A (0 , 0 , 0) ; B (−6 , 0 , 5) ; C (0 , 0 , −5) ; D (0 , 12 , 0)

$\overline{DB} = -6i - 12j + 5k$

$\therefore \quad \hat{n}_{DB} = -0.419i - 0.838j + 0.349k$

$\overline{DC} = -12j - 5k$

$\therefore \quad \hat{n}_{DC} = -0.923j - 0.3846k$

Let T_1 and T_2 be tensions in DB and DC respectively

$\therefore$ $\mathbf{T_1} = T_1\,(-0.419\mathbf{i} - 0.838\mathbf{j} + 0.349\mathbf{k})$ ..... (1)

$\mathbf{T_2} = T_2\,(-0.923\mathbf{j} - 0.3846\mathbf{k})$ ...... (2)

Taking moment of all forces about A

$$\Sigma M_A = T_1 \begin{vmatrix} i & j & k \\ 0 & 12 & 0 \\ -0.419 & -0.838 & 0.349 \end{vmatrix}$$

$$+ T_2 \begin{vmatrix} i & j & k \\ 0 & 12 & 0 \\ 0 & -0.923 & -0.3846 \end{vmatrix} + \begin{vmatrix} i & j & k \\ 0 & 6 & 0 \\ 600 & 0 & 0 \end{vmatrix} + \begin{vmatrix} i & j & k \\ 0 & 6 & 0 \\ 0 & 0 & 400 \end{vmatrix}$$

$$= (4.188\,T_1 - 4.6152\,T_2 + 2400)\,\mathbf{i} + (5.028\,T_1 - 3600)\,\mathbf{k} = 0$$

$\therefore$ $\Sigma M_x = 4.188\,T_1 - 4.6152\,T_2 + 2400 = 0$ ..... (3)

$\Sigma M_z = 5.028\,T_1 - 3600 = 0$ ..... (4)

Solving equations (3) and (4) ,

$T_1 = 716$ lb and $T_2 = 1169.74$ lb

From equations (1) & (2),

$\mathbf{T_1} = -300\mathbf{i} - 600\mathbf{j} + 250\mathbf{k}$

$\mathbf{T_2} = -1079.67\mathbf{j} - 450\mathbf{k}$

$\mathbf{F} = 600\,\mathbf{i}$

$\overline{P} = 400\,\mathbf{k}$

Let reaction at A $= a_1\,\mathbf{i} + a_2\,\mathbf{j} + a_3\,\mathbf{k}$

For Equilibrium $\Sigma F = 0$ i.e. $\mathbf{T_1} + \mathbf{T_2} + \mathbf{F} + \overline{P} + \overline{A} = 0$

$\therefore$ $a_1 = -300$; $a_2 = 1679.67$; $a_3 = -200$

$\therefore$ $A = [300^2 + 1679.67^2 + 200^2]^{1/2} = 1718$ lb

Problem 34.: The boom of the derrick in fig. is rotated back 30° from the XY plane. Determine the forces in BC and BD and the components of the bearing reaction at the socket of the mast AB.

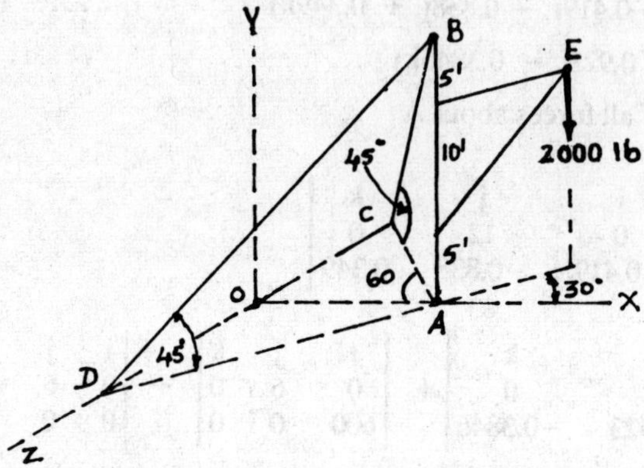

Solution : Let us find the co–ordinates of points in figure

Note that $\angle BAC = 90^\circ$ & $l(AB) = l(AC) = 20$

$\angle COA = 90^\circ$

$l(OA) = AC \cos 60 = 20 \cos 60 = 10$

$l(OC) = AC \sin 60 = 20 \sin 60 = 17.32$

$l(DA) = 20$

$\angle AOD = 90 \;;\; l(OD) = 20 \sin 60 = 17.32$

x co–ordinate of E $= 10 + 10 \cos 30^\circ = 18.66$

y co–ordinate of E $= 15$

z co–ordinate of E $= -10 \sin 30^\circ = -5$

$\therefore$ A (10 , 0 , 0) ; B (10 , 20 , 0) ; C (0 , 0 , −17.32) ;

D (0 , 0 , 17.32) ; E (18.66 , 15 , −5)

$\overline{BD} = -10i - 20j + 17.32k$

$$\therefore \quad \hat{n}_{BD} = \frac{1}{28.28}(-10i - 20j + 17.32k)$$

$\overline{BC} = -10i - 20j - 17.32k$

$$\therefore\ \hat{n}_{BC} = \frac{1}{28.28}(-10i - 20j - 17.32k)$$

$$\bar{F}_{BD} = \frac{F_{BD}}{28.28}(-10i - 20j + 17.32k)$$

$$\bar{F}_{BC} = \frac{F_{BC}}{28.28}(-10i - 20j - 17.32k)]$$

$$F_{EF} = -2000j$$

$$\overline{BA} = -20j\ ;\ \overline{EA} = -8.66i - 15j + 5k$$

Taking moment of all forces about A

$$\Sigma M_A = \begin{vmatrix} i & j & k \\ 0 & -20 & 0 \\ -10 & -20 & 17.32 \end{vmatrix} \frac{F_{BD}}{28.28}$$

$$+ \begin{vmatrix} i & j & k \\ 0 & -20 & 0 \\ -10 & -20 & -17.32 \end{vmatrix} \frac{F_{BC}}{28.28} + \begin{vmatrix} i & j & k \\ -8.66 & -15 & 5 \\ 0 & -2000 & 0 \end{vmatrix}$$

$$= \frac{F_{BD}}{28.28}(-346.4i - 200k) + \frac{F_{BC}}{28.28}(346.4i - 200k)$$
$$+ 10000i + 17320k = 0$$

$$\therefore\ -12.25\,F_{BD} + 12.25\,F_{BC} + 10000 = 0$$

$$F_{BD} - F_{BC} = 816.32 \qquad \ldots\ldots(1)$$

$$-7.07\,F_{BD} - 7.07\,F_{BC} + 17320 = 0$$

$$\therefore\ F_{BD} + F_{BC} = 2449.78 \qquad \ldots\ldots(2)$$

Solving equations (1) and (2), we get

$F_{BC} = 816.72$ lb & $F_{BD} = 1633$ lb

Problem 35. : In previous problem , determine the position of the boom which will cause a maximum force in BC. What is this maximum force ?

Solution : From the previous example we ge co–ordinates as :

A (10, 0, 0) ; B (10, 20, 0) ; C (0, 0, −17.32) ; D (0, 0, 17.32)

Let boom makes an angle θ with x axis

$$\therefore\ E(10 + 10\cos\theta\ ,\ 15\ ,\ 10\sin\theta)$$

$$\overline{BD} = -10i - 20j + 17.32k$$

$$\hat{n}_{BD} = \frac{1}{28.28}(-10i - 20j + 17.32k)$$

$$\overline{BC} = -10i - 20j - 17.32k$$

$$\hat{n}_{BC} = \frac{1}{28.28}(-10i - 20j - 17.32k)$$

$$\therefore \overline{F}_{BD} = \frac{F_{BD}}{28.28}(-10i - 20j + 17.32k)$$

$$\overline{F}_{BC} = \frac{F_{BC}}{28.28}(-10i - 20j - 17.32k)$$

$$F_{EF} = -2000j$$

$$\overline{BA} = -20j \; ; \; \overline{EA} = -10\cos\theta\, i - 15j - 10\sin\theta\, k$$

Taking moment of all forces about A

$$\Sigma M_A = \begin{vmatrix} i & j & k \\ 0 & -20 & 0 \\ -10 & -20 & 17.32 \end{vmatrix} \frac{F_{BD}}{28.28} + \begin{vmatrix} i & j & k \\ 0 & -20 & 0 \\ -10 & -20 & -17.32 \end{vmatrix} \frac{F_{BC}}{28.28} + \begin{vmatrix} i & j & k \\ -10\cos\theta & -15 & -10\sin\theta \\ 0 & -2000 & 0 \end{vmatrix}$$

$$= \frac{F_{BD}}{28.28}(-346.4i - 200k) + \frac{F_{BC}}{28.28}(346.4i - 200k) + 20000\sin\theta\, i + 20000\cos\theta\, k = 0$$

$$\Rightarrow -12.25 F_{BD} + 12.25 F_{BC} - 20000\sin\theta = 0$$

$$\therefore F_{BD} = F_{BC} - 1632.8\sin\theta \qquad \ldots..(1)$$

$$-7.07 F_{BD} - 7.07 F_{BC} + 20000\cos\theta = 0$$

$$\therefore F_{BD} + F_{BC} = 2828.8\cos\theta \qquad \ldots..(2)$$

Substituting (1) in equation (2), we get

$$2F_{BC} = 1632.8\sin\theta + 2828.8\cos\theta$$

$$\therefore F_{BC} = 816.4\sin\theta + 1414.4\cos\theta$$

For F_{BC} to be maximum $\frac{dF_{BC}}{d\theta}$ should be equal to zero.

$$\therefore \quad \frac{dF_{BC}}{d\theta} = 0 = 816.4 \cos\theta - 1414.4 \sin\theta$$

$$\therefore \quad \tan\theta = \frac{816.4}{1414.4}$$

$$\theta = 30^{\circ}$$

$\therefore$ For maximum tension in BC, boom should be rotated 30° forward.

Maximum force in BC

$$= 816.4 \sin 30 + 1414.4 \cos 30 = 1633.1 \text{ lb}$$

Problem 36.: The boom in fig. supports a load W = 1400 lb. Compute the forces in the cables AE and BD and also the components exerted by the ball–and–socket joint C.

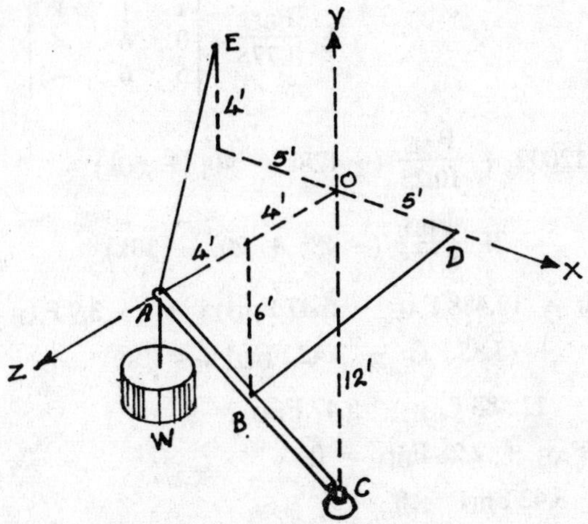

Solution : Co–ordinates of points in figure

A (0 , 0 ; 8) ; B (0 , −6 , 4) ; C (0 , −12 , 0) ; D (5 , 0 , 0) ; E (−5 , 4 , 0)

$$\overline{AE} = -5i + 4j - 8k$$

$$\therefore \quad \hat{n}_{AE} = \frac{1}{10.25}(-5i + 4j - 8k)$$

$$\overline{BD} = 5i + 6j - 4k$$

$$\therefore \quad \hat{n}_{BD} = \frac{1}{8.775}(5i + 6j - 4k)$$

$\therefore\ \bar{F}_{AE} = \dfrac{F_{AE}}{10.25}(-5i + 4j - 8k)$

$\therefore\ \bar{F}_{BD} = \dfrac{F_{BD}}{8.775}(5i + 6j - 4k)$

Let reaction at C be $F_C = C_x i + C_y j + C_z k$

$\therefore\ \overline{CA} = 12j + 8k$

$\overline{CB} = 6j + 4k$

Taking moment of all forces about C

$$\Sigma M_C = \begin{vmatrix} i & j & k \\ 0 & 12 & 8 \\ 0 & -1400 & 0 \end{vmatrix} + \frac{F_{AE}}{10.25}\begin{vmatrix} i & j & k \\ 0 & 12 & 8 \\ -5 & 4 & -8 \end{vmatrix}$$

$$+ \frac{F_{BD}}{8.775}\begin{vmatrix} i & j & k \\ 0 & 6 & 4 \\ 5 & 6 & -4 \end{vmatrix}$$

$$= 11200\, i + \frac{F_{AE}}{10.25}(-128i - 40j + 60k)$$

$$+ \frac{F_{BD}}{8.775}(-48i + 20j - 30k)$$

$$= (11200 - 12.488\, F_{AE} - 5.47\, F_{BD})\, i + (-3.9\, F_{AE} + 2.28\, F_{BD})\, j + (5.85\, F_{AE} - 3.42\, F_{BD})\, k = 0$$

$\therefore\ 11200 - 12.488\, F_{AE} - 5.47\, F_{BD} = 0$ .....(1)

$-3.9\, F_{AE} + 2.28\, F_{BD} = 0$ .....(2)

$5.85 - 3.42\, F_{BD} = 0$ .....(3)

Solving equations (1) and (2) we get,

$F_{AE} = 512.6266$ lb

$F_{BD} = 877.18$ lb

$\therefore\ \bar{F}_{AE} = -250i + 200j - 400k$

$\bar{F}_{BD} = 500i + 600j - 400k$

$\bar{W} = -1400j$

Let the reaction at C $= \bar{C} = C_x i + C_y j + C_z k$

For Equilibrium $\Sigma \bar{F} = 0$

$\therefore\ C_x = -250$ lb ; $C_y = 600$ lb ; $C_z = 800$ lb

Problem 37. : In previous problem, change the position of point D in the XY plane so that the force in BD will be a minimum. IF W = 540 lb, find this force and the x and y coordinates of D.

Solution : From Fig. co–ordinates of points are

A (0 , 0 ; 8) ; B (0 , −6 , 4) ; C (0 , −12 , 0) ; D (x , y , 0) ; E (−5 , 4 , 0)

$\overline{AE} = -5i + 4j - 8k$

$\therefore \hat{n}_{AE} = \frac{1}{10.25}(-5i + 4j - 8k)$

$\overline{BD} = xi + (y + 6)j - 4k$

$\therefore \hat{n}_{BD} = \frac{xi + (y + 6)j - 4k}{[x^2 + (y + 6)^2 + 16]^{1/2}}$

$\therefore \overline{F}_{AE} = \frac{F_{AE}}{10.25}(-5i + 4j - 8k)$

$\therefore \overline{F}_{BD} = \frac{F_{BD}}{[x^2 + (y + 6)^2 + 16]^{1/2}}\left[xi + (y + 6)j - 4k\right]$

Now for taking moments of all forces about C we need to know $\overline{CA}$ & $\overline{CB}$

$\therefore \overline{CA} = 12j + 8k$

$\overline{CB} = 6j + 4k$

Taking moment of all forces about C

$$\Sigma M_C = \begin{vmatrix} i & j & k \\ 0 & 12 & 8 \\ 0 & -540 & 0 \end{vmatrix} + \frac{F_{AE}}{10.25}\begin{vmatrix} i & j & k \\ 0 & 12 & 8 \\ -5 & 4 & -8 \end{vmatrix}$$

$$+ \frac{F_{BD}}{[x^2 + (y + 6)^2 + 16]^{1/2}}\begin{vmatrix} i & j & k \\ 0 & 6 & 4 \\ x & (y + 6) & -4 \end{vmatrix} = 0$$

$$\therefore 4320\,i + \frac{F_{AE}}{10.25}(-128i - 40j + 60k)$$

$$+ \frac{F_{BD}}{[x^2 + (y + 6)^2 + 16]^{1/2}}[(-4y - 48)i + 4x\,j - 6x\,k] = 0$$

For equilibrium $\Sigma M = 0$

$$\therefore \Sigma M_x = 4320 - 12.488 F_{AE} - \frac{(4y + 48) F_{BD}}{[x^2 + (y+6)^2 + 16]^{1/2}} = 0 \quad \ldots\ldots (1)$$

$$\& \ \Sigma M_y = -3.9 F_{AE} + \frac{4x F_{BD}}{[x^2 + (y+6)^2 + 16]^{1/2}} = 0$$

$$\therefore \ F_{AE} = \frac{1.025 x F_{BD}}{[x^2 + (y+6)^2 + 16]^{1/2}} \quad \ldots\ldots (2)$$

Substituting (2) in equation (1) we get,

$$F_{BD} = \frac{4320 \cdot [x^2 + (y+6)^2 + 16]^{1/2}}{12.8 x + 4y + 48}$$

For F_{BD} to be minimum $\dfrac{\partial F_{BD}}{\partial x}$ and $\dfrac{\partial F_{BD}}{\partial y}$ both should be equal to zero.

$$\frac{\partial F_{BD}}{\partial x} = 4320 \left[\frac{(12.8x + 4y + 48) x}{[x^2 + (y+6)^2 + 16]^{1/2}} - 12.8 [x^2 + (y+6)^2 + 16]^{1/2} \right] = 0 \quad \ldots\ldots (3)$$

$$\frac{\partial F_{BD}}{\partial y} = 4320 \left[\frac{(12.8x + 4y + 48)(y+6)}{[x^2 + (y+6)^2 + 16]^{1/2}} - 4 [x^2 + (y+6)^2 + 16]^{1/2} \right] = 0 \quad \ldots\ldots (4)$$

From equation (3) & (4),

$$(12.8x + 4y + 48) x = 12.8 \left[x^2 + (y+6)^2 + 16 \right] \quad \ldots\ldots (5)$$

$$(12.8x + 4y + 48)(y+6) = 4 \left[x^2 + (y+6)^2 + 16 \right] \ldots (6)$$

Dividing equation (5) by (6) we get,

$$\frac{x}{y+6} = \frac{12.8}{4}$$

$$\therefore \ x = 3.2 (y + 6)$$

Substituting this in equation (5) and solving, we get

$$x = 8.54 \text{ and } y = -3.33$$

Substituting this in equation of F_{BD}

$$F_{BD} = \frac{4320\,[\,(8.54)^2 + (-3.33 + 6)^2 + 16\,]^{1/2}}{12.8 \times 8.54 + 4 \times (-3.33) + 48}$$

$$= 294 \text{ lb}$$

Problem 38. : In another variation of above problem point E is to be shifted so that the force is AE will be a minimum. If W = 540 lb, find this minimum force in AE and the x and y coordinates of E.

Solution : From fig. co–ordinates of points are

A (0 , 0 ; 8) ; B (0 , −6 , 4) ; C (0 , −12 , 0) ; D (5 , 0 , 0) ; E (x , y , z)

$$\overline{AE} = xi + yj + (z - 8)\,k$$

$$\therefore \; \hat{n}_{AE} = \frac{xi + yj + (z-8)\,k}{[\,x^2 + y^2 + (z-8)^2\,]^{1/2}}$$

$$\overline{BD} = 5i + 6j - 4k$$

$$\therefore \; \hat{n}_{BD} = \frac{1}{8.775}\,(5i + 6j - 4k)$$

$$\therefore \; F_{AE} = \frac{F_{AE}\,(xi + yj + (z-8)\,k)}{[\,x^2 + y^2 + (z-8)^2\,]^{1/2}}$$

$$F_{BD} = \frac{F_{BD}\,(5i + 6j - 4k)}{8.774}$$

For taking moments of all forces about C we need to know $\overline{CA}$ & $\overline{CB}$

$$\overline{CA} = 12j + 8k$$

$$\overline{CB} = 6j + 4k$$

Taking moments of all forces about C

$$\Sigma M_C = \begin{vmatrix} i & j & k \\ 0 & 12 & 8 \\ 0 & -540 & 0 \end{vmatrix} + \frac{F_{AE}}{[\,x^2 + y^2 + (z-8)^2\,]^{1/2}}$$

$$\begin{vmatrix} i & j & k \\ 0 & 12 & 8 \\ x & y & z-8 \end{vmatrix} + \frac{F_{BD}}{8.775}\begin{vmatrix} i & j & k \\ 0 & 6 & 4 \\ 5 & 6 & -4 \end{vmatrix} = 0$$

$$\therefore \quad 4320\,i + \frac{F_{AE}\,[\,(12z - 8y - 96)\,i + 8xj - 12x\,k\,]}{[\,x^2 + y^2 + (z-8)^2\,]^{1/2}}$$

$$+ \frac{F_{BD}}{8.775}\,(-48i + 20j - 30k) = 0$$

$$\therefore \quad \Sigma M_x = 4320 + \frac{(12z - 8y - 96)\,F_{AE}}{[\,x^2 + y^2 + (z-8)^2\,]^{1/2}} + \frac{-48\,F_{BD}}{8.775} = 0 \qquad \ldots\ldots(1)$$

$$\&\ \Sigma M_y = \frac{8x\,F_{AE}}{[\,x^2 + y^2 + (z-8)^2\,]^{1/2}} + \frac{20}{8.775}\,F_{BD} = 0$$

$$\therefore \quad F_{BD} = \frac{-3.51\,x\ F_{AE}}{[\,x^2 + y^2 + (z-8)^2\,]^{1/2}} \qquad \ldots\ldots(2)$$

Substituting equation (2) in equation (1)

$$4320 + \left[\frac{12z - 8y - 96 + 19.2x}{[\,x^2 + y^2 + (z-8)^2\,]^{1/2}}\right] F_{AE} = 0$$

For AE to be minimum $\frac{\partial F_{AE}}{\partial x}$, $\frac{\partial F_{AE}}{\partial y}$, $\frac{\partial F_{AE}}{\partial z}$ should be equal to zero.

$$\therefore \quad \frac{\partial F_{AE}}{\partial z} = 0 = \frac{-4320 \cdot [\,x^2 + y^2 + 64\,]^{1/2}}{19.2x - 8y - 96}$$

$$\frac{\partial F_{AE}}{\partial x} = 0$$

$$= -4320\left[\frac{(19.2x - 8y - 96)\,x}{[\,x^2 + y^2 + 64\,]^{1/2}} - 19.2\,[\,x^2 + y^2 + 64\,]^{1/2}\right] \qquad \ldots\ldots(3)$$

$$\frac{\partial F_{AE}}{\partial y} = 0$$

$$= -4320\left[\frac{(19.2x - 8y - 96)\,y}{[\,x^2 + y^2 + 64\,]^{1/2}} + 8\,[\,x^2 + y^2 + 64\,]^{1/2}\right] \qquad \ldots\ldots(4)$$

From equations (3) and (4)

$$(19.2x - 8y - 96)\,x = 19.2\,(x^2 + y^2 + 64) \qquad \ldots\ldots(5)$$

$$(19.2\,x - 8y - 96)\,y = -8\,(x^2 + y^2 + 64) \qquad \ldots\ldots(6)$$

Dividing equation (5) by (6)

$$\frac{x}{y} = \frac{-19.2}{8}$$

$\therefore \; x = -2.4y$

Substituting this in equation (5) and solving we get

$$x = -12.8 \text{ and } y = 5.33$$

$$\therefore \; F_{AE} = \frac{-4320 \cdot [(-12.8)^2 + (5.33)^2 + 64]^{1/2}}{19.2 \times (-12.8) - 8 \times 5.33 - 96} = 180 \text{ lb}$$

$$\therefore \; \overline{F}_{AE} = \frac{180}{16}(-12.8i + 5.33j - 8k)$$

$$= -144i + 60j - 90k$$

Problem 39. : The boom AC in fig. rests in a ball–and–socket joint at A and is held in equilibrium by the cables BE and CD. The load applied at C is $P = -900\,\hat{j} - 300\,\hat{k}$ lb. Determine the coordinates of D in the YZ plane so that the force in CD will be a minimum. Neglect the weight of the boom and cables.

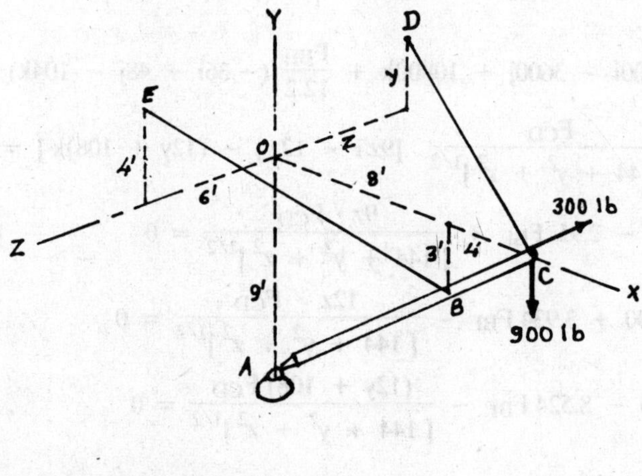

Solution : Co–ordinates of points are

A (0, −9; 0); B (8, −3, 0); C (12, 0, 0); D (0, y, −z); E (0, 4, 6)

$$\overline{BE} = -8i + 7j + 6k$$

$$\therefore\ \hat{n}_{BE} = \frac{1}{12.2}(-8i + 7j + 6k)$$

$$\overline{CD} = -12i + yj - zk$$

$$\therefore\ \hat{n}_{CD} = \frac{-12i + yj - zk}{[144 + y^2 + z^2]^{1/2}}$$

$$\therefore\ F_{BE} = \frac{F_{BE} \cdot (-8i + 7j + 6k)}{12.2}$$

$$\overline{F}_{CD} = \frac{F_{CD} \cdot (-12i + yj - zk)}{[144 + y^2 + z^2]^{1/2}}$$

$$\overline{P} = -900j - 300k$$

$$\overline{BA} = -8i - 6j \text{ and } \overline{CA} = -12i - 9j$$

Taking moment of all forces about A

$$\Sigma M_A = \begin{vmatrix} i & j & k \\ -12 & -9 & 0 \\ 0 & -900 & -300 \end{vmatrix} + \frac{F_{BE}}{12.2}\begin{vmatrix} i & j & k \\ -8 & -6 & 0 \\ -8 & 7 & 6 \end{vmatrix}$$

$$+ \frac{F_{CD}}{[144 + y^2 + z^2]^{1/2}}\begin{vmatrix} i & j & k \\ -12 & -9 & 0 \\ -12 & y & -z \end{vmatrix}$$

$$= 2700i - 3600j + 10800k + \frac{F_{BE}}{12.2}(-36i + 48j - 104k)$$

$$+ \frac{F_{CD}}{[144 + y^2 + z^2]^{1/2}}[9z\,i - 12z\,j - (12y + 108)k] = 0$$

$$\therefore\ 2700 - 2.95\,F_{BE} + \frac{9z \cdot F_{CD}}{[144 + y^2 + z^2]^{1/2}} = 0 \qquad \ldots (1)$$

$$-3600 + 3.934\,F_{BE} - \frac{12z \cdot F_{CD}}{[144 + y^2 + z^2]^{1/2}} = 0 \qquad \ldots (2)$$

$$10800 - 8.524\,F_{BE} - \frac{(12y + 108)\,F_{CD}}{[144 + y^2 + z^2]^{1/2}} = 0 \qquad \ldots (3)$$

From equation (2)

$$F_{BE} = 915 + \frac{3.05\,z \cdot F_{CD}}{[144 + y^2 + z^2]^{1/2}}$$

Substituting this in equation (3)

$$10800 - 7799.46 - \left[\frac{26z + 12y + 108}{[144 + y^2 + z^2]^{1/2}}\right] F_{CD} = 0$$

$$\therefore\ F_{CD} = \frac{3000.54\,[144 + y^2 + z^2]^{1/2}}{26z + 12y + 108}$$

For F_{CD} to be minimum $\frac{\partial F_{CD}}{\partial y}$ and $\frac{\partial F_{CD}}{\partial z}$ should be equal to zero.

$$\frac{\partial F_{CD}}{\partial y} = 3000.54\left[\frac{(26z + 12y + 108)\,y}{[144 + y^2 + z^2]^{1/2}} - 12\,[144 + y^2 + z^2]^{1/2}\right] = 0$$

$$\therefore\ (26z + 12y - 108)\,y = 12\,(144 + y^2 + z^2) \qquad \ldots.\ (4)$$

$$\frac{\partial F_{CD}}{\partial z} = 3000.54\left[\frac{(26z + 12y + 108)\,z}{[144 + y^2 + z^2]^{1/2}} - 26\,[144 + y^2 + z^2]^{1/2}\right] = 0$$

$$\therefore\ (26z + 12y - 108)\,z = 26\,(144 + y^2 + z^2) \qquad \ldots.\ (5)$$

From equation (4) and (5)

$$\frac{y}{z} = \frac{12}{26}$$

$$\therefore\ z = 2.167\,y$$

Substituting $z = 2.167\,y$ in equation (4) and solving , we get

$$y = 16 \text{ and } z = 34.7$$

Chapter Three

ANALYSIS OF STRUCTURES

Trusses :

A truss is a structure composed of members fastened together in such a way so as to resist change in shape ; it is a rigid structure.The purpose of truss is to support a larger load or span a greater distance than any individual member from which the truss may be built.

Classification :

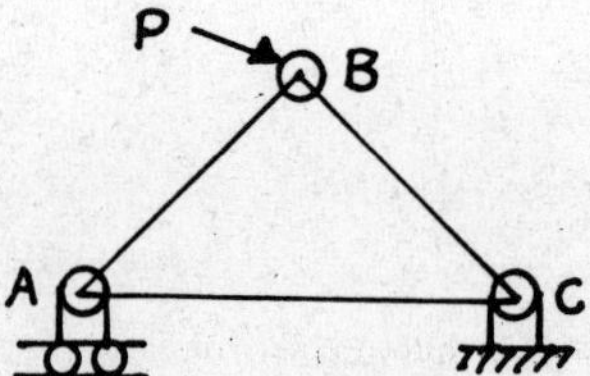

Fig. shows triangular truss. It is a rigid truss.

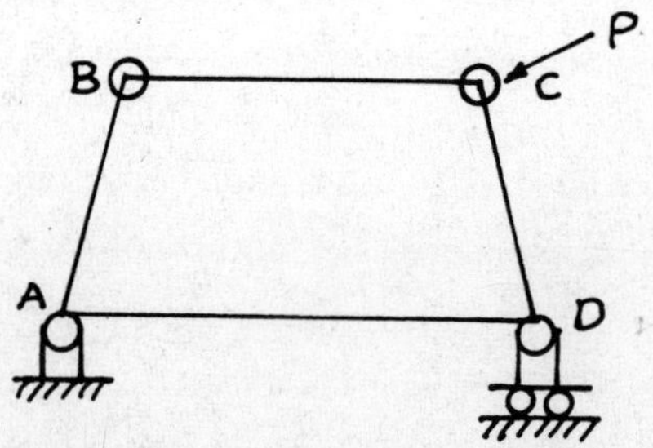

Fig. shows truss with 4-members. It is known as mechanism. It is not a stable structure.

If we continue to add bars at joints fastened to one another, additional joints can be created and a truss consisting of many members can be constructed. These types of trusses are classified as "*Simple Trusses*".

Trusses consisting of two simple trusses joined by three non-intersecting bars or by a common pin and a bar are known as "*Compound trusses*".

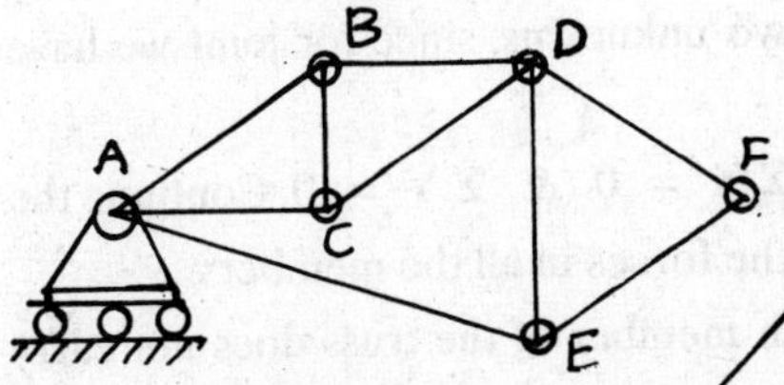

Simple truss

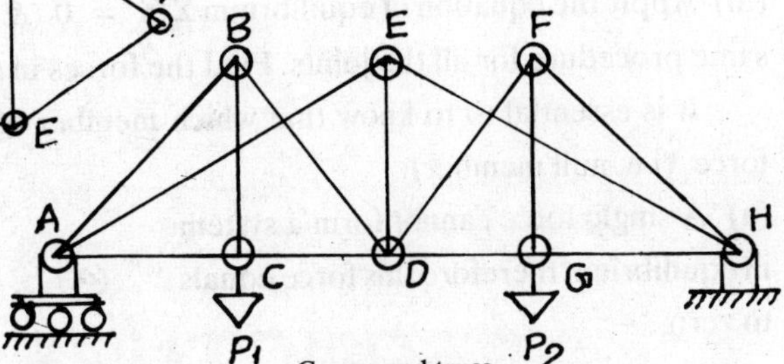

Compound truss.

For simple truss we have the relation $m = (2j - 3)$ . ... (1)

Where m = number of members

j = number of joints (including supports).

If equation (1) is satisfied then we get number of equations equal to number of unknowns and hence we say that the truss is "*Statically Determinate*".

If $m > (2j - 3)$ then number of equations becomes less than number of unknowns involved and hence it is not possible to do the complete analysis. This type of truss is known as *Statically Indeterminate Truss*.

If $m < (2j - 3)$ then truss is known as *Unstable Truss*.

Assumptions made in simple trusses :

(i) All members are assumed to be pin connected.

(ii) Applied loads acts at the ends of members i.e. at the pins.

(iii) Members carry forces along their length either tension or compression.

(iv) In FBD of joint, tension force is represented by arrow acting away from joint.

If the arrow is shown acting towards joint then it is compression force. Under the standard direction convention the tension force is taken as positive and if force of member is negative then it should be compressive force.

Method of Joints :

(i) First find the reactions at support.

(ii) Start the analysis of joint having two unknowns, since for joint we have maximum two equations available.

(iii) Apply the equation of equilibrium $\Sigma X = 0$ & $\Sigma Y = 0$. Continue the same procedure for all the joints. Find the forces in all the members.

It is essential to to know that which member of the truss does not carry force (i.e. null member).

(a) A single force cannot form a system in equilibrium therefore this force equals to zero.

(b) If two forces act at a joint then for the equilibrium of the joint these two forces should act along the same straight line, if not, then null member.

(c) If three forces act at a joint and two of them are along the same staight line then for the equilibrium of the joint the third force should be equal to zero.

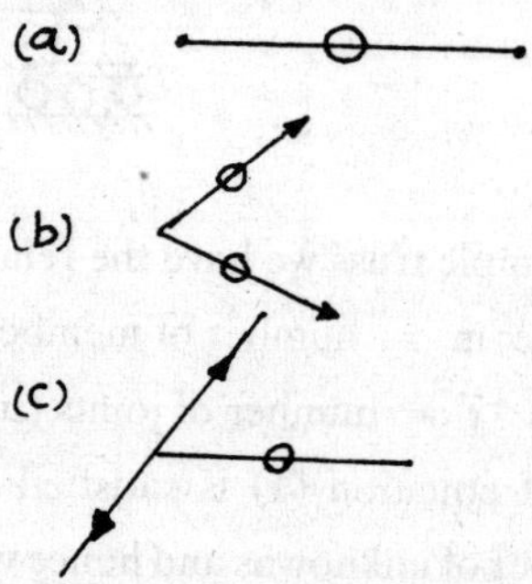

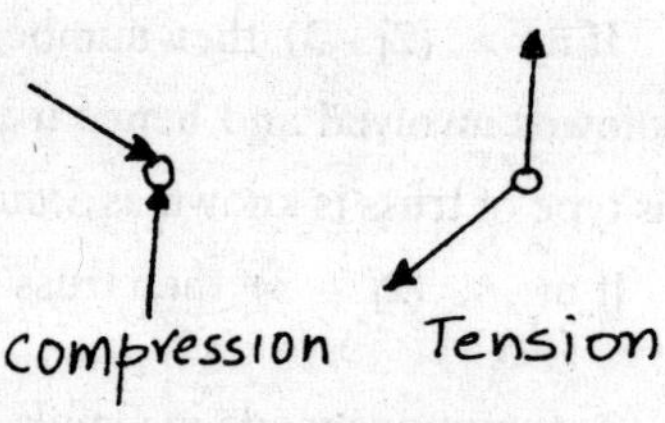

Problem 1. : Compute the force in each member of truss shown in fig.

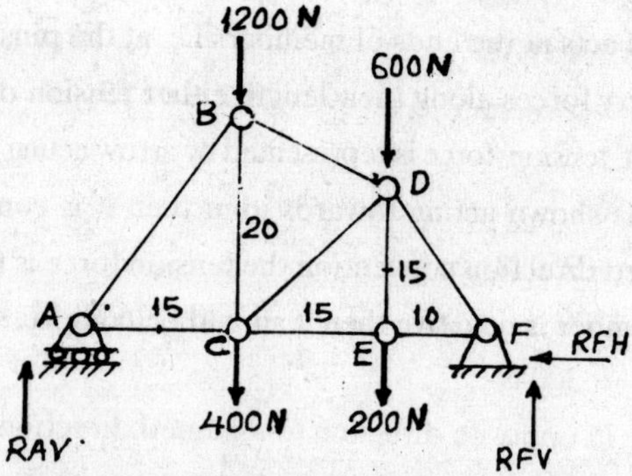

Solution : $\Sigma Y = R_{AV} + R_{FV} = 1200 + 400 + 600 + 200 = 0$

$\therefore \quad R_{AV} + R_{FV} = 2400\text{ N}$ .... (1)

$\Sigma X = R_{FH} = 0$ .... (2)

$\Sigma M_F = -R_{AV} \times 40 + 1600 \times 25 + 800 \times 10 = 0$

$R_{AV} = 1200\text{ N}$

$\therefore \quad R_{FV} = 1200\text{ N}$

Consider FBD of joint (A)

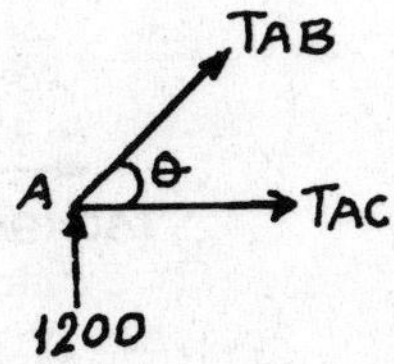

$$\theta = \tan^{-1}\frac{20}{15} = 53.13^{\circ}$$

$\Sigma X = T_{AC} + T_{AB}\cos 53.13 = 0$ ...... (3)

$\Sigma Y = 1200 + T_{AB}\sin 53.13 = 0$

i.e. $T_{AB} = -1500\text{ N}$ (Direction must be opposite to assumed).

$T_{AC} = +900\text{ N}$

Consider FBD of joint (B)

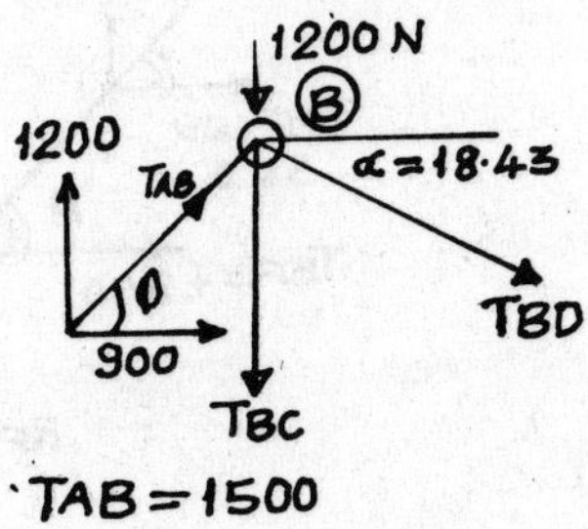

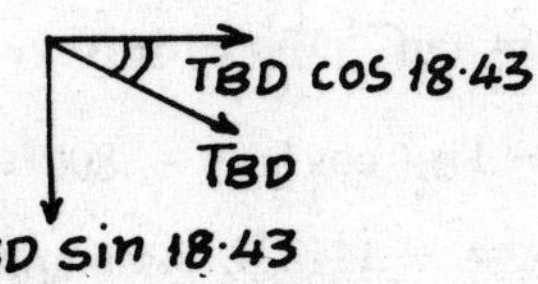

$\Sigma X = 900 + T_{BD}\cos 18.43 = 0$

$T_{BD} = -950\text{ N}$

$\Sigma Y = 1200 - 1200 - T_{BC} - T_{BD}\sin 18.43 = 0$

$T_{BC} = +300\text{ N}$

T_{BD} must be in opposite direction to assumed direction.

Consider FBD of joint (C)

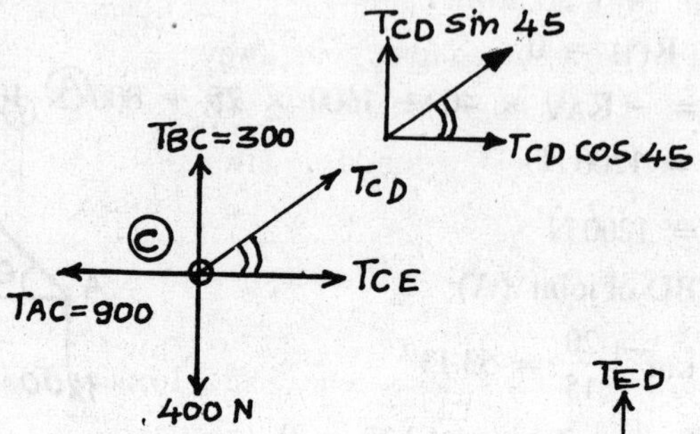

$$\Sigma Y = 300 - 400 + T_{CD} \sin 45 = 0$$

$$T_{CD} = +142 \text{ N}$$

$$\Sigma X = T_{CE} - 900 + T_{CD} \cos 45 = 0$$

$$T_{CE} = 800 \text{ N}$$

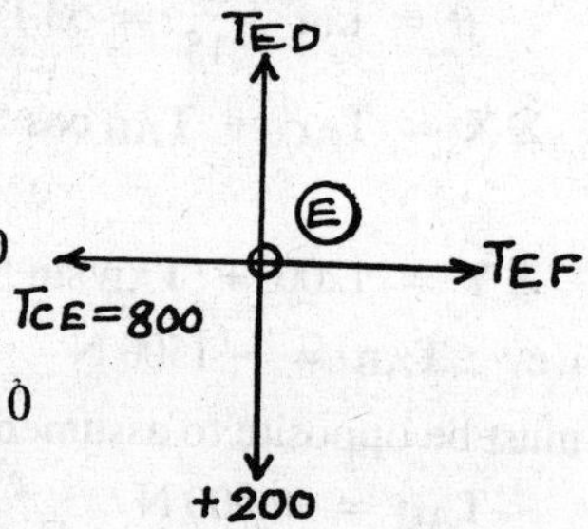

Consider FBD of joint (E)

$$\Sigma X = T_{EF} - 800 = 0$$

$$\therefore T_{EF} = 800 \text{ N}$$

$$\Sigma Y = T_{ED} - 200 = 0$$

$$\therefore T_{ED} = 200 \text{ N}$$

Consider FBD of joint (F)

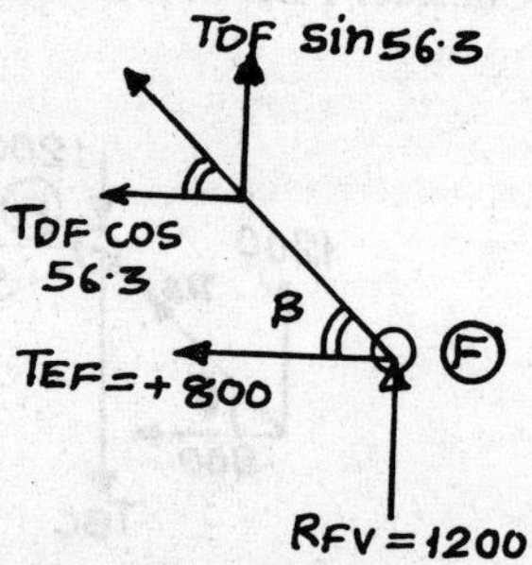

$$\angle\beta = \tan^{-1}\frac{15}{10} = 56.3^{\circ}$$

$$\Sigma X = -T_{DF} \cos 56.3 - 800 = 0$$

$\therefore$ $T_{DF} = -1442$ N. Assumption for bar DF is Tension. But since value is coming up as negative. Hence it should be compressive force for bar.

$$T_{DF} = +1442 \text{ N (Compression)}$$

To confirm answer, consider

$$\Sigma Y = T_{DF} \sin 56.3 + 1200 = 0$$

$\Rightarrow$ $T_{DF} = -1442$ N. This is cross check.

Important Note : In the above problem note the FBD of joint A (it is shown below)

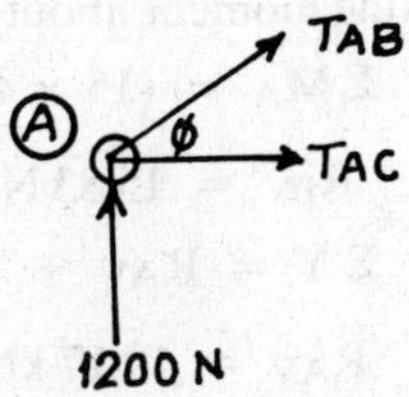

T_{AB} and T_{AC} are assumed to be going away from the joint. i.e. It is assumed that member AB and member AC are in Tension. But after analysis it is found that T_{AB} is negative and T_{AC} is positive. The meaning is that T_{AB} is in compression and T_{AC} is in tension.

This is very important, since when we have to show T_{AB} and T_{AC} in the FBD of B and C we must be careful about directions.

i.e. At B asume T_{AB} in compression and at joint C assume T_{AC} in tension as is done in above problem and only numerical value of force is used.

Problem 2. : Find the forces in all members of a pin connected truss shown in fig.

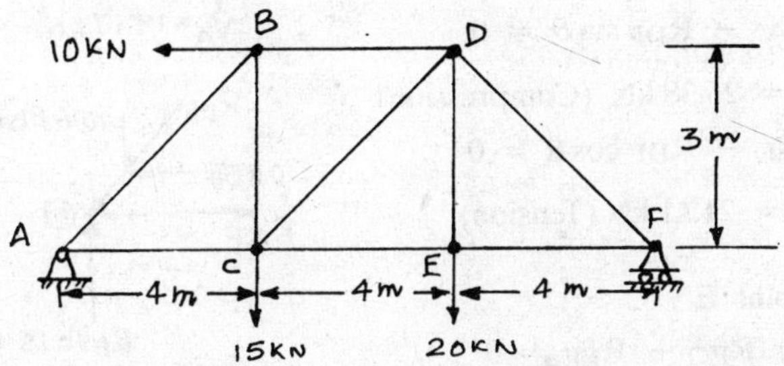

Solution :

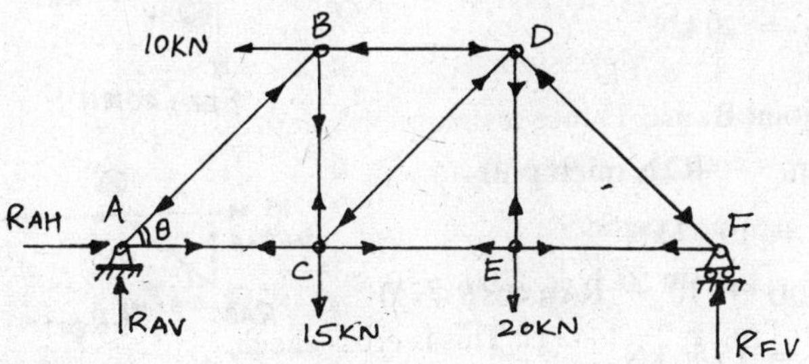

From above fig.

$$\Sigma X = R_{AH} - 10 = 0$$

$\therefore \quad R_{AH} = 10 \text{ kN}$

Taking moment about A

$$\Sigma M_A = (15 \times 4) + (20 \times 8) - (10 \times 3) - (12 \times R_{FV}) = 0$$

$\therefore \quad R_{FV} = 15.83 \text{ N}$

$$\Sigma Y = R_{AV} + R_{FV} - 15 - 20 = 0$$

$\therefore \quad R_{AV} = 19.17 \text{ kN}.$

Also, $\sin\theta = 0.6$, $\cos\theta = 0.8$

Consider joint A :

$$\Sigma Y = R_{AV} - R_{AB}\sin\theta = 0$$

$\therefore \quad R_{AB} = 31.95 \text{ kN}.$

$$\Sigma X = R_{AH} + R_{AC} - R_{AB}\cos\theta = 0$$

$\therefore \quad R_{AC} = 15.56 \text{ kN}$

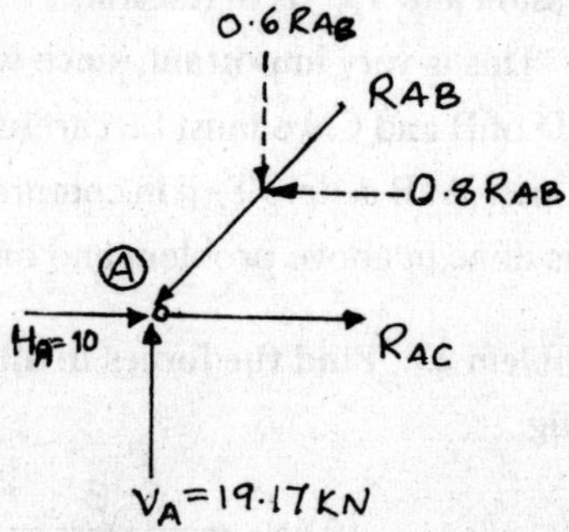

Consider joint F

$$\Sigma Y = R_{FV} - R_{DF}\sin\theta = 0$$

$\therefore \quad R_{DF} = 26.38 \text{ kN (Compression)}$

$$\Sigma X = R_{EF} - R_{DF}\cos\theta = 0$$

$\therefore \quad R_{EF} = 21.11 \text{ kN (Tension)}$

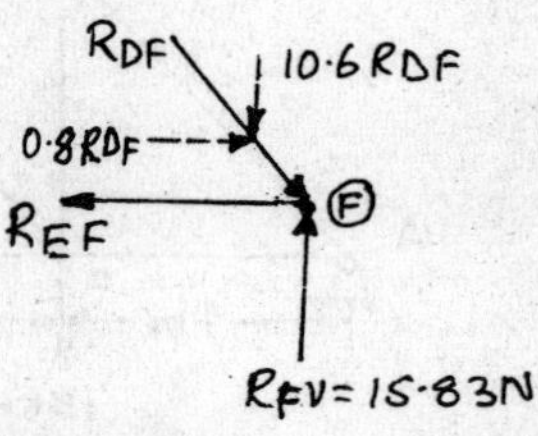

Consider joint E

$$\Sigma X = R_{EC} - R_{EF} = 0$$

$\therefore \quad R_{EC} = R_{EF} = 21.11 \text{ kN (Tension)}$

$$\Sigma Y = R_{ED} - 20 = 0$$

$\therefore \quad R_{ED} = 20 \text{ kN}$

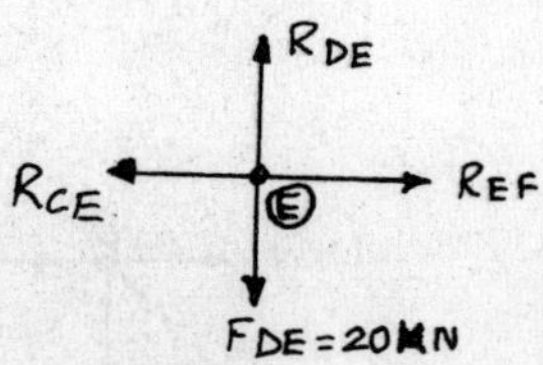

Consider joint B

$$\Sigma Y = R_{BC} - R_{AB}\sin\theta = 0$$

$\therefore \quad R_{BC} = 19.17 \text{ kN}$

$$\Sigma X = R_{BD} + 10 - R_{AB}\cos\theta = 0$$

$\therefore \quad R_{BD} = 15.56 \text{ kN}$

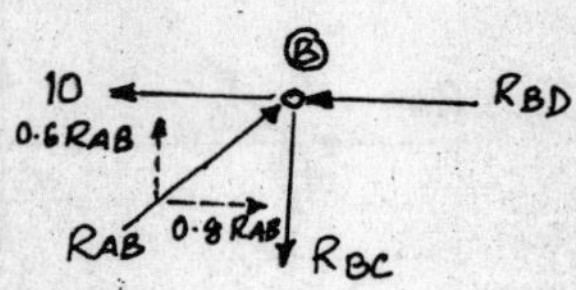

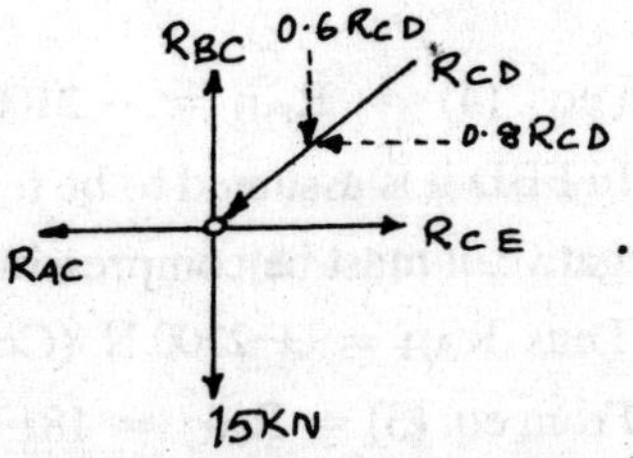

Consider joint C

$\Sigma X = R_{AC} - R_{CE} + R_{CD} \cos \theta$

$= 0$

$\therefore \quad R_{CD} = 6.94 \text{ kN}$.

Problem 3.: Determine the force in members AB, BD, BE and DE of the Howe roof truss shown in fig.

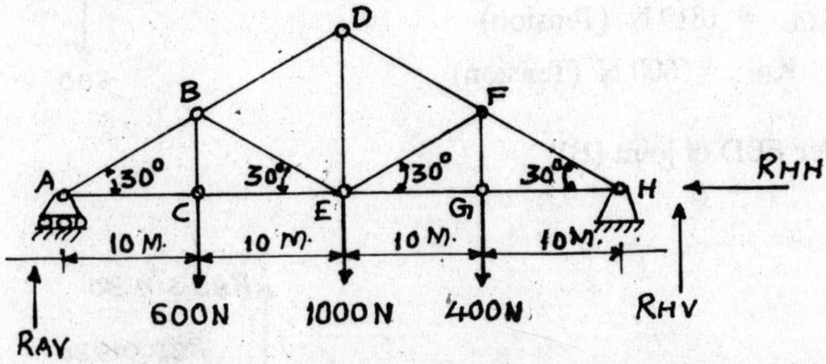

Solution : There will be only one vertically upward reaction at A, since it is a roller support and there will be two reactions at H, R_{HA} and R_{HV}, since it is a hinge support.

$$\Sigma X = R_{HH} = 0 \qquad \ldots\ldots (1)$$

$$\Sigma Y = R_{AV} + R_{HV} - 600 - 1000 - 400 = 0$$

$$\therefore \quad R_{AV} + R_{HV} = 2000 \text{ N} \qquad \ldots\ldots (2)$$

Taking moment about joint H, we get

$$\Sigma M_H = -R_{AV} \times 40 + 600 \times 30 + 1000 \times 20 + 400 \times 10 = 0$$

$$R_{AV} = 1050 \text{ N}$$

From equation (2) we get, $\quad R_{HV} = 950$ N

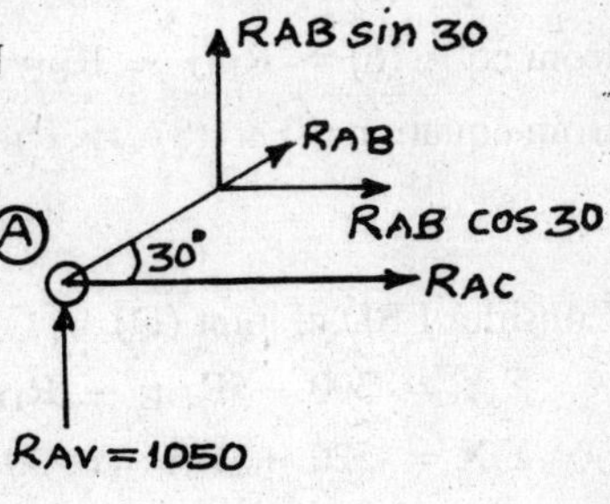

Consider FBD of joint (A)

$$\Sigma X = R_{AB} \cos 30 + R_{AC} = 0 \qquad \ldots\ldots (3)$$

$$\Sigma Y = 1050 + R_{AB} \sin 30 = 0 \qquad \ldots\ldots (4)$$

From eq^n. (4) ⇒ $R_{AB} = -2100$ N

In FBD it is assumed to be tension for bar AB. Since the value is coming as negative, it must be compressive force.

Thus $R_{AB} = +2100$ N (Compression)

∴ From eq. (3) ⇒ $R_{AC} = 1819$ N (Tension)

Consider FBD of joint (C)

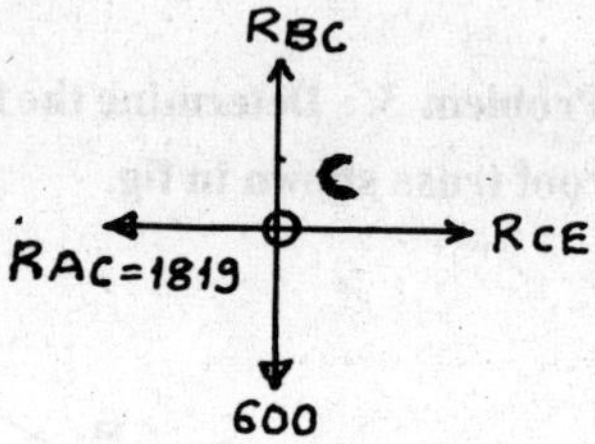

$\Sigma X = R_{CE} - 1819 = 0$

∴ $R_{CE} = 1819$ N (Tension)

& $R_{BC} = 600$ N (Tension)

Consider FBD of joint (B)

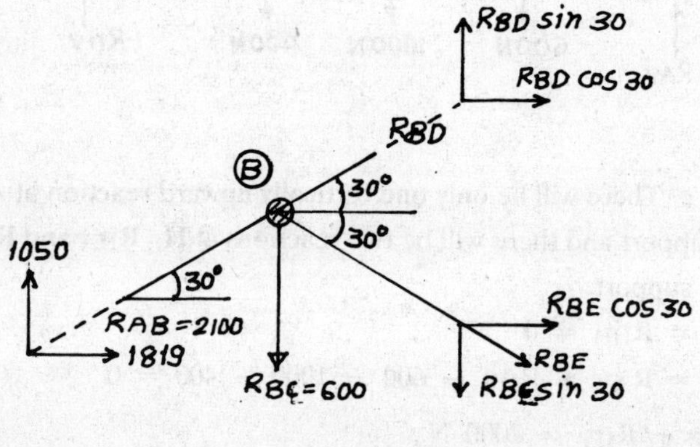

$\Sigma X = 1819 + R_{BD} \cos 30 + R_{BE} \cos 30 = 0$ ...... (5)

$\Sigma Y = 1050 - 600 + R_{BD} \sin 30 - R_{BE} \sin 30 = 0$ .... (6)

from eq^n. (5) ⇒ $R_{BD} + R_{BE} = -2100$ .....(7)

from eq^n. (6) ⇒ $R_{BD} - R_{BE} = 900$ ......(8)

from equation (7) & (8) ⇒ $R_{BD} = -600$ N (Compression)

$R_{BE} = -1500$ N (Compression)

Consider FBD of joint (D)

$\Sigma Y = 300 - R_{DE} - R_{DF} \sin 30 = 0$

$\Sigma X = 520 + R_{DF} \cos 30 = 0$

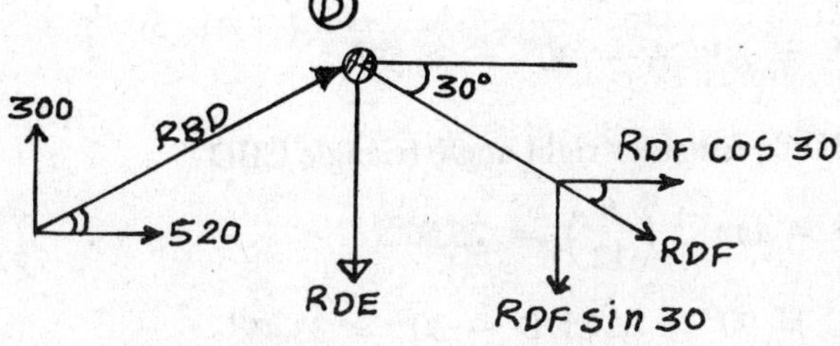

$\therefore$ R_{DF} = − 600 N (Compression)

R_{DE} = 600 N (Tension)

Problem 4. : Determine the force in each member of the crane truss shown in fig.

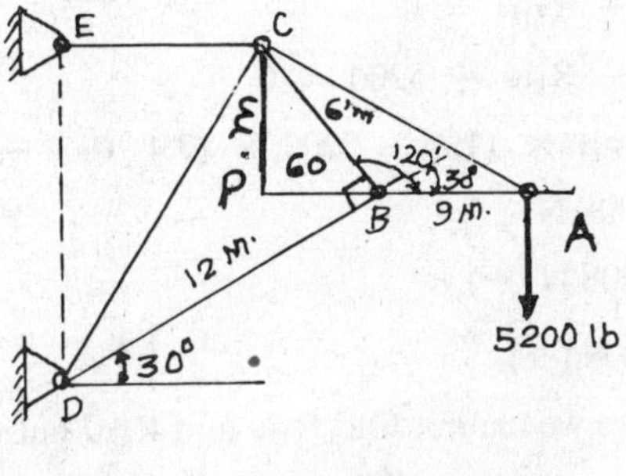

Solution : To find ∠CAB, consider following fig.

In triangle CPB

$$\angle CBP = 60^{o} \ \& \ l(CB) = 6 \text{ m}$$

$$\therefore \quad CP = CB \sin 60 = 5.2 \text{ m}$$

$$PB = CB \cos 60 = 3 \text{ m}$$

Now consider Δ CPA .

$$\therefore \quad \angle CAB = \tan^{-1}\left(\frac{CP}{PA}\right)$$

$$= \tan^{-1}\left(\frac{5.2}{12}\right) = 23.42$$

$$\& \quad \angle PCA = \tan^{-1}\left(\frac{PA}{PC}\right) = \tan^{-1}\left(\frac{12}{5.2}\right) = 66.58$$

$$\therefore \quad \angle BCA = \angle PCA - 30^\circ = 36.58^\circ$$

To find $\angle EDC$, consider right angle triangle CBD

$$\angle CDB = \tan^{-1}\left(\frac{6}{12}\right) = 26.56^\circ$$

$$\therefore \quad \angle EDC = 90^\circ - \angle CDB - 30^\circ = 33.43^\circ$$

Also $l(CD) = (6^2 + 12)^{1/2} = 13.41$ m

$$\therefore \quad l(ED) = 13.41 \cos 33.43 = 11.2 \text{ m}$$

$$l(EC) = 13.41 \sin 33.43 = 7.4 \text{ m}$$

E and D being hinge joints there will be two reactions at each joint. Applying equilibrium conditions,

$$\Sigma X = R_{EH} + R_{DH} = 0$$

$$\Sigma Y = R_{DV} - R_{EV} - 5200 = 0$$

$$\Sigma M_D = -R_{EH} \times 11.2 - 5200 \times (7.4 + 3 + 9) = 0$$

$$R_{EH} = -9008 \text{ N}$$

$$= +9008 \text{ N} (\leftarrow)$$

$$\therefore \quad R_{DH} = 9008 \text{ N} (\rightarrow)$$

Now in this case we cannot find R_{EV} and R_{DV} but we can start for joint analysis from joint A since two unknowns will be involved i.e. R_{AC} and R_{AB}

Consider FBD at joint A

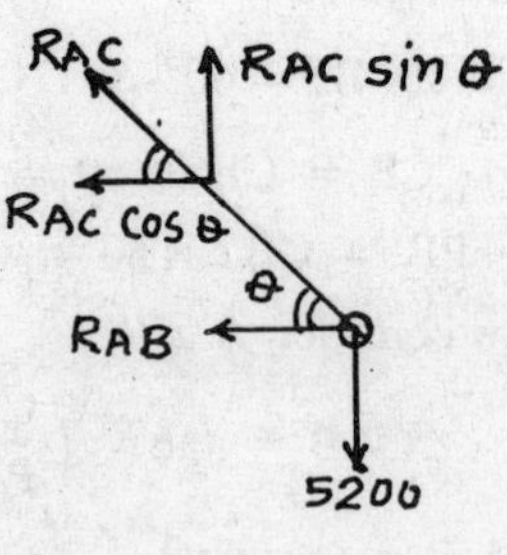

$$\Sigma Y = R_{AC} \sin 23.41 - 5200 = 0$$

$$\therefore \quad R_{AC} = \frac{5200}{\sin 23.41} = 13088 \text{ N}$$

$$\Sigma X = -R_{AB} - R_{AC} \cos 23.41 = 0$$

$$R_{AB} = -12010 \text{ N}$$

(Compression. Since negative value.)

Consider FBD at joint B

$$\Sigma X = -R_{BC} \cos 60 - R_{BD} \cos 30 - 12010 = 0 \quad \ldots\ldots (1)$$

$\Sigma Y = R_{BC} \sin 60 - R_{BD} \sin 30$

$= 0$

i.e. $R_{BC} = 0.577 R_{BD}$

$\ldots\ldots (2)$

from eq^n. (1) & (2)

$R_{BD} = -10402$ N (Compression)

$R_{BC} = -6002$ N (Compression)

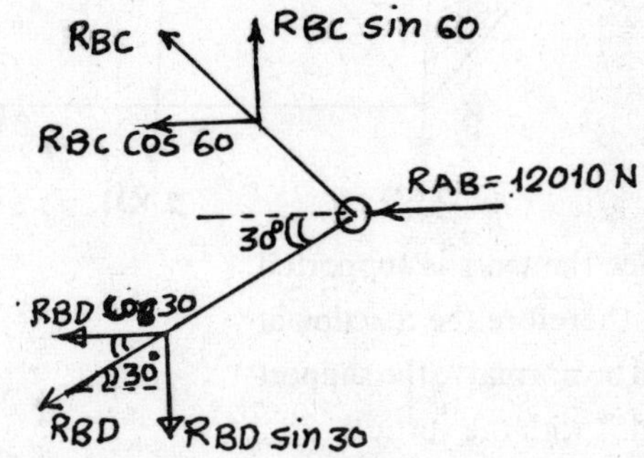

Consider FBD of joint D (or support D)

Components of R_{BD} will be, $R_{BD} \cos 30 = 10402 \cos 30 = 9008$ N

$R_{BD} \sin 30 = 10402 \sin 30 = 5201$ N

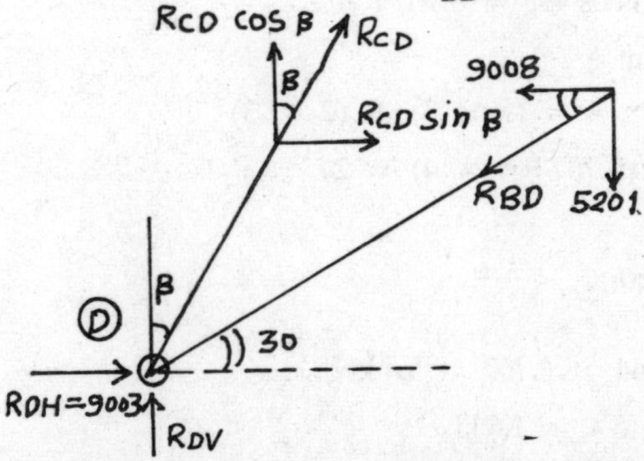

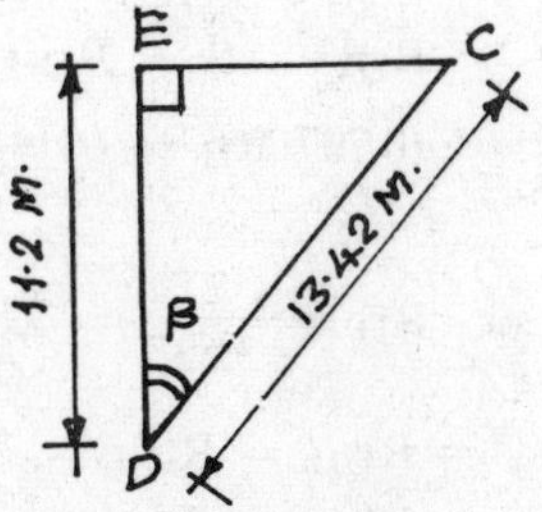

$\Sigma X = 9008 - R_{CD} \sin 33.43 - 9008 = 0 \quad \ldots\ldots (3)$

$\Sigma Y = R_{DV} + R_{CD} \cos 33.43 - 5201 = 0 \quad \ldots\ldots (4)$

from eq^n. (3) & (4) $\Rightarrow R_{CD} = 0$, $R_{DV} = 5201$ N

Problem 5. : For the truss shown in fig. find the forces in all the members of the truss and mention the nature of the force.

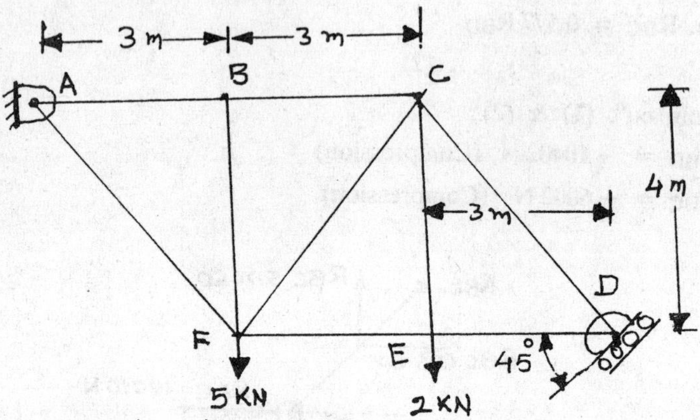

Solution : Since the truss is supported on rollers at D, therefore the reaction at this support will be normal to the support i.e. inclined at 45° with the vertical.

Consider FBD of joint D.

$\therefore$ Horizontal component of reaction at D,

$$R_{DH} = R_{DV} = R_D \cos 45^o = 0.707\, R_D$$

Now taking moments about A

$$R_{DV} \times 9 - R_{DH} \times 4 = (5 \times 3) + (2 \times 3)$$

$$(0.707\, R_D \times 9) - (0.707\, R_D \times 4) = 27$$

$$\text{or} \quad R_D = \frac{27}{3.535} = 7.64 \text{ kN}$$

$$\therefore \quad R_{DH} = R_{DV} = 7.64 \times 0.707 = 5.4 \text{ kN}$$

$$R_{AV} = (5 + 2) - 5.4 = 1.6 \text{ kN}$$

$$\text{and } R_{AH} = R_{DH} = 5.4 \text{ kN}$$

$$\therefore \quad R_A = [\,(1.6)^2 + (5.4)^2\,]^{1/2} = 5.63 \text{ kN}.$$

Consider FBD of joint A.

$$\tan\theta = \frac{4}{3} = 1.3333 \text{ or } \theta = 53.12^\circ$$

Resolving the forces vertically and equating the same,

$$R_{AF} \times \sin 53.12^\circ = 1.6$$

$$R_{AF} = \frac{1.6}{\sin 53.12^\circ} = \frac{1.6}{0.8} = 2.0 \text{ kN (Tension)}$$

and now resolving the forces horizontally and equationg the same,

$$R_{AB} = 5.4 + 2\cos 53.12^\circ$$
$$= 5.4 + (2 \times 0.6)$$
$$= 6.6 \text{ kN (Compression)}$$

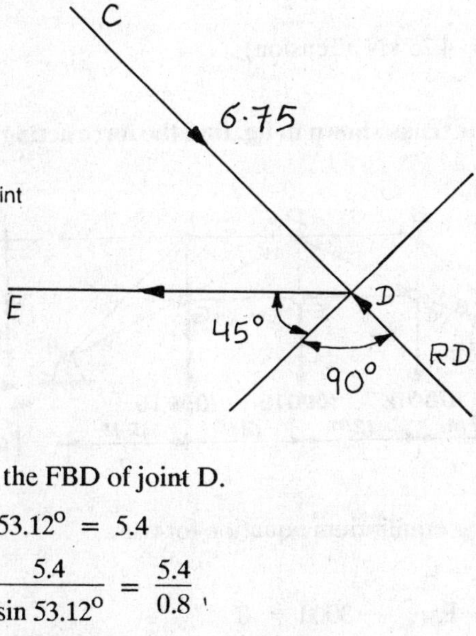

FBD of joint

Now consider the FBD of joint D.

$$R_{CD} \sin 53.12^\circ = 5.4$$

$$R_{CD} = \frac{5.4}{\sin 53.12^\circ} = \frac{5.4}{0.8}$$

$$= 6.75 \text{ kN (Compression)}$$

and now resolving the forces horizontally and equating the same,

$$R_{DE} = 5.4 - 6.75\cos 53.12^\circ$$
$$= 5.4 - (6.75 \times 0.6) \text{ kN}$$
$$= 1.35 \text{ kN (Compression)}$$

It is clear that value of R_{BC} will be equal to R_{AB} i.e. 6.6. kN (Compression). Moreover, R_{BF} will be zero, as there is no other member at joint B to balance it.

Consider the FBD of joint E.
A little consideration will show that the value of R_{CE} will be equal to the load at E. i.e. 2 kN (Tension). Similarly R_{EF} will be equal to R_{DE} i.e. 1.35 kN (Compression) as shown in fig.

Now consider the FBD of joint F.

$$R_{CF} \sin 53.12^{o} + 2.0 \sin 53.12^{o} = 5.0$$

$$\therefore \quad R_{CF} \times 0.8 = 5.0 - (2.0 \times 0.8) = 3.4$$

$$\text{or} \quad R_{CF} = \frac{3.4}{0.8} = 4.25 \text{ kN (Tension)}.$$

Problem 6. : For the truss shown in fig. find the force acting in each member.

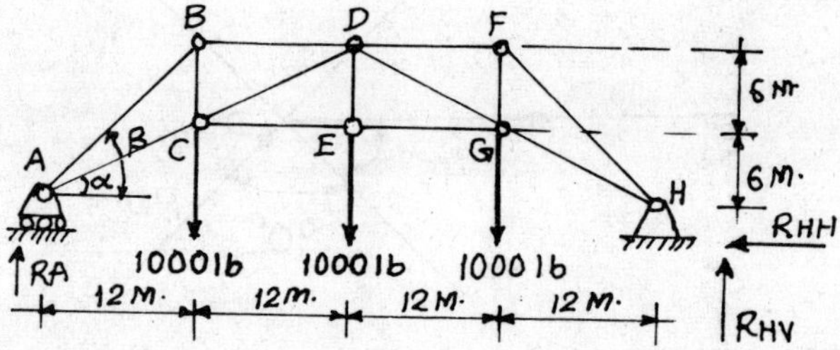

Solution : Applying equilibrium equation for truss.

$$\Sigma X = R_{HH} = 0$$

$$\Sigma Y = R_A + R_{HV} - 3000 = 0$$

$$\Sigma M_H = -R_A \times 48 + 1000 \times 36 + 1000 \times 24 + 1000 \times 12 = 0$$

$$\Rightarrow R_A = 1500 \text{ N}$$

$$\therefore \quad R_{HV} = 1500 \text{ N}$$

$$\angle\alpha = \tan^{-1}\left(\frac{6}{12}\right) = 26.56^{\circ}$$

$$\angle\beta = \tan^{-1}\left(\frac{12}{12}\right) = 45^{\circ}$$

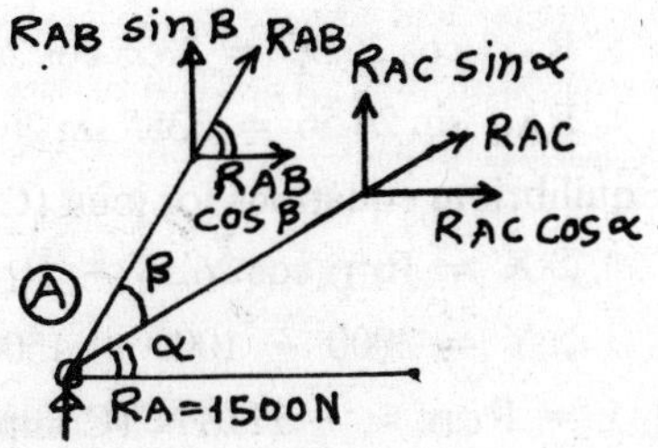

Consider FBD of joint A

Equilibrium equations for joint (A)

$$\Sigma X = R_{AB} \cos 45 + R_{AC} \cos 26.56 = 0$$

$$\Sigma Y = R_{AB} \sin 45 + R_{AC} \sin 26.56 + 1500 = 0$$

Solving above equations we get,

$$R_{AC} = 3355 \text{ N (Tension)}$$

$$R_{AB} = -4241 \text{ N (Compression)}$$

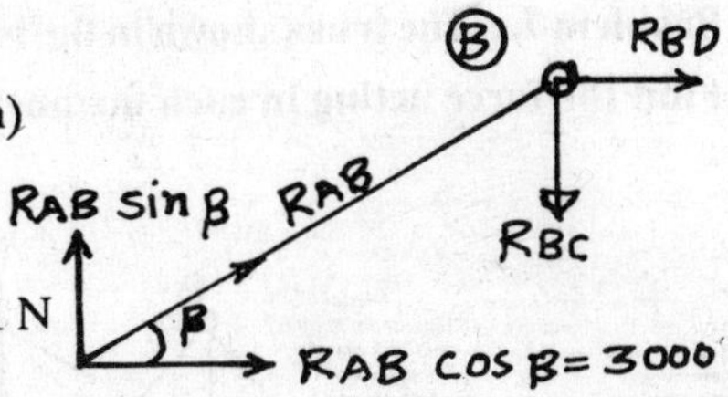

Consider FBD of joint B

We have, $\beta = 45^{\circ}$ & $R_{AB} = 4241$ N

$$\therefore \quad R_{AB} \cos\beta = 3000 \text{ N}$$

$$R_{AB} \sin\beta = 3000 \text{ N}$$

Applying equilibrium equations to joint B

$$\Sigma X = R_{BD} + 3000 = 0$$

$$\Sigma Y = 3000 - R_{BC} = 0$$

$$\therefore \quad R_{BC} = 3000 \text{ N (Tension)}$$

$$\& \quad R_{BD} = -3000 \text{ N (Compression)}$$

Consider FBD of joint C

Components of R_{AC} will be

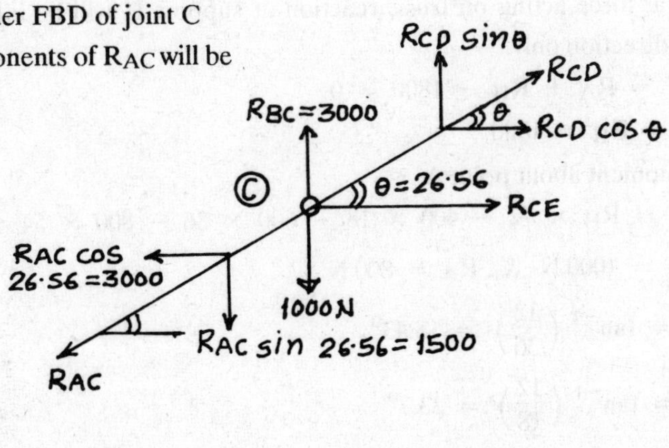

$$R_{AC}\cos 26.56 = 3355\cos 26.56 = 3000\text{ N}$$
$$R_{AC}\sin 26.56 = 3355\sin 26.56 = 1500\text{ N}$$

Equilibrium equations for joint (C) will give

$$\Sigma X = R_{CD}\cos 26.56 + R_{CE} - 3000 = 0$$
$$\Sigma Y = 3000 - 1000 - 1500 + R_{CD}\sin 26.56 = 0$$
$$\Rightarrow R_{CD} = -1118\text{ N (Compression)}$$
$$\Rightarrow R_{CE} = +4000\text{ N (Tension)}$$

Similarly we can get the forces in all the members.

Problem 7.: The truss shown in fig. is pin connected only at the lettered joints. Find the force acting in each member.

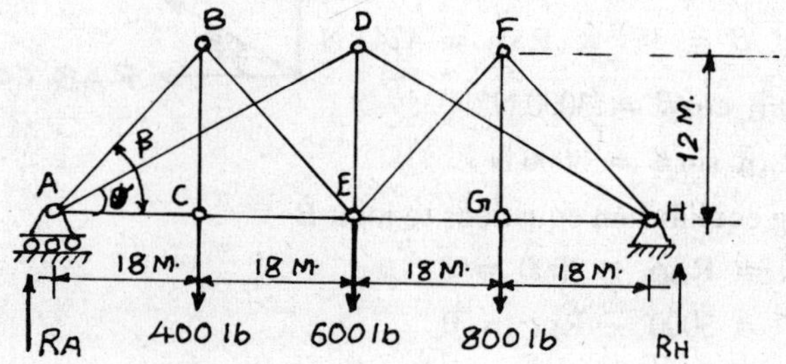

Solution : At support A only vertical reaction will be present. As there is no horizontal force acting on truss, reaction at support H will be in vertically upward direction only.

$$\Sigma Y = R_A + R_H - 1800 = 0$$

i.e. $R_A + R_H = 1800$

Taking moment about point A

$$\Sigma M_A = R_H \times 72 - 400 \times 18 - 600 \times 36 - 800 \times 54 = 0$$

$$\therefore \; R_H = 1000\text{ N} \;\&\; R_A = 800\text{ N}$$

$$\text{Now } \theta = \tan^{-1}\left(\frac{12}{36}\right) = 18.43^\circ$$

$$\text{and } \beta = \tan^{-1}\left(\frac{12}{18}\right) = 33.7^\circ$$

Consider FBD of joint and support A

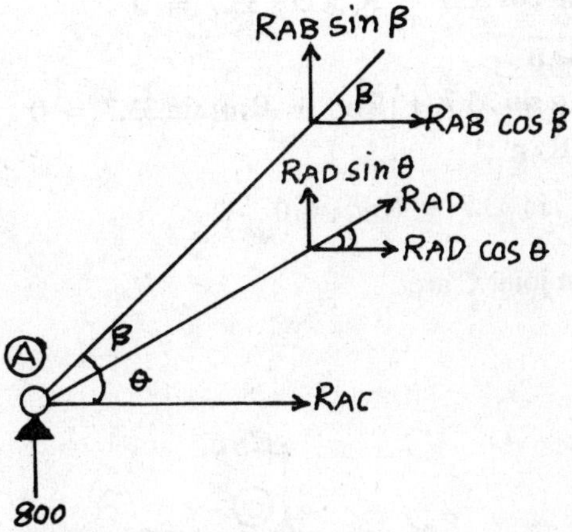

Equilibrium equations are

$$\Sigma Y = 800 + R_{AB} \sin 33.7 + R_{AD} \sin 18.43 = 0 \quad \ldots\ldots (1)$$

$$\Sigma X = R_{AB} \cos 33.7 + R_{AD} \cos 18.43 + R_{AC} = 0 \quad \ldots\ldots (2)$$

from eq^n. (1) $\Rightarrow R_{AB} + 0.569\, R_{AD} = -1442 \quad \ldots\ldots (3)$

from eq^n. (2) $\Rightarrow R_{AB} + 1.14\, R_{AD} + 1.2\, R_{AC} = 0 \quad \ldots\ldots (4)$

As number of unknowns are three and equations available are two so we cannot solve it now.

If we carefully see joint A, B & C then number of unknowns involved with these joints are six and we will have six equations. Thus we can get unknowns after writing equilibrium equations for joint B and C.

So consider FBD of joint B

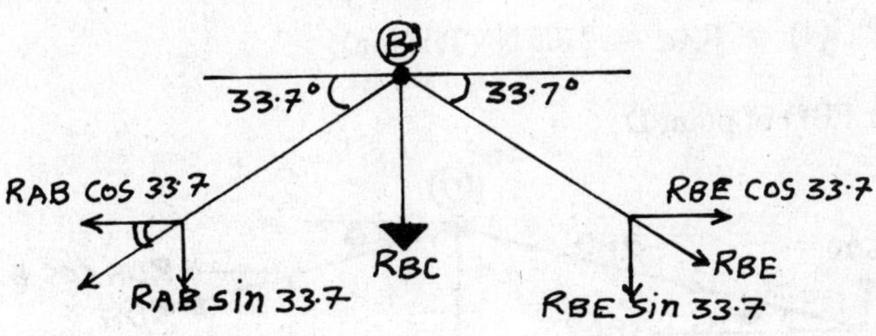

Equilibrium equations are

$\Sigma X = R_{BE} \cos 33.7 - R_{AB} \cos 33.7 = 0$

$\therefore \quad R_{BE} = R_{AB}$ ......(5)

$\Sigma Y = R_{AB} \sin 33.7 + R_{BC} + R_{BE} \sin 33.7 = 0$

Since $R_{BE} = R_{AB}$;

$\therefore \quad 2 \times R_{AB} \sin 33.7 + R_{BC} = 0$ ..... (6)

Consider FBD at joint C are

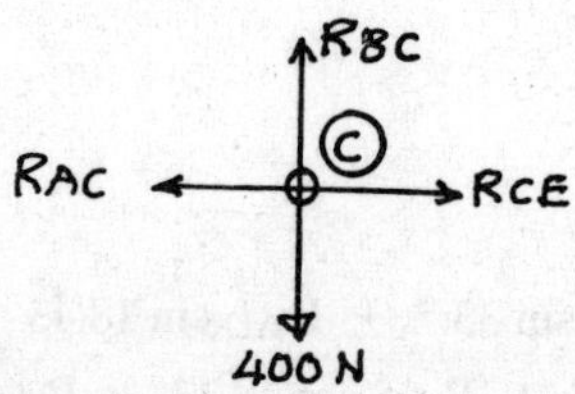

Equilibrium equations for joint (C)

$\Sigma X = R_{CE} - R_{AC} = 0$

$\therefore \quad R_{CE} = R_{AC}$ ......(7)

$\Sigma Y = R_{BC} - 400 = 0$

$\therefore \quad R_{BC} = 400$ N .... (8)

from eq^n. (6) $\Rightarrow R_{AB} = -360$ N (Compression)

from eq^n. (5) $\Rightarrow R_{BE} = -360$ N (Compression)

from eq^n. (3) $\Rightarrow R_{AD} = -1900$ N (Compression)

from eq^n. (4) $\Rightarrow R_{AC} = 2105$ N (Tension)

Consider FBD of point D

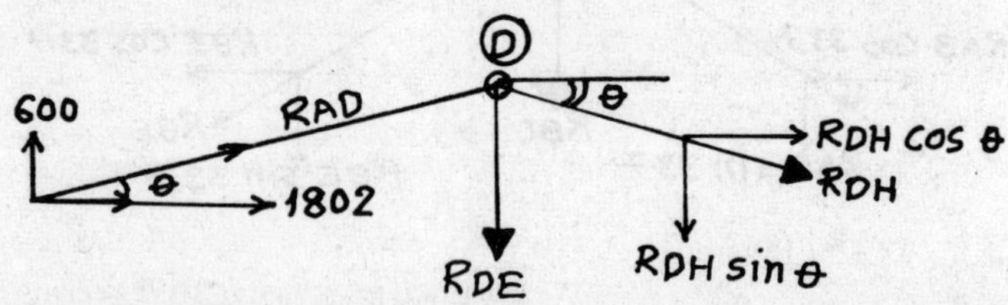

We have, $\theta = 18.43^\circ$

Equilibrium equations for joint (D) will be

$$\Sigma X = 1802 + R_{DH} \cos 18.43 = 0$$

$$\Rightarrow R_{DH} = -1900 \text{ N} \quad \ldots\ldots(9)$$

$$\Sigma Y = 600 - R_{DE} - R_{DH} \sin 18.43 = 0 \quad \ldots\ldots(10)$$

from eq^n. (9) (10) $\Rightarrow R_{DE} = 1200$ N (Tension)

and $R_{DH} = -1900$ N (Compression)

Similarly we can proceed for further joints E, F, G and H.

Further analysis is left as an assignment for students. Force in member GH = 2400 N (Tension).

Problem 8. : In the cantilever truss shown in fig. compute the force in each member.

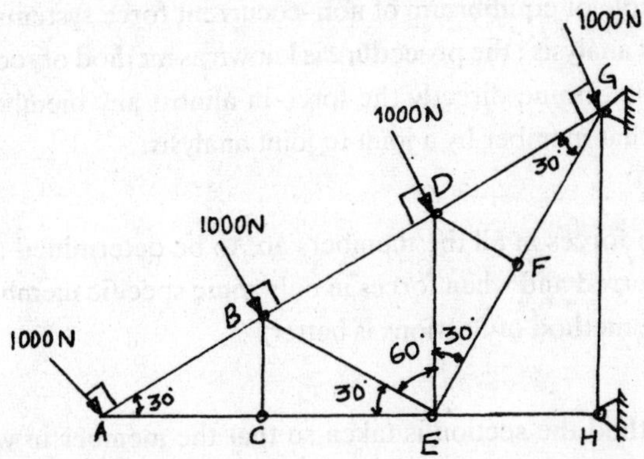

Solution : Inspection of joint C and F shows that member BC and DF are null member. i.e. $R_{BC} = R_{DF} = 0$

Consider FBD of joint A :

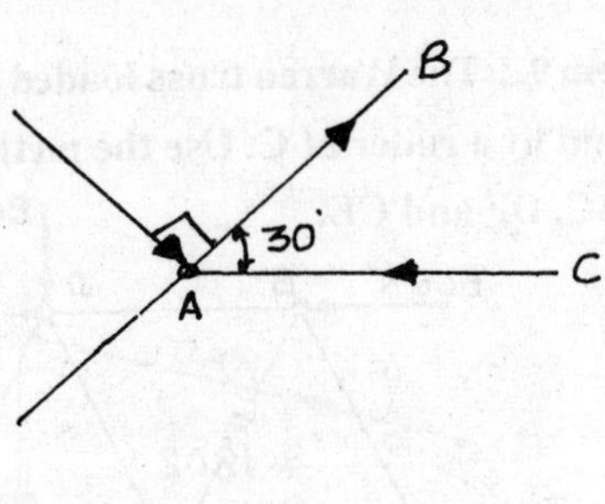

Let X–axis be along AB

$\therefore \; R_{AC} \sin 30 = 1000$ N

$\Rightarrow R_{AC} = 2000$ N (Compression)

and $R_{AC} \cos 30 = R_{AB}$

$\therefore \; R_{AB} = 1732$ (Tension)

Consider FBD of joint B :

Taking X–axis along BD.

$R_{BE} \sin 60 = 1000$ N

$\therefore \quad R_{BE} = 1154$ N (Compression)

$R_{BD} = R_{BE} \cos 60 = 577$ N

Consider FBD of joint D :

$\therefore \quad R_{DE} \sin 60 = 1000$ N

$\Rightarrow R_{DE} = 1154$ N (Compression)

Similarly we can find for other members.

Method of Sections :

The principle of equilibrium of non–cocurrent force systems may also be applied to truss analysis ; the procedure is known as method of sections. Its use permits us to determine directly the force in almost any member instead of proceeding to that member by a joint to joint analysis.

Conclusion :

Thus when forces in all the members are to be determined ; the method of joints is preferred and when forces in only some specific members are to be known then the method of Sections is better.

Procedure :

In this method the section is taken so that the member in which we are interested to find the force is cut and then equilibrium equations are applied for half of the truss.

Equations of equilibrium are $\Sigma X = 0$, $\Sigma Y = 0$ and $\Sigma M = 0$.

Problem 9.: The Warren truss loaded as shown in fig. is supported by a hinge at G and by a roiler at C. Use the method of sections to compute the force in bars BC, DF and CE.

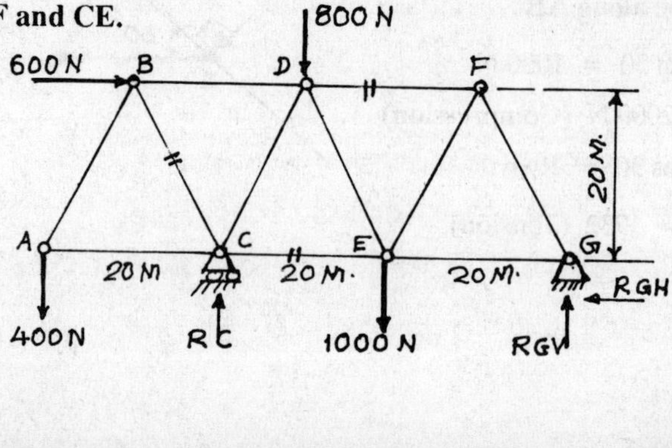

Solution : First find the reactions at C and G.

Taking moment about G we get,

$$\Sigma M_G = 400 \times 60 - 600 \times 20 - R_C \times 40 + 800 \times 30 + 1000 \times 20 = 0$$

$\Rightarrow R_C = 1400$ N

$\Sigma Y = R_C + R_{GV} - 2200 = 0$

$\Rightarrow R_{GV} = 800$ N

Also we have $R_{GH} = 600$ N

We want to find force in member BC therefore we take section which will cut BD, BC and AC as shown below.

Section for member BC :

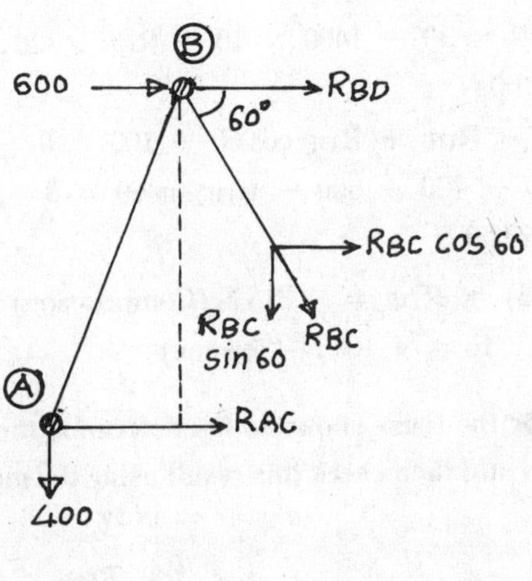

Equilibrium equations can be applied to above FBD.

Taking moment about B

$\Sigma M_B = R_{AC} \times 20 + 400 \times 10 = 0$

$\Rightarrow R_{AC} = -200$ N (Compression) ..... (1)

$\Sigma X = R_{AC} + R_{BD} + 600 + R_{BC} \cos 60 = 0$ ..... (2)

$\Sigma Y = -400 - R_{BC} \sin 60 = 0$

$\Rightarrow R_{BC} = -462$ N (Compression)

from eq^n. (2) $\Rightarrow R_{BD} = -169$ N (Compression)

Section for member DF :

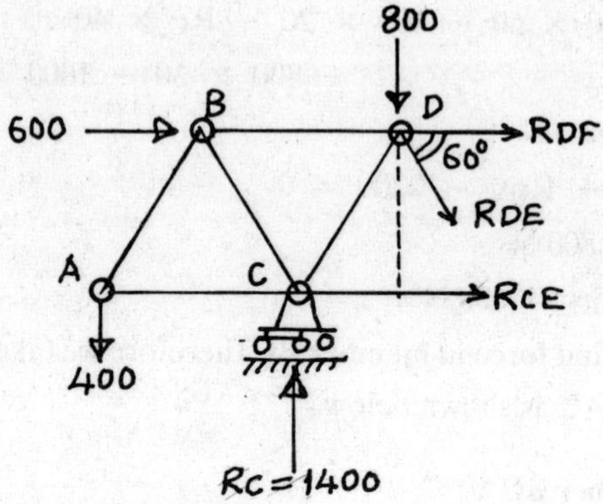

$$\Sigma M_D = 400 \times 30 - 1400 \times 10 + R_{CE} \times 20 = 0$$

$$\Rightarrow R_{CE} = 100 \text{ N}$$

$$\Sigma X = 600 + R_{DF} + R_{DE} \cos 60 + 100 = 0 \quad \ldots\ldots (3)$$

$$\Sigma Y = 1400 - 400 - 800 - R_{DE} \sin 60 = 0$$

$$\Rightarrow R_{DE} = 231.0 \text{ N}$$

and from eq^n. (3) $\Rightarrow R_{DF} = -815$ N (Compression)

Force in member CE = + 100 N (Tension) .

Problem 10. : For the truss shown in fig. determine the force in BF by the method of section and then check this result using the method of joints.

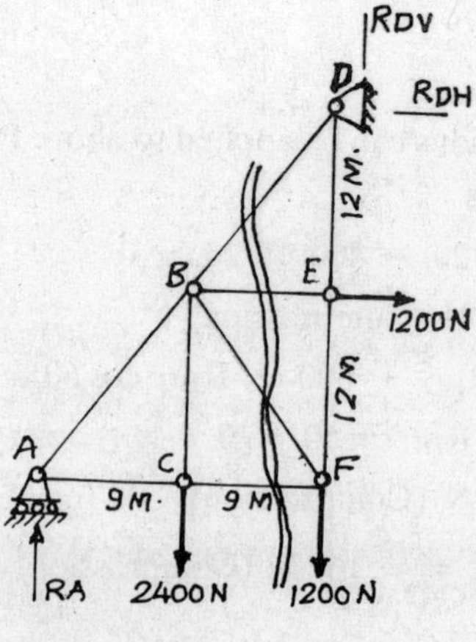

Solution : To find the force in member BF we have to take the section as shown in fig. by cutting members BD, BE, BF and CF.

Now nunmber of unknowns will be four. Due to this reason we have to apply intution to find one of the unknowns. Now if we draw FBD of joint (E) we will get $R_{BE} = +1200$ N (Tension)

Now we can apply equilibrium equations to the section.

Also we have $R_A = +2000$ N

$R_{DV} = -1600$ N & $R_{DH} = +1200$ N

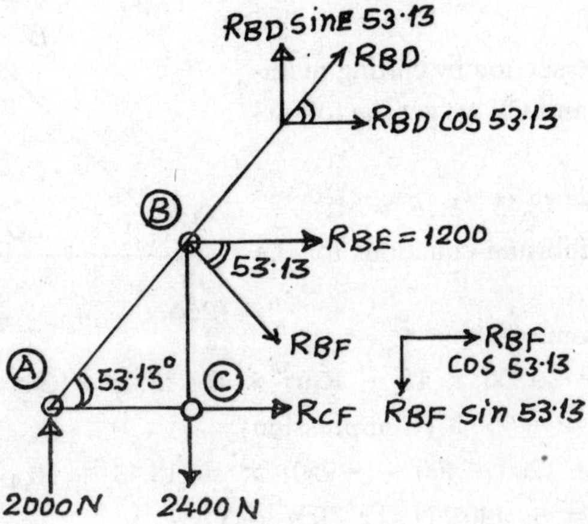

Taking moment about B

$\Sigma M_B = R_{CF} \times 12 - 2000 \times 9 = 0$

$\Rightarrow R_{CF} = +1500$ N

$\Sigma X = R_{CF} + R_{BF} \cos 53.13 + 1200 + R_{BD} \cos 53.13 = 0$

$R_{BF} + R_{BD} = -4500$ N ...... (1)

$\Sigma Y = 2000 - 2400 - R_{BF} \sin 53.13 + R_{BD} \sin 53.13 = 0$

$\Rightarrow -R_{BF} + R_{BD} = 500$ N ...... (2)

Adding (1) and (2) we get

$R_{BD} = -2000$ N (Compression)

$\therefore$ $R_{BF} = -2500$ N (Compression).

Solution using method of joints is left as an exercise for the students.

Problem 11.: For the truss shown in fig. determine each of the forces in bars BD, CD and CE by means of an equilibrium equation that determine it independently of the others.

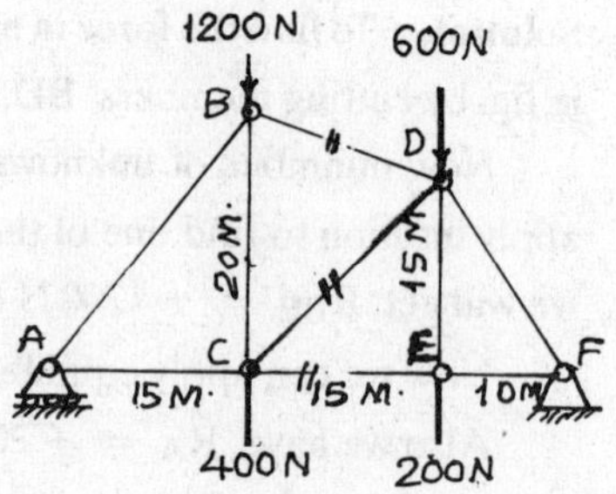

Solution : Using the reactions determined in problem (1).

$R_{AV} = 1200$ N

$R_{FV} = 1200$ N

$R_{FH} = 0$

If we take section by cutting members BD, CD and CE we get the following FBD.

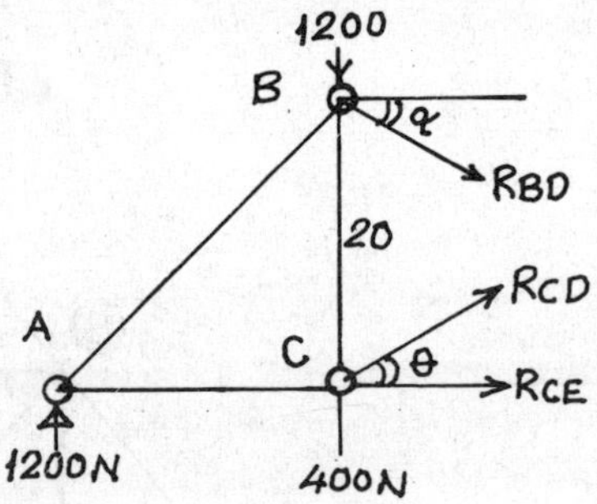

We have, $\alpha = 18.43^\circ$ & $\theta = 45^\circ$.

Applying equilibrium equations for the truss we get

Taking moment about C

$\Sigma M_C = -1200 \times 15 - R_{BD} \times \cos 18.43 \times 20 = 0$

$\Rightarrow R_{BD} = -950$ N (Compression)

$\Sigma Y = 1200 - 1200 - 400 - (-950) \times \sin 18.43 + R_{CD} \sin 45 = 0$

$\Rightarrow R_{CD} = +140.9$ N (i.e. Tension)

$\Sigma X = -950 \cos 18.43 + 140.9 \cos 45 + R_{CE} = 0$

$\Rightarrow R_{CE} = +801$ N (Tension)

Problem 12.: Use method of sections to determine independently the force in bars AD, BE & CE of truss shown in fig. in problem no. 7.

Solution : Referring to the same fig and reactions of problem no.7 we get the desired section and FBD of truss as shown below.

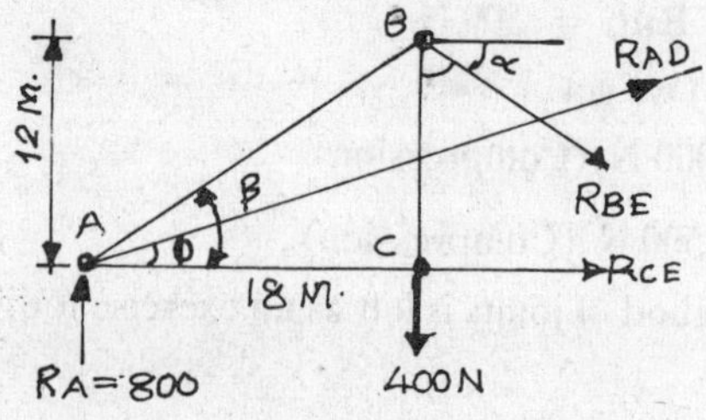

where $\theta = 18.43^o$; $\beta = 33.7^o$; & $\alpha = 33.7^o$

Taking moment about B we get

$\Sigma M_B = -800 \times 18 - R_{AD} \sin\theta \times 18$

$+ R_{AD} \cos\theta \times 12 + R_{CE} \times 12 = 0$ ..... (1)

$\Sigma Y = 800 - 400 + R_{AD} \sin\theta - R_{BE} \sin\alpha = 0$ ..... (2)

$\Sigma X = R_{AD} \cos\theta + R_{BE} \cos\alpha + R_{CE} = 0$ ..... (3)

from eq^n. (2) $\Rightarrow R_{AD} - 1.755\ R_{BE} = -1265$ ..... (4)

from eq^n. (1) $\Rightarrow 5.69\ R_{AD} + 12\ R_{CE} = 14400$ ..... (5)

from eq^n. (3) $\Rightarrow 0.948\ R_{AD} + 0.831\ R_{BE} + R_{CE} = 0$ ..... (6)

Solving equation (4), (5), & (6) we get

$R_{BE} = -360$ N (Compression)

$R_{AD} = -1900$ N (Compression)

$R_{CE} = +2105$ N (Tension)

METHOD OF MEMBERS :

Frames :

If some or all the members of a pin connected structure are subjected to bending action because of loads applied along their length in addition to loads at their ends, the structure is called a 'Frame'.

To analyse frames, we consider the separate Free Body Diagram of each member subjected to bending. The procedure is called the method of members.

Problem 13 : The structure shown in fig. is hinged at A and C. Find the horizontal and vertical components of the hinge force at A, B & C.

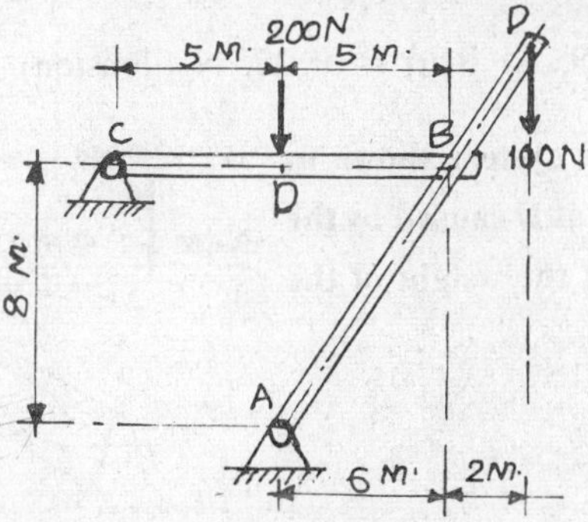

Solution : Consider the FBD of whole system,

Taking moment about C, we get

$\Sigma M_C = -200 \times 5 - 100 \times 12 + R_{AV} \times 4 + R_{AH} \times 8 = 0$

$\Rightarrow R_{AV} + 2R_{AH} = 550$ ..... (1)

$\Sigma Y = R_{AV} + R_{CV} - 300 = 0$ ..... (2)

$\Sigma X = R_{CH} + R_{AH} = 0$ ..... (3)

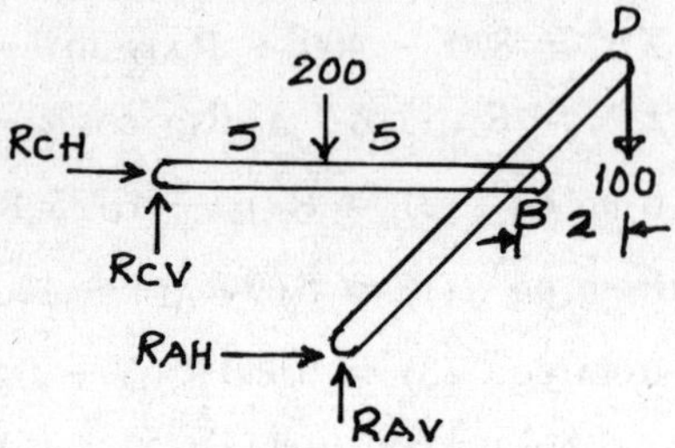

Now we have three equations and four unknowns. Hence to get more equations we shall split the frame.

Considering the FBD of member CB we get

(Note that directions of R_{CH} and R_{CV} are maintained same).

$\Sigma X = R_{CH} - R_{BH} = 0$

$\Rightarrow R_{CH} = R_{BH}$ ..... (4)

$\Sigma Y = R_{CV} + R_{BV} - 200 = 0$

$\Rightarrow R_{CV} + R_{BV} = 200$ ..... (5)

Taking moment about B

$\Sigma M_B = -R_{CV} \times 10 + 200 \times 5 = 0$

$\Rightarrow R_{CV} = +100$ N

from eq^n. (5) $\Rightarrow R_{BV} = +100$ N

from eq^n. (2) $\Rightarrow R_{AV} = +200$ N

from eq^n. (1) $\Rightarrow R_{AH} = +175$ N

from eq^n. (3) $\Rightarrow R_{CH} = -175$ N

from eq^n. (4) $\Rightarrow R_{BH} = -175$ N

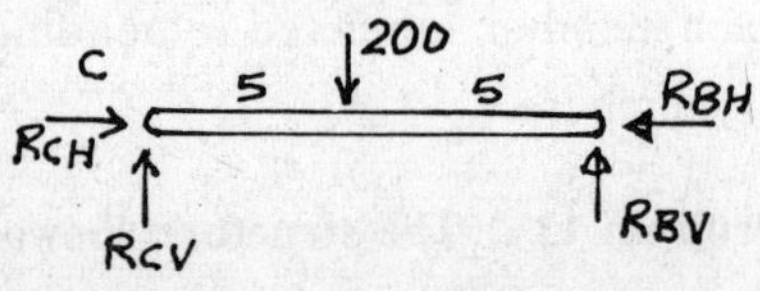

$R_{BV} = +100$ N & $R_{BH} = +175$ N (Tension).

Problem 14. : In the system shown in fig. find the reaction at D caused by the 1200 N load. Neglect the weight of the member.

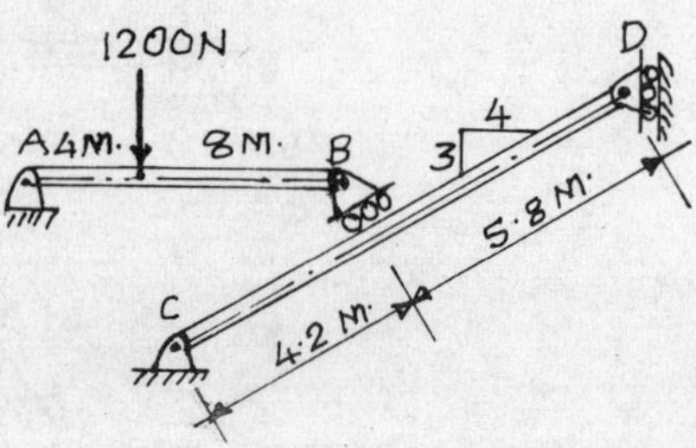

Solution : Consider FBD of given system :

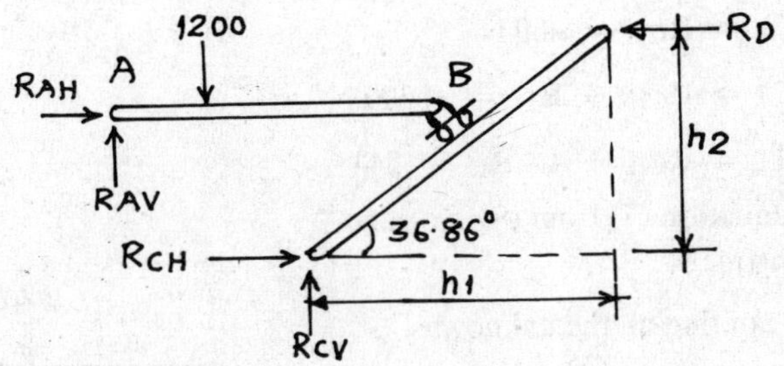

First we find distances h_2 and h_1

Now $\sin 36.86 = \dfrac{h_2}{10}$ and $\cos 36.86 = \dfrac{h_1}{10}$

$\therefore\ h_2 = 6m\ ;\ h_1 = 8m$

From Equilibrium equations for above FBD we get

$\Sigma M_A = -1200 \times 4 + R_{CV} \times (12 - 4.2 \cos 36.86)$
$\quad + R_{CH} \times (6 - 5.8 \sin 36.86) + R_D \times 5.8 \sin 36.86 = 0$

$\therefore\ 8.6\, R_{CV} + 2.5\, R_{CH} + 3.5\, R_D = 4800$ ..... (1)

$\Sigma X = R_{AH} + R_{CH} - R_D = 0$ ..... (2)

$\Sigma Y = R_{AV} + R_{CV} - 1200 = 0$

$\Rightarrow R_{AV} + R_{CV} = 1200$ ..... (3)

Again considering the FBD of member AB alone to get 3 more equations, so that number of unknowns becomes equal to number of equations.

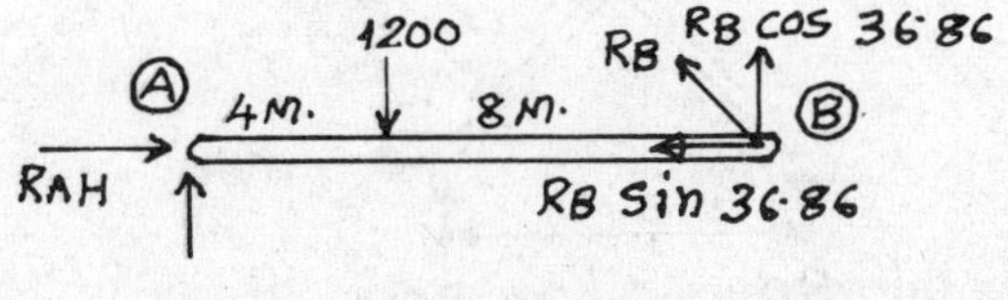

$\Sigma M_A = -1200 \times 4 + R_B \cos 36.86 \times 12 = 0$

$\Rightarrow R_B = 500\ N$ ..... (4)

$\Sigma X = 0\ ;\ R_{AH} = 300\ N$ ..... (5)

$$\Sigma Y = R_{AV} - 1200 + 500 \cos 36.86 = 0$$

$$R_{AV} = 800 \text{ N} \quad \ldots\ldots (6)$$

from eq^n. (3) $\Rightarrow R_{CV} = 400$ N

from eq^n. (2) $\Rightarrow R_{CH} - R_D = -300$ N ..... (7)

from eq^n. (1) $\Rightarrow R_{CH} + 1.4 R_D = 544$ ..... (8)

Subtracting equation (7) from (8) we get

$$R_D = 350 \text{ N}$$

Similarly we can find all the unknowns.

Problem 15.: The frame shown in fig. is supported by a hinge at E and a roller at D. Determine the horizontal and vertical components of the hinge force at C as it acts upon member BD.

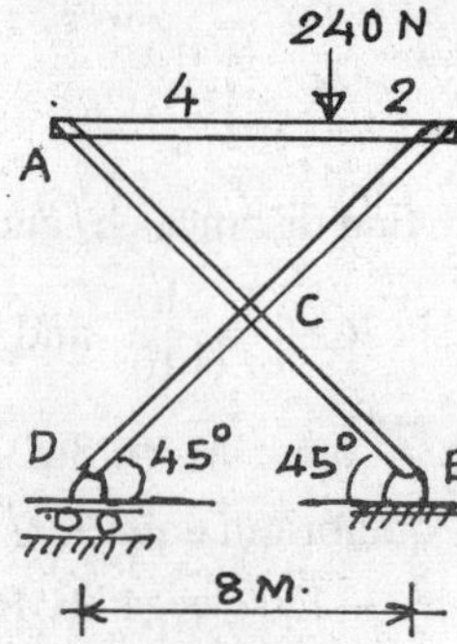

Solution : Applying Equilibrium Equations

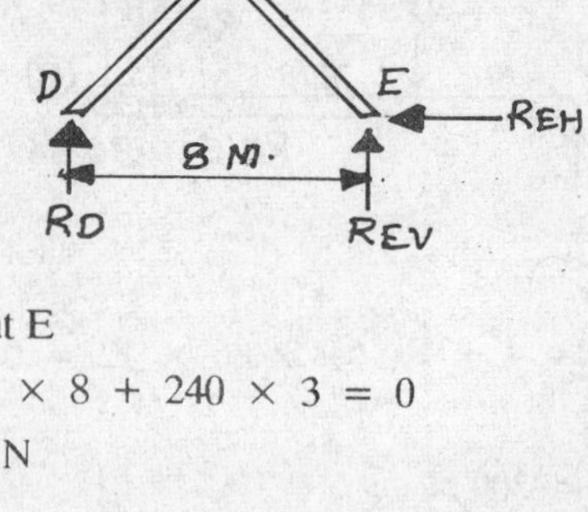

Taking moment about E

$$\Sigma M_E = - R_D \times 8 + 240 \times 3 = 0$$

$$\Rightarrow R_D = + 90 \text{ N} \quad \ldots\ldots (1)$$

$\Sigma Y = R_D + R_{EV} - 240 = 0$

$\therefore \quad R_{EV} = +150 \text{ N}$ ......(2)

$\Sigma X = R_{EH} = 0$ ......(3)

Consider FBD of DCB

$\Sigma M_B = -R_D \times 7 - R_{CV} \times 3 + R_{CH} \times 3 = 0$

$\therefore \quad R_{CH} - R_{CV} = 210$ ..... (4)

$\Sigma Y = 90 + R_{CV} + R_{BV} = 0$ .....(5)

$\Sigma X = R_{BH} + R_{CH} = 0$

$\Rightarrow R_{BH} = -R_{CH}$ ..... (6)

Consider FBD of member AB.

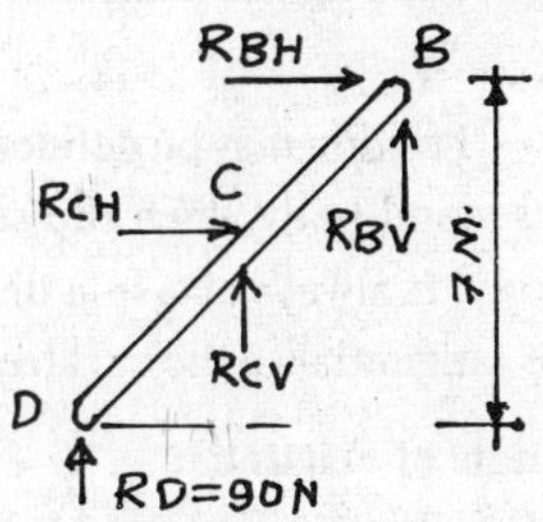

$\Sigma M_A = -240 \times 4 - R_{BV} \times 6 = 0$

$\Rightarrow R_{BV} = -160 \text{ N}$

$\Sigma Y = R_{AV} - 240 - R_{BV} = 0$

$\Rightarrow R_{AV} = +80 \text{ N}$

$\Sigma X = R_{AH} - R_{BH} = 0$

......(7)

from eq^n. (5) $\Rightarrow R_{CV} = 70 \text{ N}$

from eq^n. (4) $\Rightarrow R_{CH} = 280 \text{ N}$

from eq^n. (6) $\Rightarrow R_{BH} = -280 \text{ N}$

from eq^n. (7) $\Rightarrow R_{AH} = -280 \text{ N}$

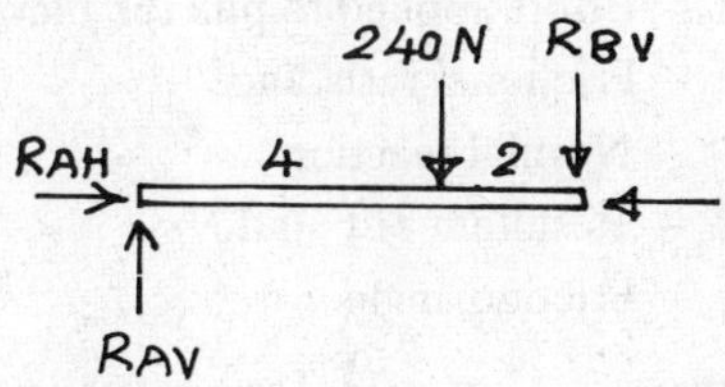

Chapter Four

FRICTION

Friction may be defined as the contact resistance exerted by one body upon a second body when the second body moves or tends to move past the first body. It always acts in a direction opposite to that of relative motion and will be tangential to the surface of two bodies at point of contact.

Angle of Friction :

W – Weight of block

P – Effort applied to pull the block

F – Frictional resistance

N – Normal reaction

R – Resultant of F and N

φ – Friction angle

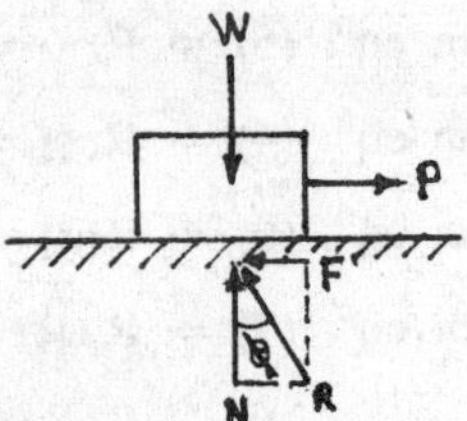

Thus $\tan\varphi = \left(\frac{F}{N}\right)$

Laws of Friction : (Coulomb's laws of friction)

(1) The force of friction always acts in a direction opposite to that in which the body tends to move.

(2) The magnitude of frictional force is directly proportional to a normal reaction between the contact surfaces. i.e. $F \alpha N$ or $F_S = f_S N$.

The proportionality constant is called coefficient of static friction.

i.e. coefficient of static friction (f_s) = F_s/N = maximum static friction when motion is impending.

(3) If friction is neglected, reactions are always normal to surface of contact.

(4) If static friction is acting the value of friction force may vary from zero to its maximum available value F_s adjusting itself to resultant force tending to cause motion.
(5) If motion occurs, the kinectic friction force always acts at a constant value of f_k. N, where f_k is coefficient of kinectic friction.
(6) The coefficient of static friction is always greater than that of kinetic friction i.e $f_s > f_k$
(7) The angle between the total reaction (R) and its normal component (N) in limiting condition is called the angle of friction (ϕ) . The tangent of this angle is equal to the coefficient of friction i.e. $\tan\phi = f$.

Static Friction and Kinetic Friction :

Static friction is defined as the friction experienced by a body, when it is at rest. Or in otherwords it is the friction when the body tends to move. Kinetic friction is the friction experienced by the body when it is in motion. It is also called as dynamic friction.

Results from experiments are plotted in the graph. Thus it is clear that static friction is always more than kinetic friction.

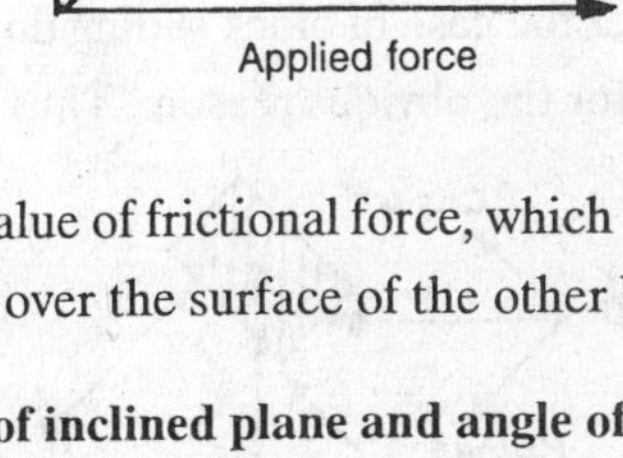

Limiting Friction :

It is the maximum value of frictional force, which comes into play, when a body just begins to slide over the surface of the other body.

Relation between angle of inclined plane and angle of friction :

If θ = angle of inclined plane and φ = angle of friction.
(i) If $\varphi > \theta$ then block will not move
(ii) If $\varphi = \theta$ Block just starts sliding down
(iii) If $\varphi < \theta$ Block starts moving down on its own weight.

Problem 1. : The 500 N block shown is in contact with 45° incline. The coefficient of friction is 0.25. Compute the value of horizontal force P necessary to

(a) just start the block up the incline.

(b) just prevent motion down the incline.

(c) If P = 400 N, what is the amount and direction of the friction force ?

Solution : (a) If the case under consideration is block moving upwards, then in that case friction will be acting downwards.

FBD will be as shown .

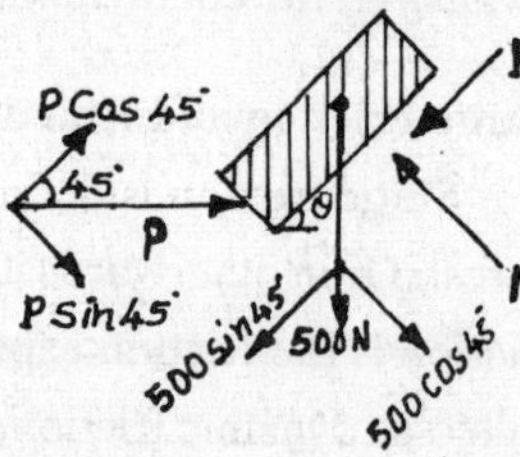

Applying equilibrium equations , we have

$$\Sigma X = P \cos 45 - F - 500 \sin 45 = 0 \quad \ldots\ldots (1)$$

$$\Sigma Y = N - 500 \cos 45 - P \sin 45 = 0 \quad \ldots\ldots (2)$$

$$\text{Also we have } \mu = \tan \varphi = \frac{F}{N} = 0.25 \quad \ldots\ldots (3)$$

where, μ is coefficient of friction.

from eq^n. (1) $\Rightarrow$ P cos 45 $-$ F = 354

from eq^n. (2) $\Rightarrow$ 4F $-$ P sin 45 = 354 $\quad (\because \quad F = 0.25\,N)$

Solving above equations we get,

F = 236 N ; P = 834 N

(b) If we have to consider the case of block sliding down, then the friction will act in upward direction for the obvious reason. Thus FBD will be ;

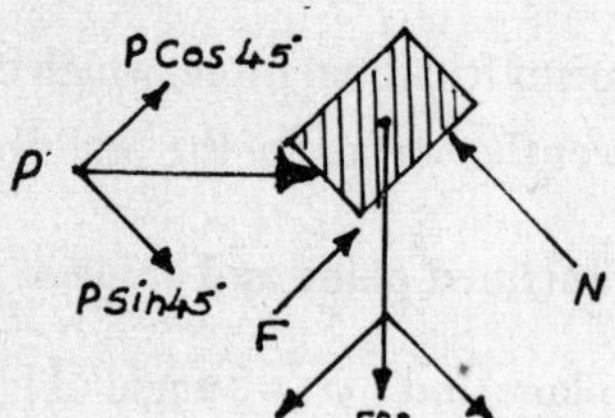

Applying equilibrium equations.

$$\Sigma X = F - 354 + P \cos 45 = 0 \quad \ldots\ldots (4)$$

$\Sigma Y = N - 354 - P \sin 45 = 0$ ..... (5)

from eq^n. (4) $\Rightarrow$ $F + P \cos 45 = 354$

from eq^n. (5) $\Rightarrow$ $4F - P \sin 45 = 354$ $(\because \; F = 0.25\, N)$

Solving above two equations, we get

$F = 141.6\, N \quad \& \quad P = 300\, N$

(c) If $P = 400$ then we have to determine magnitude and direction of friction force. In this case we shall assume the direction of F arbitrarily and proceed for FBD.

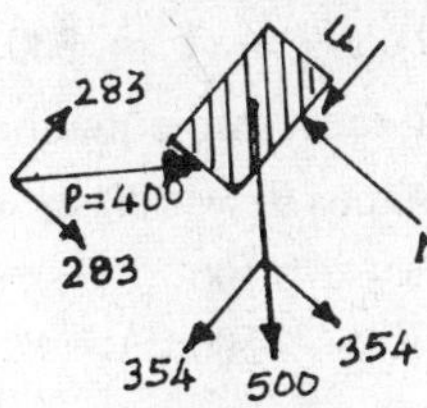

$\Sigma X = 283 - 354 - F = 0$

$F = -71\, N$

Thus F should be acting upwards.

$F = 71\, N$ (upwards)

Problem 2.: Bodies A and B are joined by a cord parallel to the inclined plane shown in fig. Under body A which weighs 200 N, f = 0.20, while f = 0.50 under body B which weighs 300 N. Determine the angle θ at which motion impends. What is then the tension in the cord ?

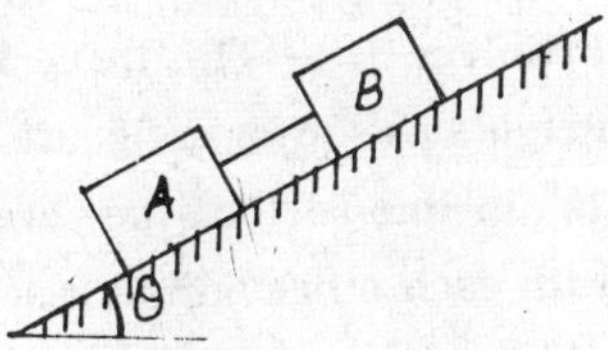

Solution : Consider FBD of block A

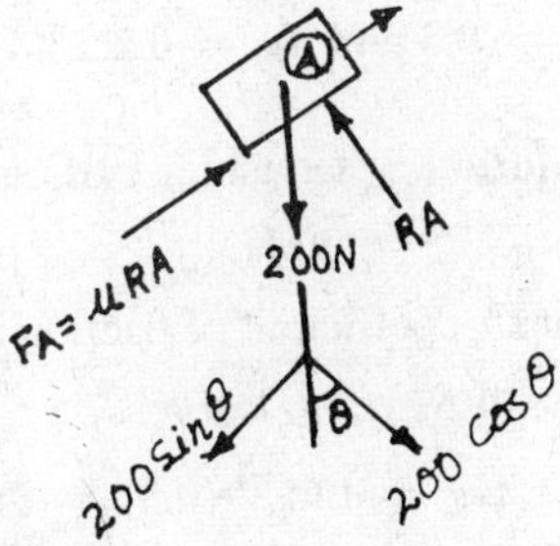

Applying equilibrium equations

$\Sigma Y = R_A - 200 \cos\theta = 0$ ...... (1)

$\Sigma X = \mu R_A + T - 200 \sin\theta = 0$ ...... (2)

$\mu = 0.2$ (given)

from eq^n. (1) $\Rightarrow$ $R_A = 200 \cos\theta$ ..... (3)

from eq^n. (2) $\Rightarrow$ $0.2 R_A + T = 200 \sin\theta$ ..... (4)

Consider FBD of block B

Applying equilibrium equations :

$$\Sigma X = 0.5\,R_B - 300 \sin\theta - T = 0 \quad \ldots\ldots (5)$$

$$\Sigma Y = R_B - 300 \cos\theta = 0 \quad \ldots\ldots (6)$$

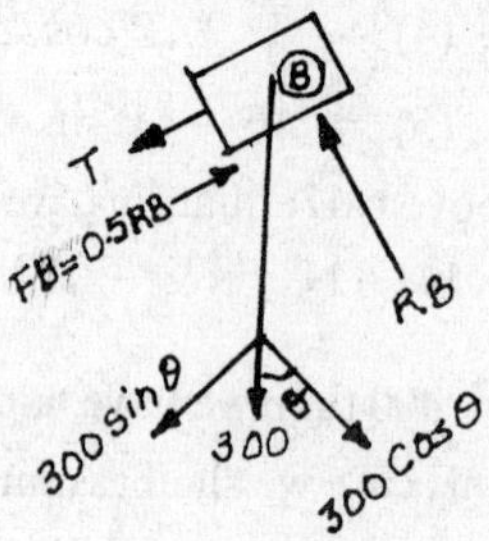

from eq^n. (3) & (4)

$$40 \cos\theta + T = 200 \sin\theta$$

from eq^n. (5) & (6)

$$\Rightarrow 150 \cos\theta - T = 300 \sin\theta$$

Adding these two equations

$$190 \cos\theta = 500 \sin\theta$$

$$\tan\theta = \frac{190}{500}$$

$\therefore\ \theta = 20.80^o$ & therefore $T = 34$ N

Problem 3. : The three blocks with weights as shown in fig. are placed on a 20° incline so that they are in contact with each other and at rest. Determine which, if any, of the blocks will move and the friction force acting under each. Assume that under blocks A & C, the coefficients of friction $f_s = 0.5$ and $f_k = 0.4$ while under B they are $f_s = 0.3$ and $f_k = 0.2$.

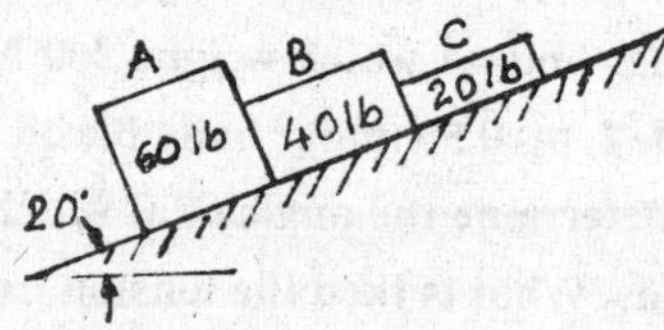

Solution : Consider FBD of (C)

$W_C = 20$ N

For C coefficient of friction is 0.5

We know $\mu = \tan\varphi$

$\therefore\ \varphi_C = \tan^{-1}(0.5) = 26.56^o$

and angle of inclination is 20°.

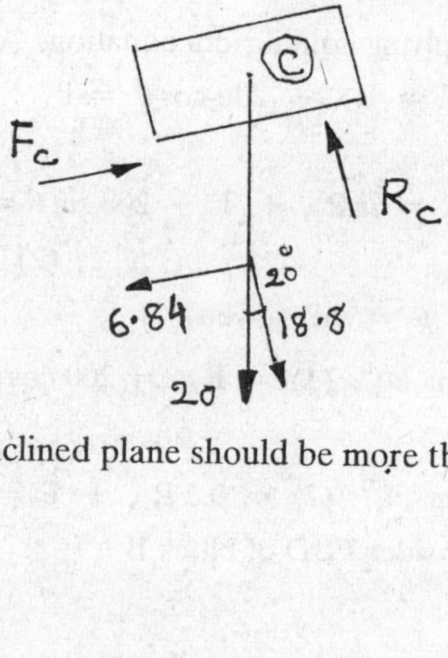

For body to move on its own the angle of inclined plane should be more than friction angle. Since $20^o < 26.56^o$.

The block C will not move.

From FBD of block C , $F_C = 6.84$ N

For block B the coefficient of friction is 0.3

$\therefore \quad \varphi_B = \tan^{-1}(0.3) = 16.69^o$

Now φ_B is less than 20°

$\therefore$ Block B will tend to move down.

$F_B = 0.3\, R_B$

$= 0.3\,(37.6) = 11.28$ N

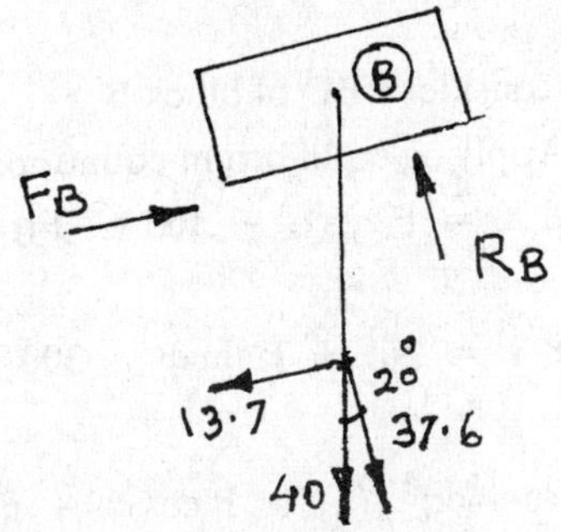

For block A

$\varphi_A = 26.56^o$

$\therefore$ Block A will not move.

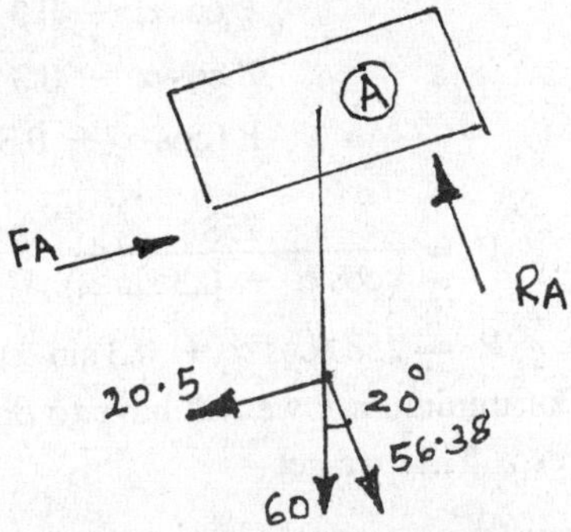

Consider FBD of block A

$R_A = 56.38$ N

$F_A = (20.5 + R_{AB}) = 22.9$ N

Thus no block will move.

Problem 4. : Find the least value of P that will just start the system of blocks shown moving to the right. The coefficient of friction under each block is 0.3

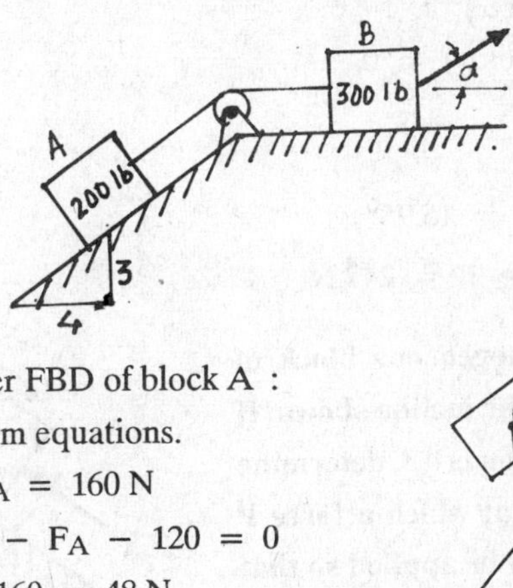

Solution : Consider FBD of block A :

Applying equilibrium equations.

$\Sigma Y = 0 \; ; \; N_A = 160$ N

$\Sigma X = 0 \; ; \; T - F_A - 120 = 0$

$F_A = 0.3 \times 160 = 48$ N

$\therefore \quad T = 168$ N

Consider FBD of block B :

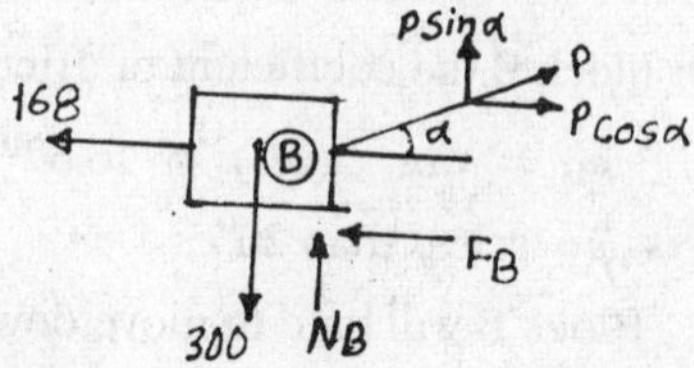

Applying equilibrium equations.

$$\Sigma X = P\cos\alpha - 168 - F_B = 0 \quad \ldots\ldots (1)$$

$$\Sigma Y = N_B + P\sin\alpha - 300 = 0 \quad \ldots\ldots (2)$$

from eq^n. (1) ⇒ $P\cos\alpha - F_B = 168$

$$P\cos\alpha - 0.3\,N_B = 168$$

$$P\cos\alpha - 0.3\,(300 - P\sin\alpha) = 168$$

$$P(\cos\alpha + 0.3\sin\alpha) = 258 \quad \ldots\ldots (3)$$

$$P = \frac{258}{(\cos\alpha + 0.3\sin\alpha)}$$

$$P = 258\,(\cos\alpha + 0.3\sin\alpha)^{-1}$$

For minimum P we will have to differentiate above equation and equate it to zero. Thus we get

$$\frac{dP}{d\alpha} = 258\,(\cos\alpha + 0.3\sin\alpha)^{-2}\,(-\sin\alpha + 0.3\cos\alpha) = 0$$

i.e. $\dfrac{(-\sin\alpha + 0.3\cos\alpha)}{(\cos\alpha + 0.3\sin\alpha)^2} = 0$ (Neglect the value of denominator)

i.e. $(-\sin\alpha + 0.3\cos\alpha) = 0$

$$\sin\alpha = 0.3\cos\alpha$$

$$\tan\alpha = 0.3$$

$$\alpha = \tan^{-1}(0.3) = 16.95^\circ$$

∴ from eq^n. (3) ⇒ $P = 247\ \text{N}$

Problem 5. : A homogeneous block of weight W rests upon the incline shown. If the coefficient of friction is 0.3, determine the greatest height h at which a force P parallel to incline may be applied so that block will slide up the incline without tipping over.

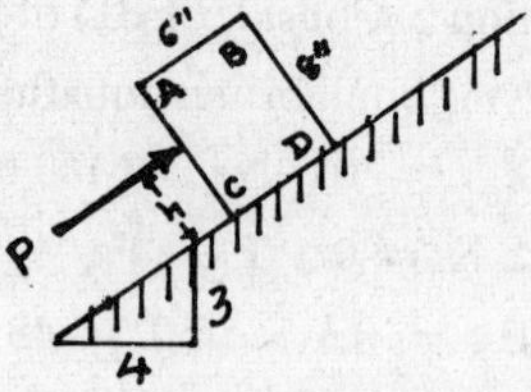

Solution : The FBD of block will be as shown in fig.:

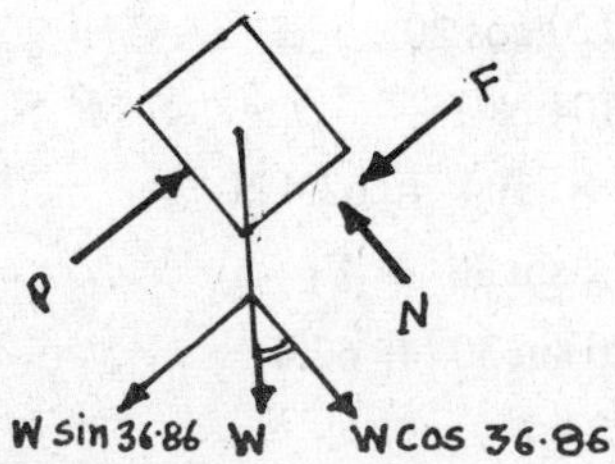

Applying equilibrium equations to above FBD

$\Sigma X = P - F - W \sin 36.86 = 0$ ...... (1)

$\Sigma Y = N - W \cos 36.86 = 0$ ...... (2)

Also $F = \mu N$ where $\mu = 0.3$ ..... (3)

From equation (1) (2) & (3)

$P = W (\sin 36.86 + 0.3 \times \cos 36.86)$

or $P = 0.83 W$ ......(4)

To satisfy the given condition we must have moment of all forces about point D as zero. Moment equation will be

$P \times h - (W \cos 36.86 \times 3) - (W \sin 36.86 \times 4) = 0$

$P \times h = W (4.79)$ ..... (5)

From (4) and (5) we get

$h = 5.78$ m .

Problem 6. : Block 'A' in fig. weighs 120 N & block 'B' weighs 200 N, and the cord is parallel to incline. If the coefficient of friction is $\mu = 0.60$ and $\theta = 30^o$, what force P applied to B acting down and parallel to the incline will start motion ? What will be the tension in the cord attached to A ?

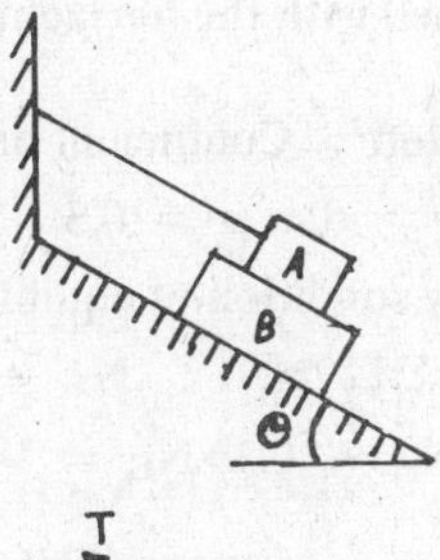

Solution : Given $\theta = 30^o$;

$\mu = 0.60$

Consider FBD of Block A

Apply condition of equilibrium,

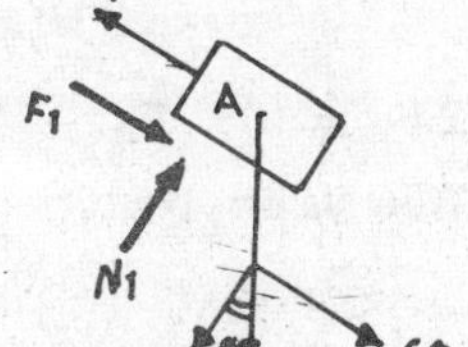

$\Sigma F_y = 0$; $N_1 = W_A \cos 30$

$= 120 \cos 30$

$= 104$ N

Then $F_1 = \mu N_1 = 0.6 \times 104 = 62.4$ N

$\Sigma F_x = 0$; $T = W_A \sin 30 + F_1$

$= 120 \sin 30 + 62.4$

$= 122.4$ N

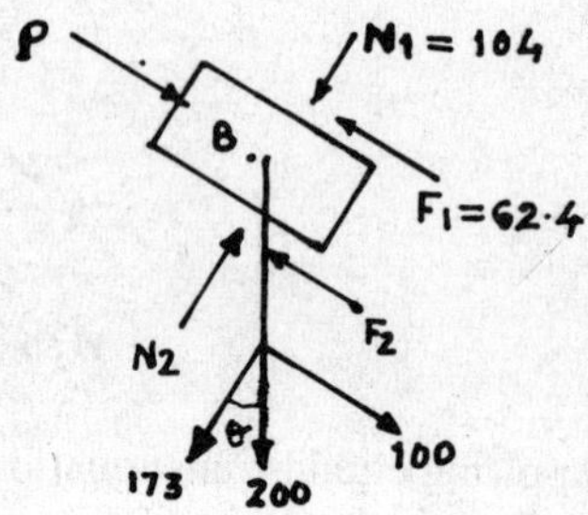

Consider FBD of Block B :

$\Sigma F_y = 0$; $N_2 = W_B \cos 30 + N_1$

$= 200 \cos 30 + 104$

$= 277.2$ N

& $F_2 = \mu N_2$

$= 0.60 \times 277.2 = 166.32$ N

$\Sigma F_x = 0$; $P = F_1 + F_2 - W_B \sin 30$

$= 62.4 + 166.32 - 200 \sin 30$

$= 128.63$ N & $T = 122.41$ N

Problem 7. : A uniform ladder of length l and weight W is placed with one end on the ground and the other against a vertical wall. The angle of friction at all contact surfaces is 20°. Find the minimum angle θ at which the ladder can be inclined with the horizontal before slipping occurs.

Solution : Coefficient of friction.

$\mu = \tan \varphi = 0.36$

Apply condition of equilibrium.

$\Sigma F_y = 0$; $N_A + F_B = W$

$N_A + 0.36 N_B = W$

..... (1)

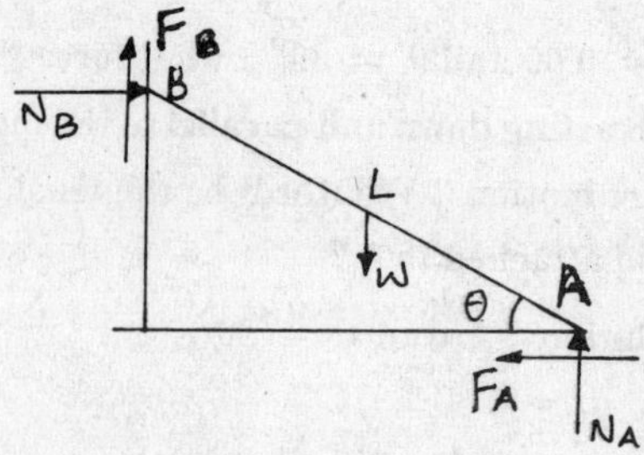

$\Sigma F_x = 0$; $F_A - N_B = 0$

$0.36 N_A = N_B$

...... (2)

Taking moment about A.

$$\Sigma M_A = 0$$

$$W \times \frac{L}{2} \cos\theta = F_B \times L\cos\theta + N_B\, L \sin\theta$$

Substitute the value of W from eq^n. (1) & (2),

$$\left(0.36\, N_B + \frac{N_B}{0.36}\right)\frac{\cos\theta}{2} = 0.36\, N_B \cos\theta + N_B \sin\theta$$

$$\Rightarrow \left(0.36 + \frac{1}{0.36}\right)\frac{\cos\theta}{2} = 0.36 \cos\theta + \sin\theta$$

$$\Rightarrow 1.56 \cos\theta = 0.36\cos\theta + \sin\theta$$

$$\therefore \quad \tan\theta = 1.2$$

$$\theta = 50.5^\circ.$$

Problem 8. : A uniform ladder 24ft. long weighs 40 N. It is placed against a vertical wall at an angle of 60° with the ground. How far along the ladder can a 160–N man climb before the ladder is on the verge of slipping ? The angle of friction at all contact surface is 15°.

Solution : Coefficient of friction $\mu = \tan\varphi = 0.26$

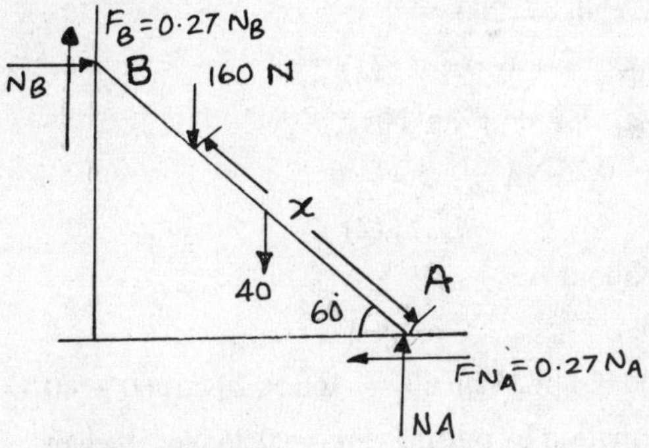

Applying conditions of equilibrium

$$\Sigma F_y = 0; \quad N_A + F_B - 40 - 160 = 0$$

$$N_A + F_B = 200$$

$$\text{or} \quad N_A + 0.26\, N_B = 200 \qquad \ldots\ldots (1)$$

$$\Sigma F_x = 0; \quad F_A - N_B = 0$$

$$0.26\, N_A = N_B \qquad \ldots\ldots (2)$$

Solving equations (1) and (2), we get

$$N_A + 0.26\ (0.26\ N_A) = 200$$

$$\therefore \quad N_A = 187.33\ \text{N}\ ;\ \&\ N_B = 48.70\ \text{N}$$

Now $F_B = \mu\, N_B$

$$= 48.70 \times 0.26 = 12.662\ \text{N}.$$

Taking momen about A.

$$\Sigma M_A = 0\ ;$$

$$40 \times 12 \cos 60 + 160\ (x \cos 60) = F_B\ (24\ \cos 60) + N_B\ (24 \sin 60)$$

Solving we get

$$x = 14.25\ \text{m}$$

Problem 9. : With what horizontal force must an assistant brace the bottom of the ladder in previous problem when the man has climbed 20 ft along the ladder ?

Solution : Applying equilibrium condition

$\Sigma F_y = 0\ ;$

$N_A + F_B - 160 - 40 = 0$

$$\therefore \quad N_A + 0.26\ N_B = 200 \qquad \ldots\ldots (1)$$

$$\Sigma F_x = 0\ ; \quad F_A + P = N_B$$

or $P = N_B - 0.26\ N_A \qquad \ldots\ldots (2)$

Taking moment about A.

$$\Sigma M_A = 0\ ;$$

$$F_B \times 24 \cos 60 + N_B\ 24 \sin 60 = 160 \times 20 \cos 60 + 40 \times 12 \cos 60$$

Solving above equation by putting $F_B = 0.26\ N_B$, we get

$$N_B = 76.665\ \text{N}$$

From equation (1) & (2) we get ,

$$N_A = 179.45\ \text{N}$$

& $P = N_B - 0.26 \times N_A$

$$= 28.5\ \text{N}$$

Problem 10. : A horizontal bar, 10 m long and of negligible weight, rests on rough inclined plane as shown in fig. If the angle of friction is 15°, how close to B may the 200 N force be applied before motion impends.

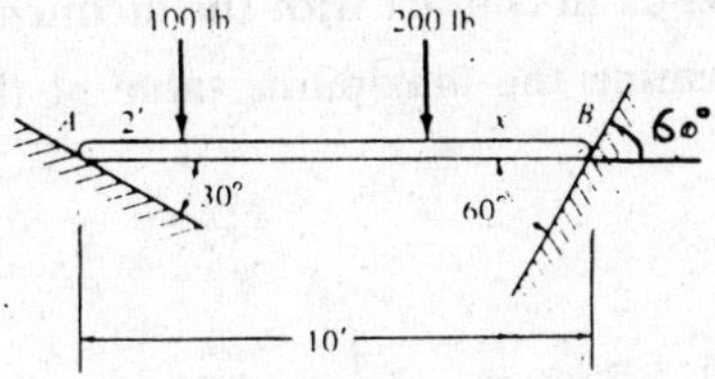

Solution : Consider FBD of horizontal bar

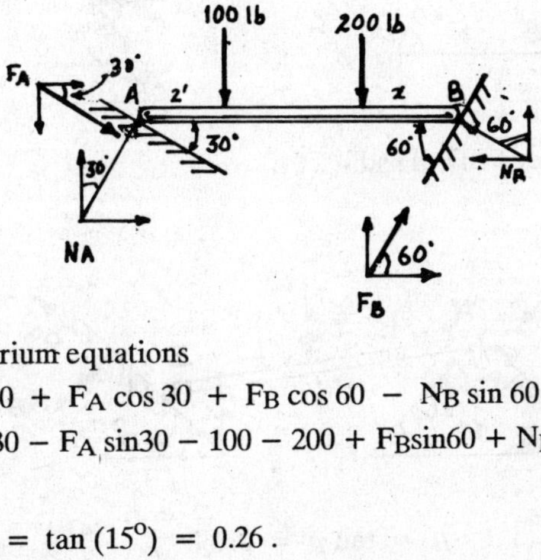

Applying equilibrium equations

$\Sigma X = N_A \sin 30 + F_A \cos 30 + F_B \cos 60 - N_B \sin 60 = 0$ ..... (1)

$\Sigma Y = N_A \cos 30 - F_A \sin 30 - 100 - 200 + F_B \sin 60 + N_B \cos 60 = 0$ ..... (2)

As $\mu = \tan \varphi = \tan (15^\circ) = 0.26$.

$\therefore F_A = 0.26\, N_A$ ..... (3)

$F_B = 0.26\, N_B$ ..... (4)

Solving (1) (2) (3) & (4) we get

$N_A = N_B = 205$ N

and $F_A = F_B = 55$ N

To satisfy the given condition we must have moment of all forces about point B as zero.

So taking moment about B we get ,

$\Sigma M_B = (100 \times 8) + (200 \times x) - (N_A \cos 30 \times 10)$

$+ (F_A \sin 30 \times 10) = 0$

Solving above equation for x, we get $x = 3.5$ m

Problem 11. : A uniform plank of weight W and total length 2L is placed as shown in fig. with its ends in contact with the inclined planes. The angle of frictioon is 15°. Determine the maximum value of the angle α at which slipping impends.

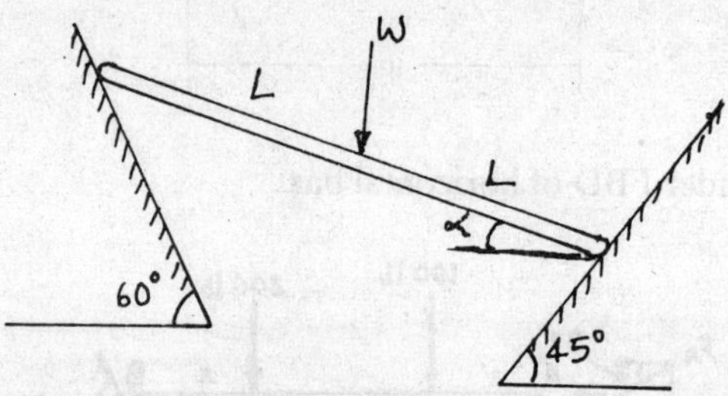

Solution : Consider FBD of rod

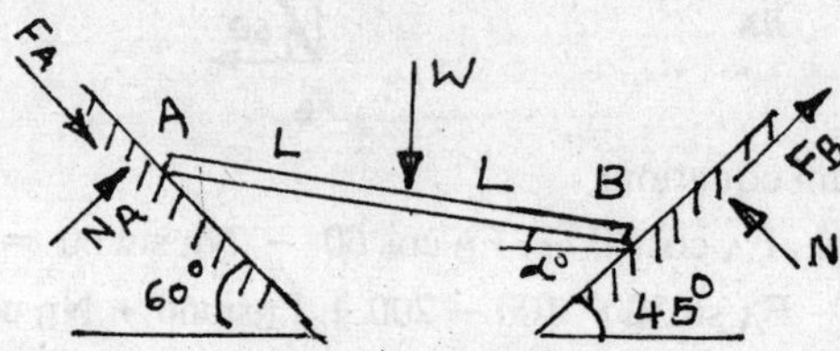

Given $\varphi = 15^{o}$ $\therefore$ $\mu = \tan\varphi = 0.268$

Applying equilibrium equations, we get

$\Sigma X = F_A \cos 60 + N_A \sin 60 + F_B \cos 45 - N_B \sin 45 = 0$

also we know that

$F_A = 0.268\, N_A$ & $F_B = 0.268\, N_B$

Solving above equations we get

$N_A = 0.518\, N_B$

$\Sigma Y = N_A \cos 60 - F_A \sin 60 + N_B \cos 45 + F_B \sin 45 = W$

Solving these equations we get

$N_B = 0.966$ W & $F = 0.259$ W

Taking moment about A and equationg it to zero, we get

$$\Sigma M_A = (W \times L \cos\alpha) - (N_B \cos 45 \times 2L \cos\alpha)$$
$$+ (N_B \sin 45 \times 2L \sin\alpha) - (F_B \sin 45 \times 2L \cos\alpha)$$
$$- (F_B \cos 45 \times 2L \times \sin\alpha) = 0$$

By putting the values of known quantities in above equation we get

$$\alpha = 36.2^o.$$

Problem 12. : A plank, 12ft long and of negligible weight, is supported and carries the horizontal and vertical loads shown in fig. The coefficient of friction is 0.30. If T = 300 N, determine the value of P to start motion downward.

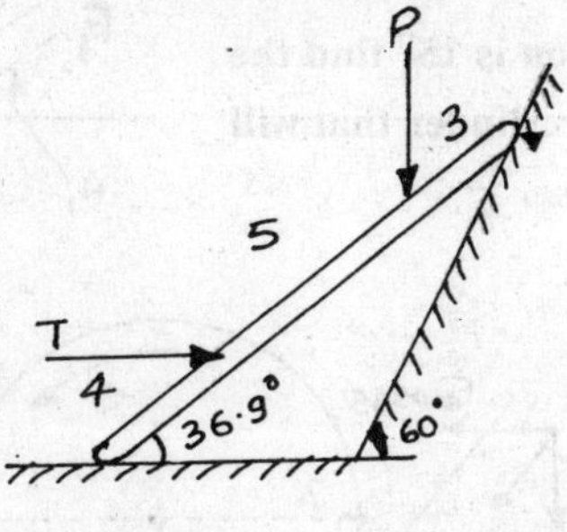

Solution :

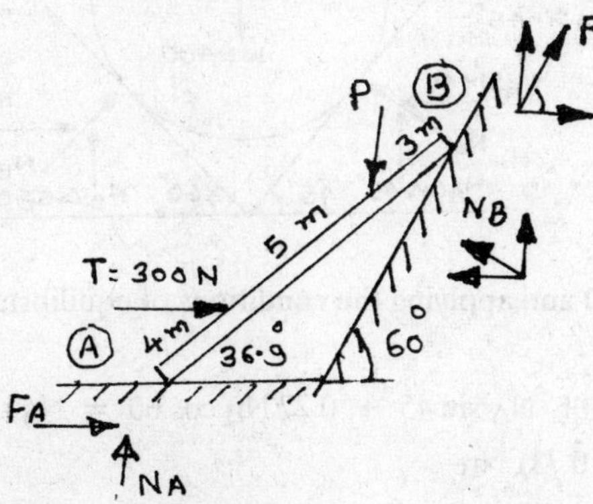

Applying Equilibrium Equations :

$$\Sigma X = T + F_A + F_B \cos 60 - N_B \sin 60 = 0$$

$$F_A = 0.3\, N_A \quad \& \quad F_B = 0.3\, N_B$$

Solving above we get

$$N_A = 2.387\ N_B - 1000 \qquad \ldots\ldots (1)$$

$$\Sigma Y = N_A + N_B \cos 60 + F_B \sin 60 - P = 0$$

Solving above equations we get

$N_A + 0.759\,N_B - P = 0$ ..... (2)

from eq^n. (1) & (2) $\Rightarrow N_B = 0.317\,P + 317.8$ ..... (3)

from eq^n. (1) & (3) $\Rightarrow N_A = 0.758\,P - 242$ ..... (4)

$\Sigma M_B = F_A \sin 36.9 \times 12 + (P \cos 36.9 \times 3)$

$- (N_A \cdot \cos 36.9 \times 12) + (300 \sin 36.9 \times 8) = 0$ .... (5)

Solving this, we get P = 995 N

Problem 13. : A homogeneous cylinder 3 ft. in diameter and weighing 400 N resting on two rough inclined surface as shown in fig. If the angle of friction is 15° find the couple C applied to the cylinder that will start it rotating clockwise.

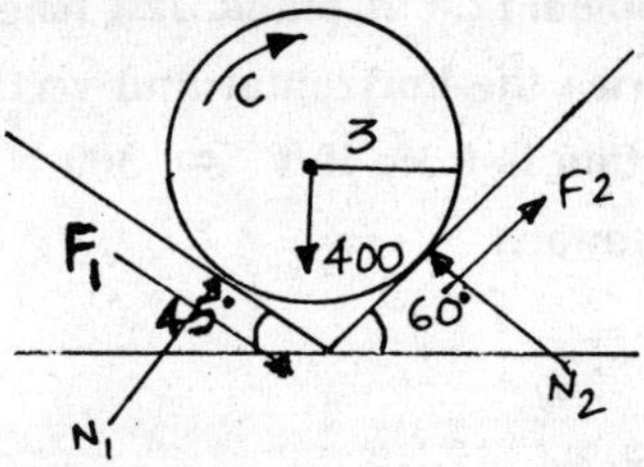

Solution :

Coefficient of friction

$\mu = \tan \varphi = 0.27$

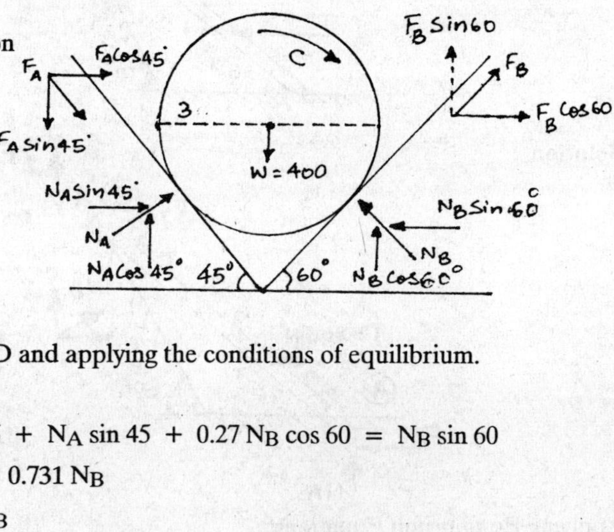

Considering the FBD and applying the conditions of equilibrium.

$\Sigma F_x = 0$;

i.e. $0.27\,N_A \cos 45 + N_A \sin 45 + 0.27\,N_B \cos 60 = N_B \sin 60$

$\Rightarrow 0.897\,N_A = 0.731\,N_B$

$N_A = 0.814\,N_B$

$\Sigma F_y = 0$;

$0.27\,N_A \sin 45 + 400 = N_A \cos 45 + N_B \cos 60 + 0.27\,N_B \sin 60$

$\therefore\ N_B = 347.5$ N

then, $N_A = 282.86$N

Taking moment about O. $\Sigma M_o = 0$;

Note that normal reaction N_A & N_B pass through the centre so only frictional forces are to be considered.

Couple $C = 0.27\, N_A \times 1.5 + 0.27\, N_B \times 1.5$

$= 254.9$ N–ft.

Problem 14. : Find the couple applied to the cylinder of above problem that will start counterclock wise rotation.

Solution :

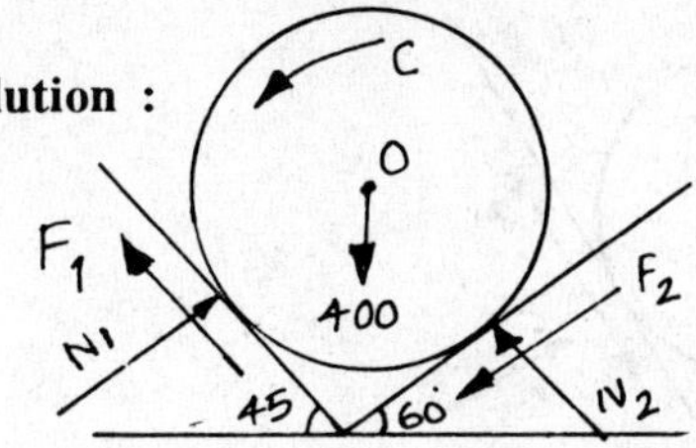

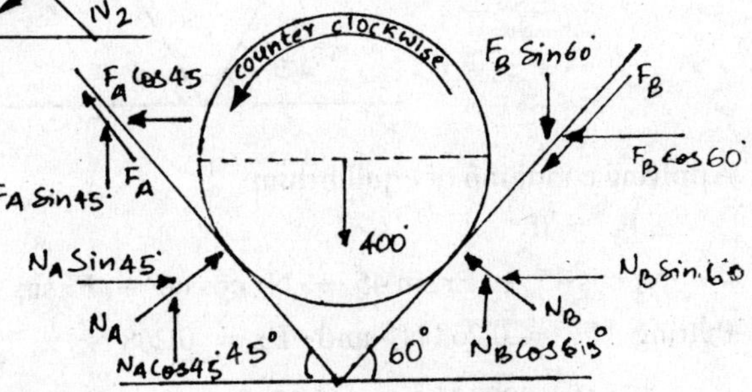

$\mu = 0.27$

Applying condition of equilibrium

$\Sigma F_x = 0$

$-\,0.27\, N_A \cos 45 + N_A \sin 45 = 0.27\, N_B \cos 60 + N_B \sin 60$

$\therefore\; N_A = 1.6856\, N_B$

$\Sigma F_y = 0$

$0.27\, N_B \sin 60 + 400 = N_A \cos 45 + N_B \cos 60 + 0.27\, N_A \sin 45$

$400 + 0.23\, N_B = 1.19\, N_B + 0.5\, N_B + 0.32\, N_B$

$400 = 1.77\, N_B$

$N_B = 225.20$ N & $N_A = 378.34$ N

$\Sigma M_o = 0$

$C = 0.27\, N_A \times 1.5 + 0.27\, N_B \times 1.5$

$= 0.27 \times 378.34 \times 1.5 + 0.27 \times 1.5 \times 225.20$

$= 244.43$ N−ft.

Problem 15. : For the cylinder described in above problem, determine the mininum horizontal force P applied tangentially to the left at the top that will start counter clockwise rotation.

Solution : $\mu = \tan 15^o = 0.268$

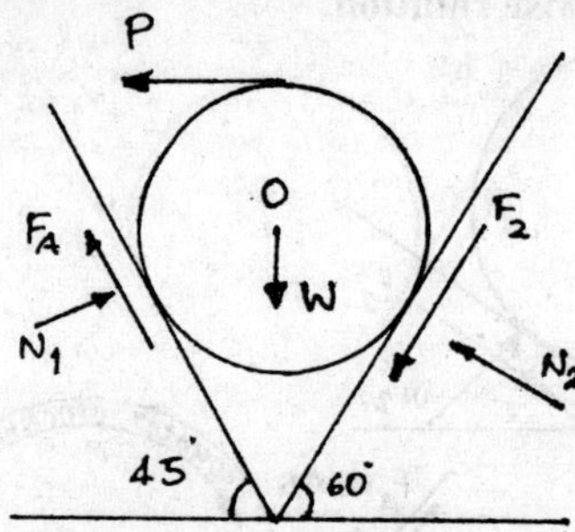

Applying condition of equilibrium

$\Sigma F_y = 0$

$N_1 \cos 45 + F_1 \sin 45 + N_2 \cos 60 - F_2 \sin 60 = W$

Putting $F_1 = 0.268\, N_1$ and $F_2 = 0.268\, N_2$

$\therefore \quad N_1 + 0.2988\, N_2 = 446.1298$ ..... (1)

$\Sigma F_x = 0$

$N_1 \sin 45 - F_1 \cos 45 - N_2 \sin 60 - F_2 \cos 60\, P = 0$

$N_1 - 1.932\, N_2 = 1.932\, P$ ..... (2)

Taking moment about O

$\Sigma M_o = 0$

$P \times 1.5 - (F_1 \times 1.5) - F_2 \times 1.5 = 0$

$\therefore \quad P = F_1 + F_2 = 0.268\,(N_1 + N_2)$ .... (3)

From equation (2) and (3),

$N_1 = 5.0804\, N_2$

∴ From equation (1) and (2)

$N_1 = 421.348$ N

$N_2 = 82.9359$ N

From equaton (3),

$P = 135.148$ N

Problem 16. : A uniform bar AB, 10 ft long and weighing 280 N, is hinged at B and rests upon a 400–N block at A as shown in fig. If the coefficient of friction is 0.40 at all contact surfaces, find the horizontal force P required to start moving the 400–N block.

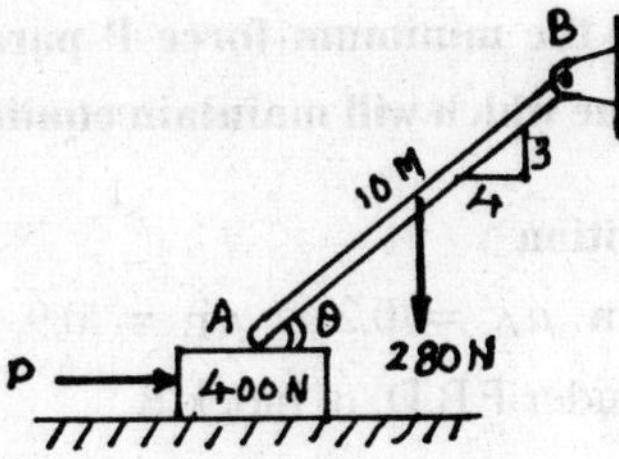

Solution : Consider FBD of rod

Applying equilibrium equations,

$\Sigma X = F_A = B_H = 0$ ......(1)

$\Sigma Y = N_A - B_V = 280 = 0$ ..... (2)

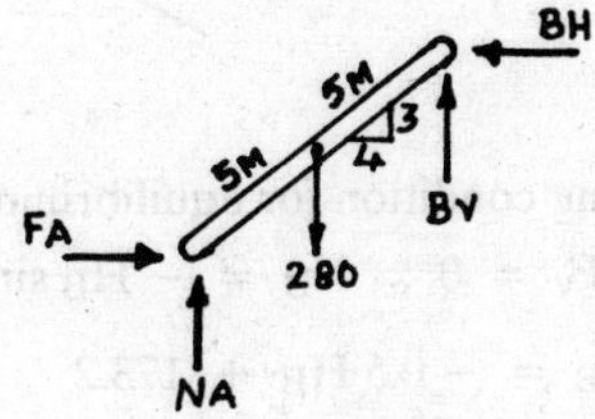

$F_A = 0.4\, N_A$;

given $\mu = 0.4$ (at all surfaces)

Also $\theta = \tan^{-1} 3/4 = 36.86^o$

Taking moment about B and equation it to zero

$$\Sigma M_B = 280 \times 5 \cos 36.86 + F_A \times 10 \sin 36.86 - N_A \times 10 \cos 36.86 = 0$$

Solving above equation, we get

$N_A = 200$ N ; $F_A = 80$ N

Consider FBD of block

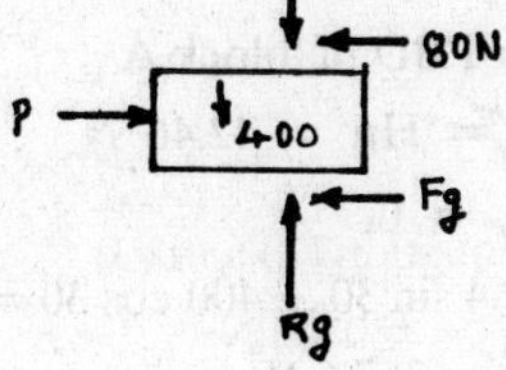

$F_g = 0.4\, R_g$

Applying equilibrium equation.

$\Sigma X = P - 80 - F_g = 0$

& $R_g = 600$ N

$\therefore$ $F_g = 240$ N & $P = 320$ N.

Problem 17. : The two crates shown in fig. pin connected by the horizontal strut AB. Find the minimum force P parallel to the incline which will maintain equilibrium.

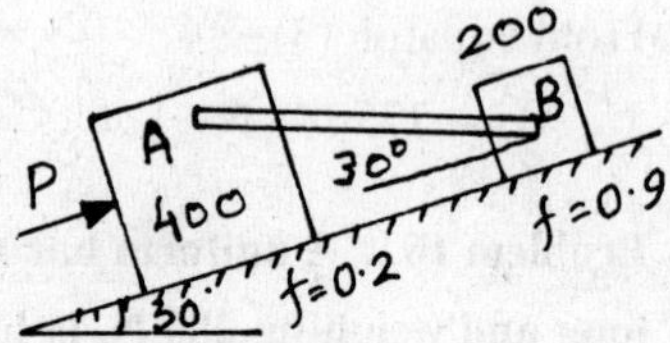

Soluition :

Given $\mu_A = 0.2$ & $\mu_B = 0.9$

Consider F.B.D. of block B

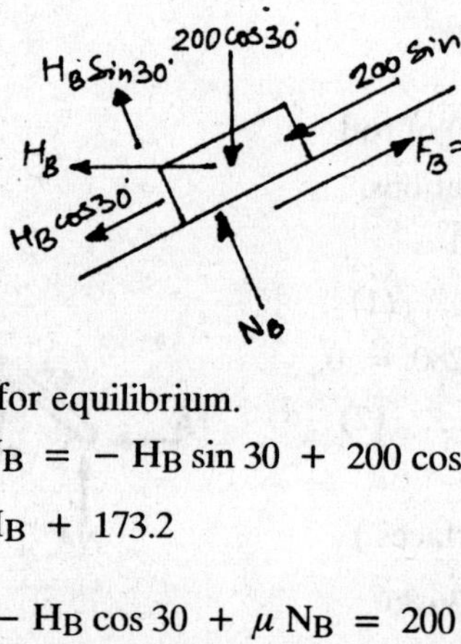

Applying condition for equilibrium.

$\Sigma F_y = 0$; $N_B = -H_B \sin 30 + 200 \cos 30$

$\therefore$ $N_B = -0.5 H_B + 173.2$ ..... (1)

$\Sigma F_x = 0$; $-H_B \cos 30 + \mu N_B = 200 \sin 30$

$\therefore$ $-0.866 H_B + 0.9 N_B = 100$

Put the value of N_B in this equation.

$-0.866 H_B + 0.9 [-0.5 H_B + 173.2] = 0$

$\therefore$ $H_B = 42.46$ N

Consider FBD of block A

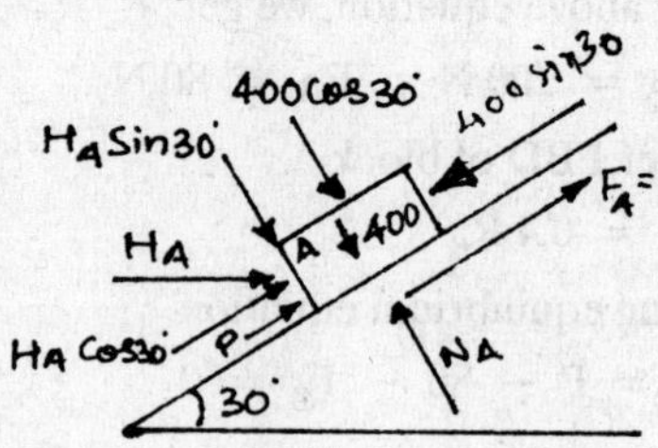

$H_A = H_B = 42.46$ N

$\Sigma F_y = 0$

$N_A - 42.4 \sin 30 + 400 \cos 30 = 0$

$N_A = 367.6$ N

Now $\Sigma F_x = 0$

$$P + F_A + H_A \cos 30 = 400 \sin 30$$

or $\quad P + 0.2 \times 367.6 + 42.46 \cos 30 = 400 \sin 30$

$\therefore \quad P = 89.71$ N.

Problem 18. : In fig. two blocks are separated by a uniform strut attached to each block with frictionless pins. Block A weighs 400 N, block B weighs 300 N, and the strut AB weighs 200 N. If f = 0.25 under B, determine the minimum coefficient of friction under A to prevent motion

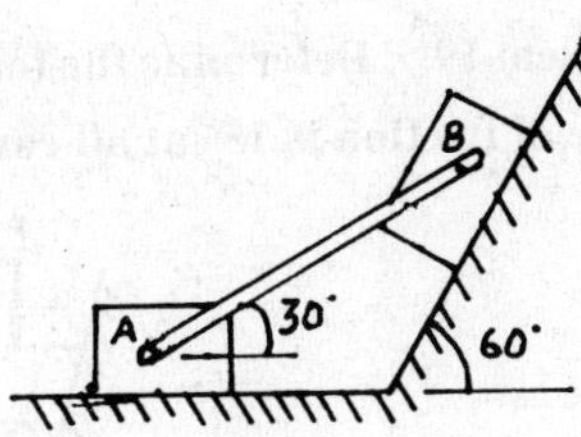

Solution : Consider FBD of structure.

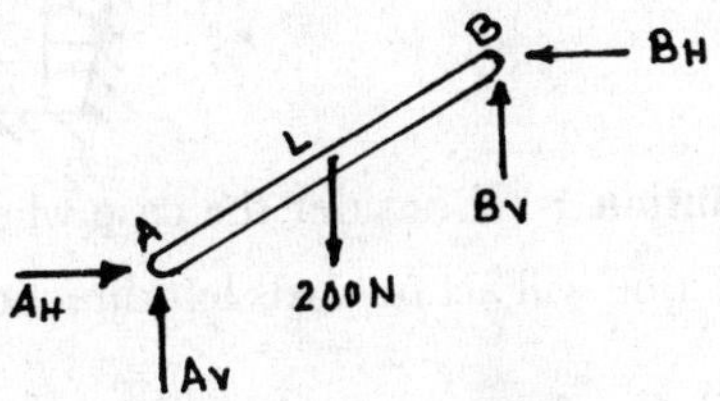

Applying equilibrium equations

$$A_V + B_V = 200 \text{ N} \qquad \ldots\ldots (1)$$

$$A_H = B_H \qquad \ldots\ldots (2)$$

From FBD of Block B ,

$$B_H + F_B \cos 60 - N_B \sin 60 = 0$$

$$N_B \cos 60 - B_V - 300 + F_B \sin 60 = 0$$

$$F_B = 0.25\, N_B$$

$$B_H - 0.74\, N_B = 0 \qquad \ldots\ldots (3)$$

$$- B_V + 0.71\, N_B = 300 \qquad \ldots\ldots (4)$$

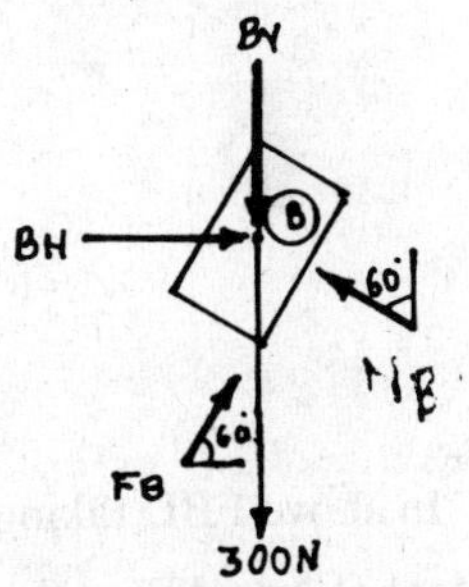

FBD of Block A

$$F_A - A_H = 0$$

$$N_A - A_V = 400 \qquad \ldots\ldots (5)$$

$$F_A = \mu_A N_A$$

$$\therefore \quad \mu_A N_A - A_H = 0 \qquad \ldots\ldots (6)$$

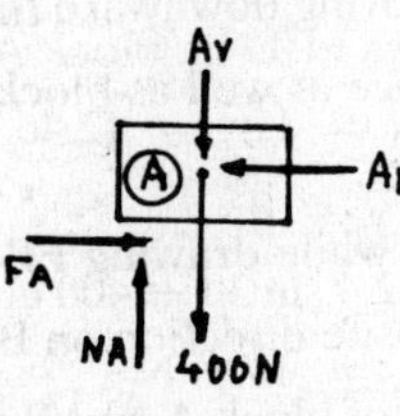

On solving above equations we get

$$N_A = 650 \text{ N}$$

$$F_A = 260 \text{ N}$$

$$F_A = \mu_A N_A$$

$$\therefore \quad \mu_A = \frac{260}{650} = 0.4 .$$

Problem 19. : Determine the force P required to start wedge shown in fig. The angle of friction is 15° at all contact surfaces.

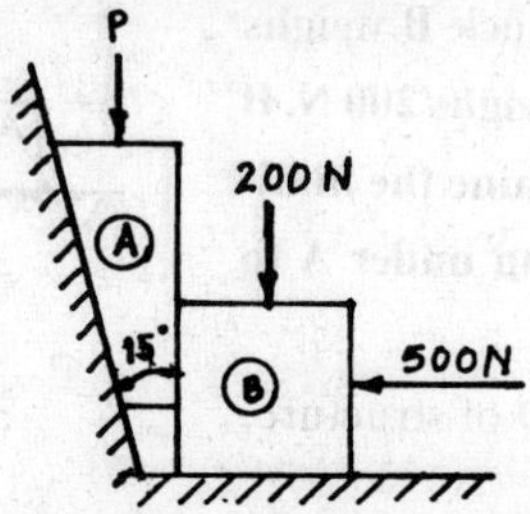

Solution : Consider the case when B starts moving in rightward direction (friction will act towards left direction).

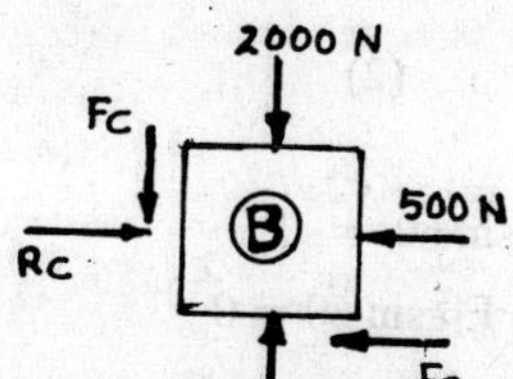

In above FBD taking about 'B' the direction of frictional force at point of contact of A and B is taken in downward direction, since if we consider wedge A moving downward then in that case both the surfaces will resist i.e. inclined surface as well as block B. Thus we get direction of friction of forces on A as shown.

Now while drawing FBD of B we have to consider the contact friction F_C in opposite direction on B i.e. downward.

FBD of block A and B will be as shown

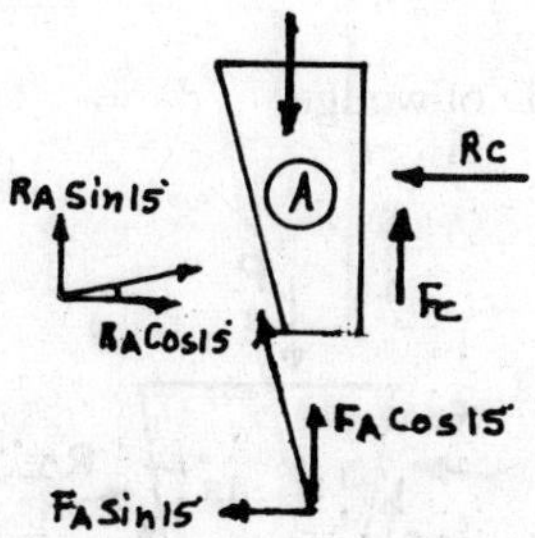

Applying equilibrium equations to FBD of B we get

$R_C - F_B = 500$ ..... (1)

$R_B - F_C = 2000$ ..... (2)

$F_B = 0.27\ R_B$ ..... (3)

$F_C = 0.27\ R_C$ ..... (4)

Solving these equations, we get

$R_B = 2303$ N & $R_C = 1122$ N

$F_B = 622$ N & $F_C = 303$ N

Applying equilibrium equations to FBD of wedge A

$R_A \cos 15 - F_A \sin 15 = 1122$ ..... (5)

$R_A \sin 15 + F_A \cos 15 - P = -303$ ..... (6)

$F_A = 0.27\ R_A$ ..... (7)

$F_C = 0.27\ R_C$ ..... (8)

from eq^n. (5) & (7) $\Rightarrow R_A \cos 15 - 0.27 R_A \sin 15 = 1122$

$\therefore\ R_A = 1252$ N

$F_A = 338$ N

from eq^n. (6) $\Rightarrow P = 953$ N :

Problem 20. : As shown in fig. two blocks, each weighing 200 N, are to be pushed apart by a 30° wedge. The angle of friction is 15° for all contact surfaces. What value of P is required to start movement of the blocks ? How would this answer be changed if the weight of one of the blocks were increased to 300 N ?

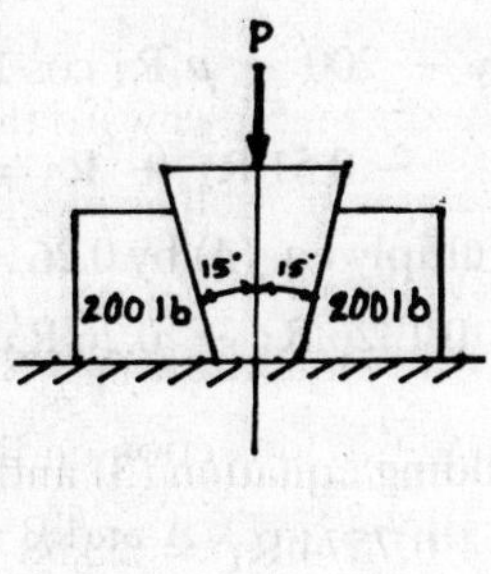

Solution : Consider FBD of wedge

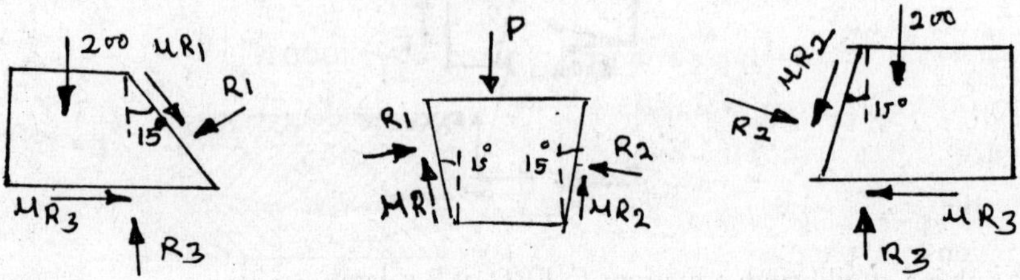

Applying equilibrium equations.

$R_2 \sin 15 + \mu R_2 \cos 15 + \mu R_1 \cos 15 + R_1 \sin 15 = P$

$\mu = 0.26$ (given)

$\therefore \quad 0.50 R_2 + 0.5 R_1 = P$

i.e. $R_1 + R_2 = 2P$ ......(1)

$R_2 \cos 15 - \mu R_2 \sin 15 + \mu R_1 \sin 15 - R_1 \cos 15 = 0$

$0.89 R_2 + (-0.89) R_1 = 0$

$\Rightarrow R_1 = R_2$ .....(2)

$\therefore$ from eq^n. (1) & (2), we get

$R_1 = P = R_2$

Consider FBD of block,

Applying equilibrium equations.

$R_1 \cos 15 - \mu R_1 \sin 15 - \mu R_3 = 0$

$\therefore \quad 0.89 R_1 - 0.26 R_3 = 0$ ......(3)

$R_3 - 200 - \mu R_1 \cos 15 - R_1 \sin 15 = 0$

$\therefore \quad -0.51 R_1 + R_3 = 200$ ......(4)

Multiply eq. (4) by 0.26

$-0.1326 R_1 + 0.26 R_3 = 53.58$ .....(5)

Adding equation (3) and (5), we get

$0.7574 R_1 = 53.58$

$\therefore\ R_1 = 71$ N

i.e. $P = 71$ N .

Problem 21. : What force P must be applied to the weightless wedges shown to start them under the 1000 N block ? The angle of friction at all contact surfaces is 10°.

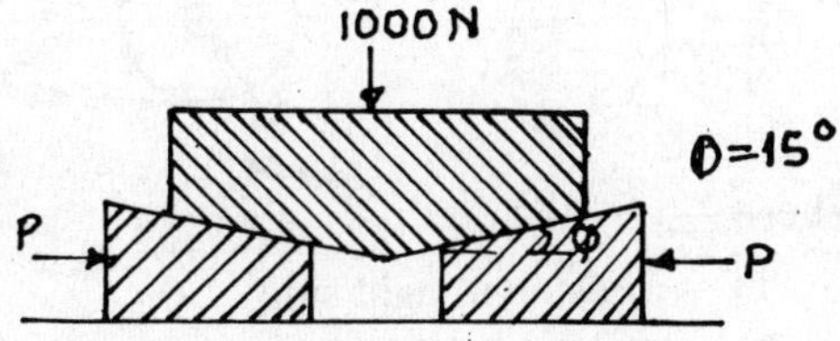

Solution : Consider FBD of block.

We can write from force polygon as follows :

Consider R_1 as resultant of F_1 & N_1 & R_2 as resultant of F_2 & N_2 then ,

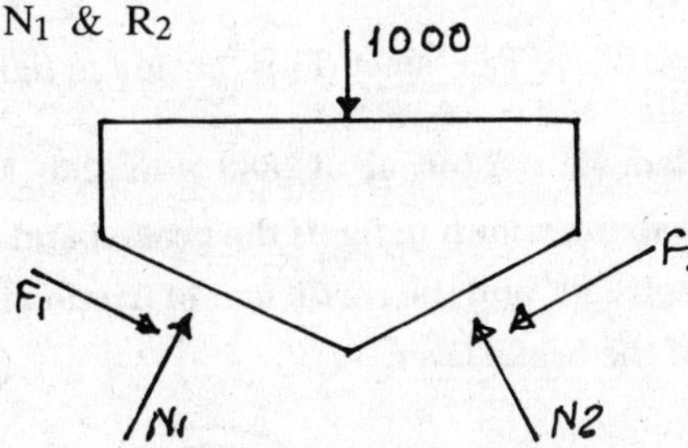

$$\frac{R_1}{\sin 25} = \frac{1000}{\sin 130}$$

$\therefore\ R_1 = R_2 = 551.7$ N

Consider FBD of wedge

$F_3 = \mu N_3$,

Resultant of F_3 & N_3 is R_3 ..

Applying equilibrium equations we get

$R_1 \cos 25 = R_3 \cos 10$

$\Rightarrow R_3 = 507.72$ N .

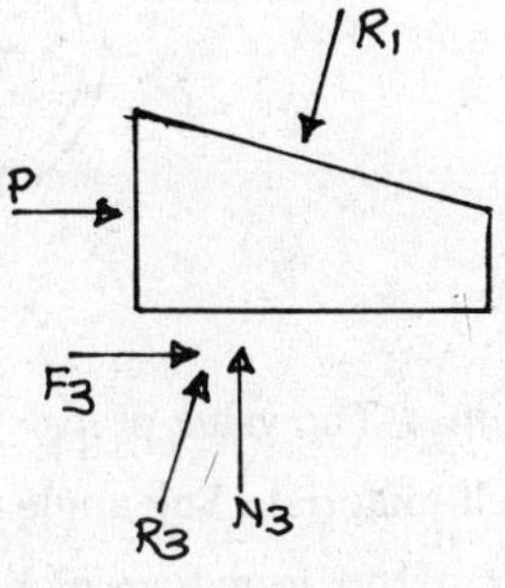

Also we can write

$P = R_1 \sin 25 + R_3 \sin 10$

$= 551.7 \times \sin 25 + 507.72 \times \sin 10$

$= 321.32$ N .

Friction in Belts :

If we have system of flat belt for transmitting power as shown below

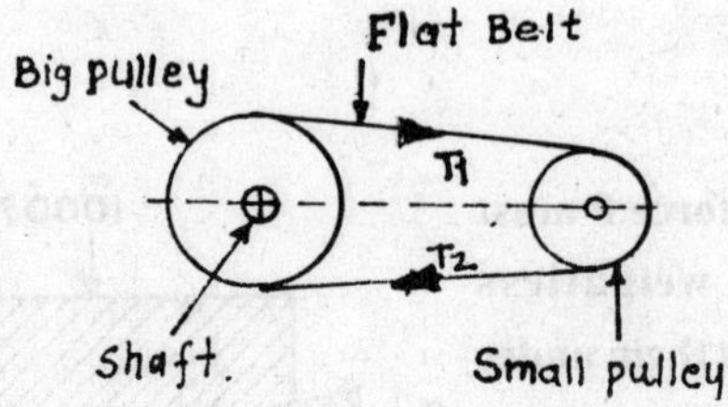

Where

T_1 – Tension in tight side

T_2 – Tension in slack side

θ – Angle of lap of belt over the pulley.

μ – Coefficient of friction between belt material & material of pulley

Then for above flat system, we have

$$\frac{T_1}{T_2} = e^{\mu\theta}$$

Where $T_1 > T_2$, since T_1 is tension in tight side.

Problem 22. : A torque of 2880 N-m (240 N/ft = 240 × 12 n−m) acts on brake drum shown in fig. If the brake band is in contact with the brake drum through 250° and the coefficient of friction is 0.3, determine the force P at the end of the brake lever.

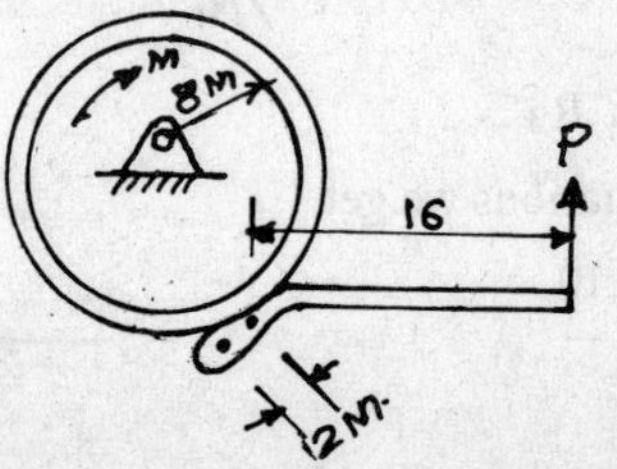

Solution : The value of $\mu = 0.3$ is the coefficient of friction between drum and belt material. The **angle of lap** is 250°, it is the angle along the circumference of the drum for which the belt is in contact with the drum.

The relationship between tensions in belts is given by

$$\frac{T_1}{T_2} = e^{\mu\theta}$$

where T_1 – Tension in tight side belt

T_2 – Tension in slack side belt

μ – Coefficient of friction between drum and belt material

θ – angle of lap (in radians)

$$\frac{T_1}{T_2} = e^{0.3 \times 250 \times \frac{\pi}{180}}$$

$$\Rightarrow T_1 = 3.70\ T_2 \qquad \ldots\ldots (1)$$

Torque is defined as relationship between tensions $= (T_1 - T_2) \times r$, where r is radius of brake drum.

Now Torque $= (T_1 - T_2) \times r$

$$\therefore \quad 2880 = (T_1 - T_2) \times 8$$

$$\Rightarrow (T_1 - T_2) = 360 \qquad \ldots\ldots (2)$$

from eq^n. (1) & (2) ;

$$T_2 = 133.4\ \text{N} \ \ \& \ \ T_1 = 493.4\ \text{N}$$

Consider FBD of brake lever

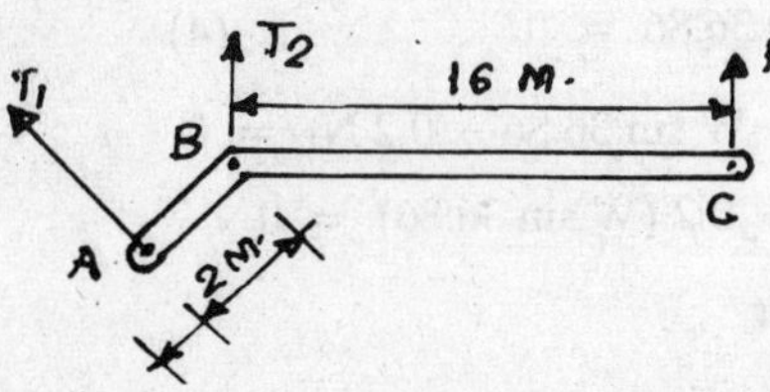

ıaking moment about B (Moment is taken about B so as to eliminate distance calculation part for T_2).

$$\Sigma M_B = T_1 \times 2 = P \times 16$$

$$P = 61.7\ \text{N}$$

Problem 23. : In fig. shown the coefficient of friction is 0.2 between the rope and the fixed drum and between all surfaces in contact. Determine the minimum weight W to prevent down plane motion off 1000 N body.

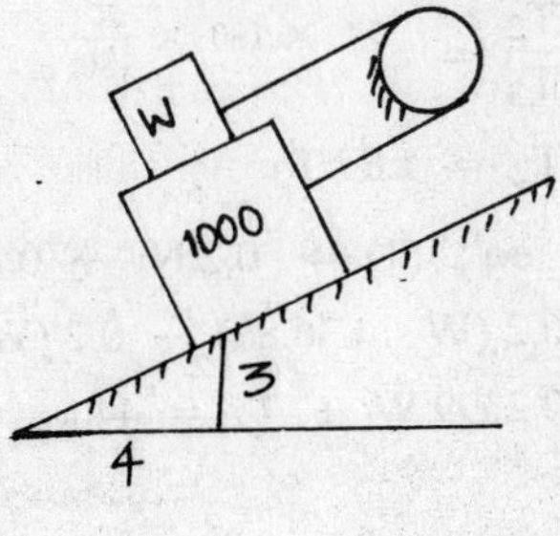

Solution :

Consider FBD of 1000 N block

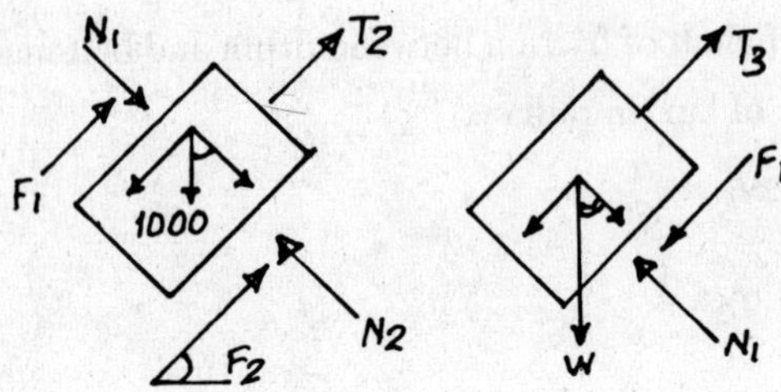

Applying equilibrium equations

$\Sigma X = F_1 + F_2 - 1000 \sin 36.8 + T_2 = 0$ .... (1)

$\Sigma Y = N_2 - N_1 - 1000 \cos 36.8 = 0$

$N_2 = (N_1 + 800)$ .....(2)

Also $F_2 = 0.2 N_2$ & $F_1 = 0.2 N_1$

Consider FBD of upper block

$\Sigma X = T_3 - W \sin 36.86 - F_1 = 0$ .... (3)

$\Sigma Y = N_1 - W \sin 36.86 = 0$ ..... (4)

from eq^n. (3) $\Rightarrow T_3 - W \sin 36.86 - 0.2 N_1 = 0$

$T_3 - W \sin 36.86 - 0.2 (W \sin 36.86) = 0$

from eq^n. (4) we get

$\therefore \; T_3 = 0.719 W$ ..... (5)

Now $\frac{T_2}{T_3} = e^{\mu\theta}$

$\frac{T_2}{T_3} = e^{0.2 \times 180 \times \frac{\pi}{180}}$

$T_2 = 1.87 T_3$

from eq^n. (1) $\Rightarrow$ $0.2 N_1 + 0.2 N_2 - 600 + T_2 = 0$

$0.2 (W \sin 36.86) + 0.2 (W \sin 36.86 + 800) - 600 + T_2 = 0$

$0.2399 W + T_2 = 440$

Problem 24. : A rope ABCD is looped over two pipes as shown knowing that the coefficient of static friction is 0.25 determine :

(a) The smallest value of mass for which equilibrium is possible.

(b) The corresponding tension in portion BC of the rope.

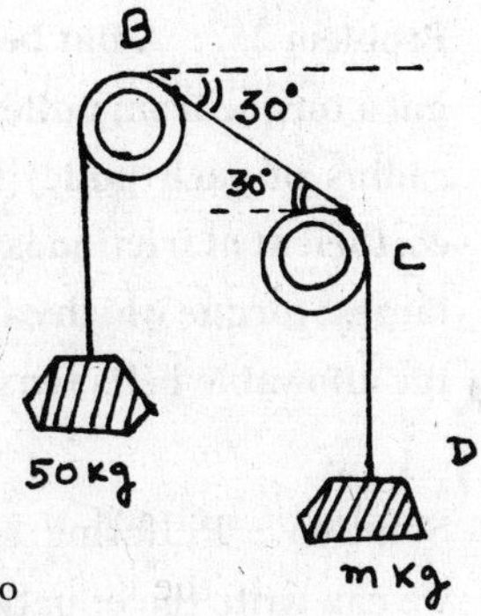

Solution : From the geometry of the fig., the angle of lap at pulley B is 120° and at pulley C is 60°

We know the relation

$$\frac{T_1}{T_2} = e^{\mu\theta} \qquad \ldots\ldots (1)$$

T_1 – Tension in tight side

T_2 – Tension in slack side

μ – Coefficient of friction

θ – angle of lap in radians.

Refer FBD of pulley (B)

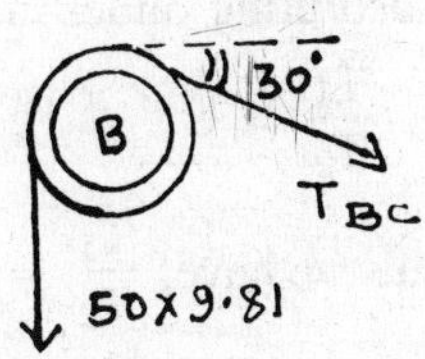

Applying equation (1) for pulley (B), we get

$$\frac{50 \times 9.81}{T_{BC}} = e^{0.25 \times 120 \times \frac{\pi}{180}}$$

Solving we get $T_{BC} = 290$ N

Now refering FBD of pulley (C)

Using equation (1) for pulley (C) we get

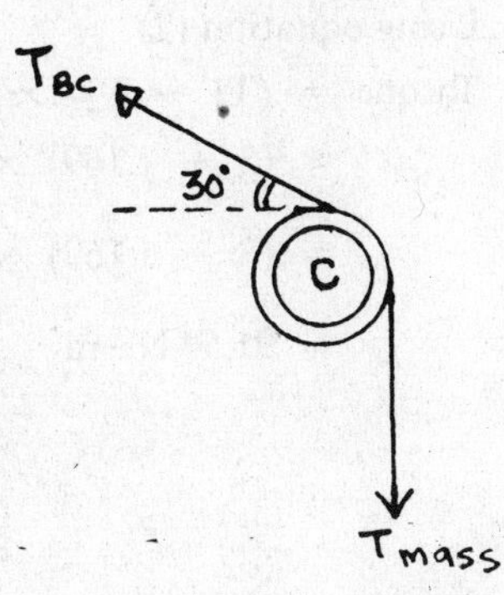

$$\frac{T_{BC}}{T_{mass}} = e^{\mu\theta}$$

$$\frac{290}{T_{mass}} = e^{0.25 \times 60 \times \frac{\pi}{180}}$$

Solving we get $T_{mass} = 223.2$ N

But we want in terms of mass.

$$\text{Hence } m = \frac{T_{mass}}{9.81} = 22.7 \text{ kg}$$

Problem 25. : A flat belt is used to transmit a torque from pulley A to pulley B. The radius of each pulley is 50 mm and the coefficient of friction is 0.3. Determine the largest torque which can be transmitted if the allowable belts tension is 3 kN.

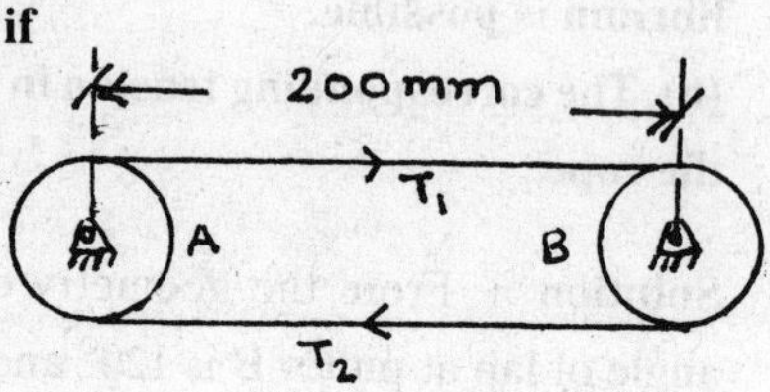

Solution : Reffering above fig.
we can write the equation

$$\frac{T_1}{T_2} = e^{\mu\theta} \qquad \ldots\ldots (1)$$

Largest torque transmitted is given by

$$\text{Torque} = (T_1 - T_2) \times \text{radius of pulley} \qquad \ldots\ldots (2)$$

Substituting values in equation (1)

$$\frac{T_1}{T_2} = e^{0.3 \times 180 \times \frac{\pi}{180}} = 2.566$$

Thus we have, $\frac{T_1}{T_2} = 2.566 \qquad \ldots\ldots (3)$

We have given that maximum allowable tension is 3kN
i.e. value of T_1 is known.

$T_1 = 3$ kN

Using equation (3)
we get $T_2 = 1.169$ kN

Using equation (2)

$$\begin{aligned} \text{Torque} &= (T_1 - T_2) \times \text{radius} \\ &= (3 - 1.169) \times 50 \text{ kN-mm} \\ &= (3 - 1.169) \times 1000 \times \frac{50}{1000} \text{ N-m} \\ &= 91.54 \text{ N-m} \end{aligned}$$

Chapter Five

CENTROIDS AND CENTRE OF GRAVITY

Centroids :

The point at which whole mass of the body is supposed to be concentrated is called centre of gravity of the body.

If we consider thin homogeneous plate then its centre of gravity is centre of area of thin plate which is known as centroid.

Co–ordinates of centroid are given by

$$\bar{x} = \frac{\Sigma a \cdot x}{A} \quad \text{and} \quad \bar{y} = \frac{\Sigma a \cdot y}{A}$$

where $\bar{x}$, $\bar{y}$ are coordinates of centroid of area

a – area of component element of given area.

x – x coordinate of centroid of component element of given area.

y – y coordinate of centroid of component element of given area

A – Total area of plate = sum of component element areas of given fig.

They are also given by

$$\bar{x} = \frac{\int x \cdot dA}{A} \quad \text{and} \quad \bar{y} = \frac{\int y \cdot dA}{A}$$

if area is confined by known curves, and where dA is area of elemental strip.

Centroid of line is given by

$$\bar{x} = \frac{\Sigma l \cdot x}{L} \quad \text{and} \quad \bar{y} = \frac{\Sigma l \cdot y}{L}$$

Where L is total length of line, l is length of part of line , x is x-coordinate of centroid of part of line and y is y-coordinate of centroid of part of line

Centroids of Common Geometric Shapes :

Shape		Area of length	$\bar{x}$	$\bar{y}$
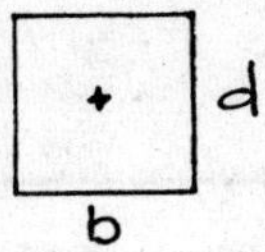	Rectangle	bd	$\frac{1}{2}b$	$\frac{1}{2}d$
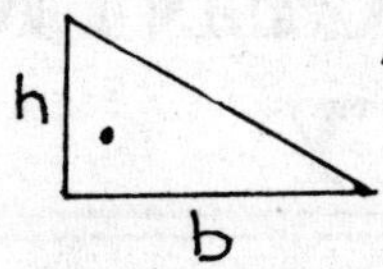	Any triangle	$\frac{1}{2}bh$		$\frac{1}{3}h$
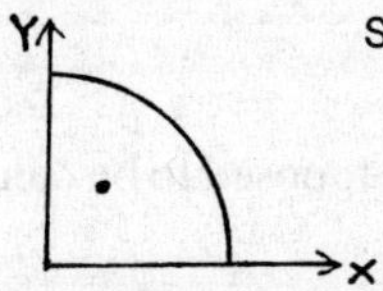	Semicircle	$\frac{\pi r^2}{2}$	0	$\frac{4r}{3\pi}$ or 0.424r
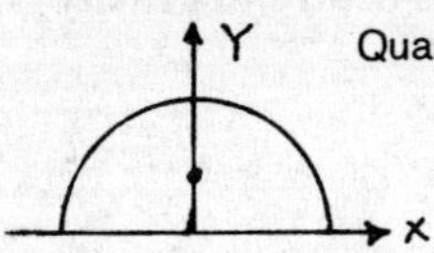	Quarter circle	$\frac{1}{4}\pi r^2$	$\frac{4r}{3\pi}$ or 0.424r	$\frac{4r}{3\pi}$ or 0.424r
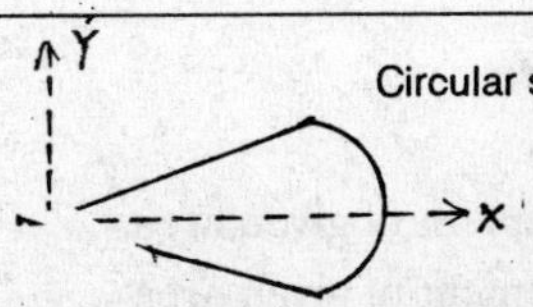	Circular sector	$r^2 \alpha$	$\frac{2}{3}\frac{r \sin \alpha}{\alpha}$	0
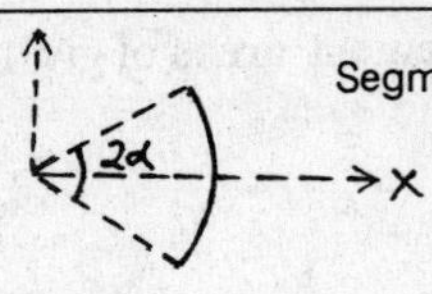	Segment of arc	$2r\alpha$	$\frac{r \sin \alpha}{\alpha}$	0
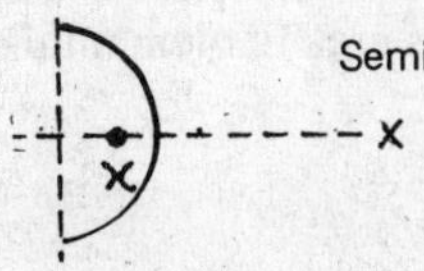	Semicircular arc	πr	$\frac{2r}{\pi}$	0

Problem 1.: Determine the centroid of the shaded area shown in fig. which is bounded by the X axis, the line x = a and the parabola $y^2 = kx$.

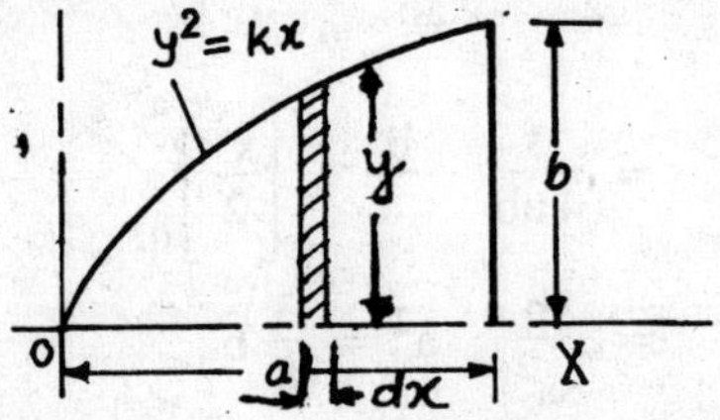

Solution : Area of elementary strip of length y and width dx is given by

$dA = y\,dx = \sqrt{kx}\,dx$

$\therefore$ Total area A is given by

$$A = \int_0^a dA = \int_0^a \sqrt{kx}\,dx = \left[\frac{2\sqrt{k}\cdot x^{3/2}}{3}\right]_0^a$$

$$= \frac{2}{3}\sqrt{k}\,a^{3/2}$$

Now, as we know (a, b) is a point on curve $y^2 = kx$,

$$b^2 = k\cdot a \Rightarrow k = \frac{b^2}{a}$$

$$\therefore \quad A = \frac{2}{3}\left[\frac{b^2}{a}\right]^{1/2}\cdot a^{3/2} = \frac{2}{3}\,a\cdot b$$

Now centroid of area is given by

$$\bar{x} = \frac{\int x\cdot dA}{A} \quad \& \quad \bar{y} = \frac{\int y\,dA}{A}$$

$$\bar{x} = \frac{\int_0^a x\cdot\sqrt{kx}\cdot dx}{2/3\cdot a\cdot b} = \frac{3}{2\,ab}\cdot\frac{b}{\sqrt{a}}\cdot\left[\frac{x^{5/2}}{5/2}\right]_0^a = \frac{3}{5}\,a$$

To determine $\bar{y}$, we use same elementary strip but each point of the element is not at the same distance from the x–axis,

In such case $\bar{y}$ is given by

$$\int \frac{1}{2}\,y\,dA \quad \& \text{ not by } \int y\,dA$$

$$\therefore \quad \bar{y} = \frac{\int_0^a \frac{1}{2} y \, dA}{A} = \frac{1}{2} \times \frac{3}{2\,ab} \int_0^a \sqrt{kx} \cdot \sqrt{kx} \cdot dx$$

$$= \frac{3}{4\,ab} \cdot \frac{b^2}{a} \cdot \left[\frac{x^2}{2}\right]_0^a$$

$$= \frac{3b}{8a^2} \cdot a^2 = \frac{3}{8}\, b.$$

Problem 2.: Determine the centroid of the quarter circle shown in fig. whose radius is r.

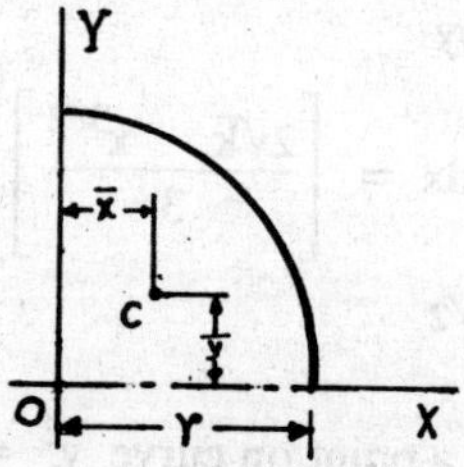

Solution : Equation of circle is $x^2 + y^2 = r^2$

$$\therefore \quad y = [r^2 - x^2]^{1/2}$$

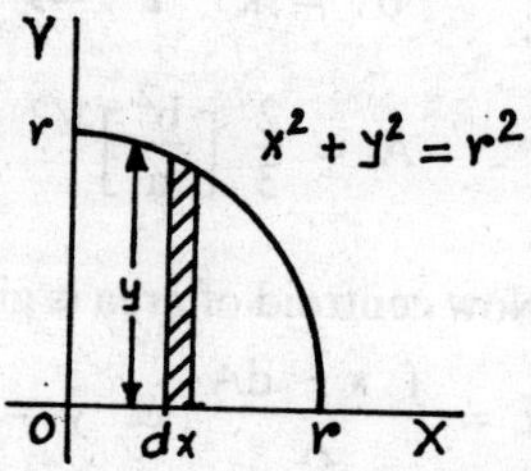

Area of element $= \int_0^r dA = \int_0^r y \cdot dx$

$$\therefore \quad A = \int_0^r [r^2 - x^2]^{1/2} \, dx$$

Solving this we get $A = \frac{\pi}{4} r^2$

$$\bar{x} = \frac{\int_0^r x \cdot dA}{A} = \frac{4}{\pi r^2} \int_0^r x \cdot [r^2 - x^2]^{1/2} \, dx$$

Substituting $x = r \sin\theta$, $dx = r\cos\theta \, d\theta$

at $x = 0$, $\theta = 0$ and when $x = r$, $\theta = \frac{\pi}{2}$

$$\therefore \quad \bar{x} = \frac{4}{\pi r^2} \int_0^{\pi/2} r\sin\theta \cdot r\cos\theta \cdot r\cos\theta \, d\theta$$

Substituting $\cos\theta = t \quad \therefore \quad -\sin\theta \, d\theta = dt$

at $\theta = 0$, $t = 1$ and when $\theta = \frac{\pi}{2}$, $t = 0$

$$\therefore \quad \bar{x} = \frac{4r}{\pi} \int_1^0 -t^2 \, dt = \frac{4r}{\pi} \left[\frac{-t^3}{3} \right]_1^0 = \frac{4r}{3\pi}$$

Øn the similar logic as in earlier problem

$$\bar{y} = \frac{\int_0^r \frac{1}{2} y \cdot y \, dx}{\frac{\pi}{4} r^2} = \frac{2}{\pi r^2} \int_0^r (r^2 - x^2) \, dx$$

Substituting $x = r\sin\theta \quad \therefore \quad dx = r\cos\theta \, d\theta$

at $x = r$, $\theta = \frac{\pi}{2}$; and when $x = 0$, $\theta = 0$

$$\therefore \quad \bar{y} = \frac{2}{\pi r^2} \cdot r^2 \int_0^{\pi/2} \cos^2\theta \, r\cos\theta \, d\theta$$

$$= \frac{2r}{\pi} \int_0^{\pi/2} \cos^3\theta \, d\theta = \frac{4r}{3\pi}$$

Problem 3. : Determine the centroid of the quadrant of the ellipse shown in fig. The equation of the ellipse is $\frac{x^2}{a^2} + \frac{y^2}{b^2} = 1$.

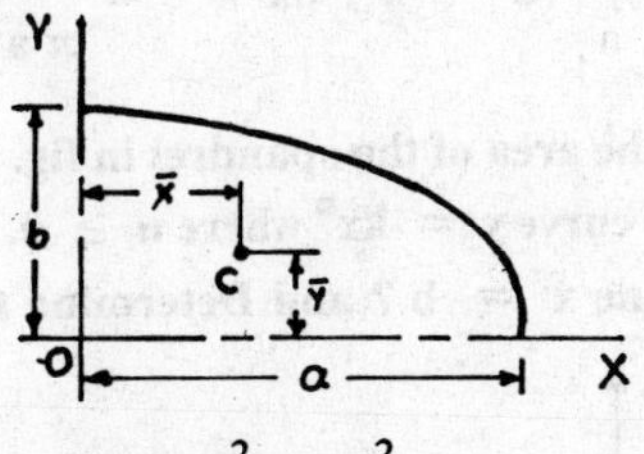

Solution : Equation of ellips is $\frac{x^2}{a^2} + \frac{y^2}{b^2} = 1$

$$\therefore \quad y = \left[b^2 \left(1 - \frac{x^2}{a^2} \right) \right]^{1/2} = \frac{b}{a} [a^2 - x^2]^{1/2}$$

Area of element $dA = y \cdot dx$

$$dA = \frac{b}{a}\,[a^2 - x^2]^{1/2} \cdot dx$$

Total area $$A = \int_0^a \frac{b}{a}\,[a^2 - x^2]^{1/2} \cdot dx$$

Solving we get $A = \dfrac{\pi ab}{4}$

Now, $$\bar{x} = \frac{\int_0^a x \cdot y\,dx}{A} = \frac{4}{\pi ab} \int_0^a x \cdot \frac{b}{a}\,[a^2 - x^2]^{1/2}\,dx$$

$$= \frac{4}{\pi ab} \times \frac{b}{a} \int_0^a x \cdot [a^2 - x^2]^{1/2}\,dx$$

$$= \frac{4}{\pi a^2} \times \frac{a^3}{3} = \frac{4a}{3\pi}$$

On the same logic, again

$$\bar{y} = \frac{\int_0^a \frac{1}{2} y \cdot y\,dx}{A} = \frac{4}{\pi ab} \int_0^a \frac{1}{2} \cdot \frac{b^2}{a^2} \cdot (a^2 - x^2)\,dx$$

$$= \frac{2b}{\pi a^3} \int_0^a (a^2 - x^2)\,dx \qquad = \frac{2b}{\pi a^3} \cdot a^3 \cdot \frac{2}{3} = \frac{4b}{3\pi}$$

Problem 4. : Compute the area of the spandrel in fig. bounded by the X axis, the line $x = b$, and the curve $y = kx^n$ where $n \geq 0$. What is the location of its centroid from the line $x = b$? and Determine the y coordinate of the spandrel.

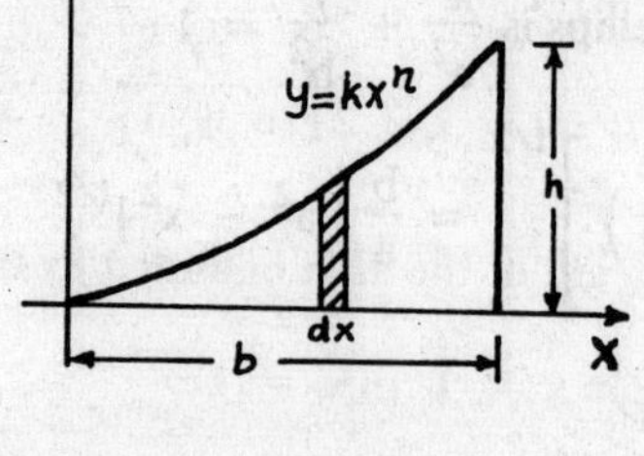

Solution : (b, h) is a point on spandrel $y = kx^n$

$\therefore \quad h = k\,b^n \qquad \therefore \quad k = \dfrac{h}{b^n}$

$\therefore$ Equation of spandrel is $y = \dfrac{h}{b^n}\, x^n$

Area of element $dA = y \cdot dx = \dfrac{h\,x^n}{b^n}\, dx$

Total area $A = \int_0^b \dfrac{h\,x^n}{b^n}\, dx$

$$= \frac{h}{b^n}\left[\frac{x^{n+1}}{(n+1)}\right]_0^b = \frac{1}{n+1}\cdot bh$$

Now $\bar{x} = \dfrac{\int_0^b x\,y\,dx}{A} = \dfrac{n+1}{bh}\int_0^b x \cdot \dfrac{h}{b^n}\, x^n\, dx$

$$= \frac{n+1}{b^{n+1}}\left[\frac{x^{n+2}}{n+2}\right]_0^b$$

$\therefore \quad \bar{x} = \dfrac{n+1}{n+2}\cdot b$ from origin

from other end $\bar{x} = \left(b - \dfrac{n+1}{n+2}\cdot b\right) = \dfrac{b}{n+2}$

Now, $\bar{y} = \dfrac{\int_0^b \frac{1}{2} y \cdot y\, dx}{A} = \dfrac{n+1}{2bh}\int_0^b \dfrac{h^2}{b^{2n}}\, x^{2n}\, dx$

$$= \frac{(n+1)\,h}{2b^{2n+1}}\left[\frac{x^{2n+1}}{2n+1}\right]_0^b = \frac{(n+1)\,h}{2\,(2n+1)} = \frac{n+1}{4n+2}\cdot h$$

Problem 5. : Locate the centroid of the area bounded by the X axis and the sine curve $y = a\dfrac{\sin \pi x}{L}$ from $x = 0$ to $x = L$.

Solution : $y = a\dfrac{\sin \pi x}{L}$

Elemental area $dA = y\,dx = a \sin\dfrac{\pi x}{L} \cdot dx$

Total area $A = \int_0^L dA = \int_0^L a \sin\dfrac{\pi x}{L}\, dx$

$$= a \cdot \frac{L}{\pi}\left[\cos\frac{\pi x}{L}\right]_0^L = a \cdot \frac{L}{\pi}[1+1] = \frac{2aL}{\pi}$$

$$\bar{x} = \frac{\int_0^L x\,y\,dx}{A} = \frac{\pi}{2aL}\int_0^L x \cdot a \sin\frac{\pi x}{L}\,dx$$

$$= \frac{\pi}{2aL} \cdot \frac{aL^2}{\pi} = \frac{L}{2}$$

$$\bar{y} = \frac{\int_0^L \frac{1}{2}y \cdot y\,dx}{A} = \frac{\pi}{2aL} \times \frac{1}{2}a^2 \int_0^L \sin^2\frac{\pi x}{L}\,dx$$

$$= \frac{\pi a}{4L} \cdot \frac{L}{2} = \frac{\pi a}{8}$$

Problem 6. : A slender homogeneous wire of uniform cross section is bent into the shape shown in fig. Determine the coordinates of its centre of gravity.

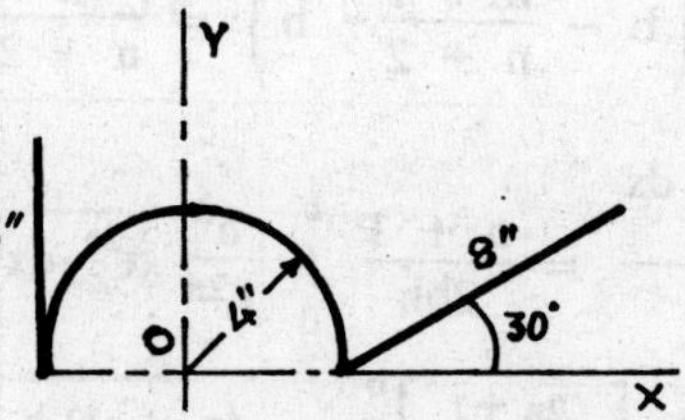

Solution :

$$\bar{x} = \frac{\Sigma x \cdot dL}{L} \quad ; \quad \bar{y} = \frac{\Sigma y \cdot dL}{L}$$

Taking 'O' as origin :

Element	Length L	$\bar{x}$	$\bar{y}$	$L\bar{x}$	$L\bar{y}$
Vertical rod	6	– 4	3	– 24	18
Semicircular rod	4π	0	$\frac{2 \times 4}{\pi}$	0	32
Inclined rod	8	4 + 4 cos 30	4 sin 30	59.7128	16
Total	26.566			35.7128	66

$$\bar{x} = \frac{35.7128}{26.566} = 1.34 \;;\; \bar{y} = \frac{66}{26.566} = 2.48$$

Problem 7. : Locate the center of gravity of the bent wire shown in fig. The wire is homogeneous and of uniform cross section.

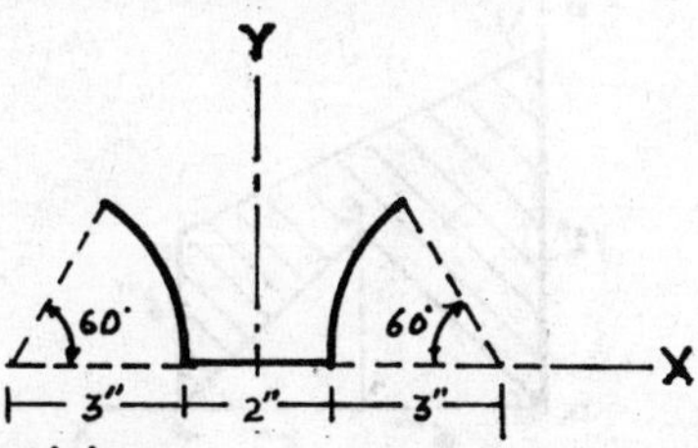

Solution : Taking 'O' origin as

Element	Length L	$\bar{x}$	$\bar{y}$	$L\bar{x}$	$L\bar{y}$
60° arc rod	$\frac{\pi}{3} \times 3$	– 1.516 *	1.4324**	– 4.772	4.5
2" rod	2	0	0	0	0
60° arc rod	$\frac{\pi}{3} \times 3$	1.516	1.4324	4.772	4.5
Total	8.283			0	9

$$\bar{x} = \frac{0}{8.283} = 0 \;;\; \bar{y} = \frac{9}{8.283} = 1.086$$

Centroid position of 60° arc is found as follows :

From the table of centroids it is clear that point P is

$$\frac{r \sin 30}{\pi / 6} = \frac{3 \sin 30}{\pi / 6} = 2.86479 \text{ away from O along OQ.}$$

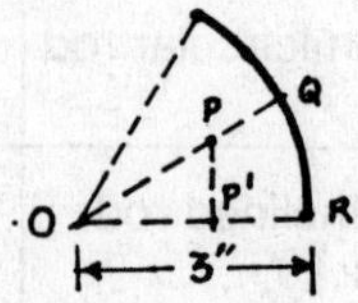

$OP = 2.86479$

$\therefore \; OP' = 2.86479 \cos 30 = 2.4833$

$\therefore \; RP' = 3 - 2.4833 = 0.5166$

$\therefore$ Distance of centroid of arc from origin along x axis is

$-1 - 0.5166 = -1.5166$

Distance of centroid of arc from origin along y axis is

$PP' = 2.86479 \sin 30 = 1.4324$

Problem 8. : Determine the centroid of the lines that form the boundary of the shaded area in fig.

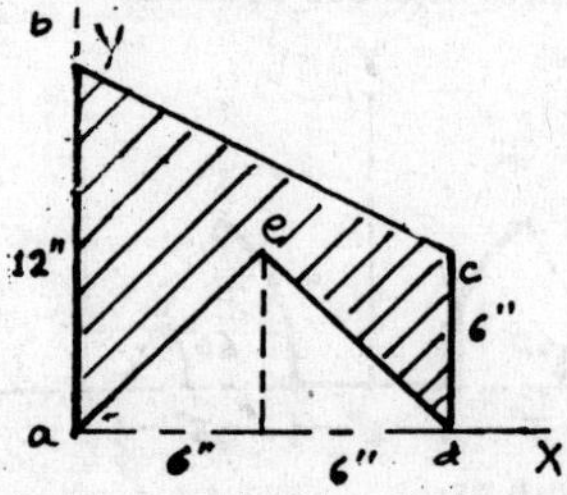

Solution : $\angle\theta$ in above fig. $= \tan^{-1}(0.5) = 26.56^{o}$

Element	Length L	$\bar{x}$	$\bar{y}$	$L\bar{x}$	$L\bar{y}$
ab	12	0	6	0	72
bc	13.416	$12 - \frac{13.416}{2} \cos\theta$	$6 + \frac{13.416}{2} \sin\theta$	80.5	120.73
cd	6	12	3	72	18
de	8.485	$\frac{8.485}{2} \cos 60$	$\frac{8.485}{2} \sin 60$	18	31.175
ea	8.485	$12 - \frac{8.485}{2} \cos 60$	$\frac{8.485}{2} \sin 60$	83.82	31.175
Total	48.386			254.322	273.08

$$\bar{x} = \frac{254.32}{48.386} = 5.26$$

$$\bar{y} = \frac{273.08}{48.386} = 5.64$$

Problem 9. : Locate the centroid of the shaded are in fig. created by cutting a semicircleof diameter r from a quarter which of radius r.

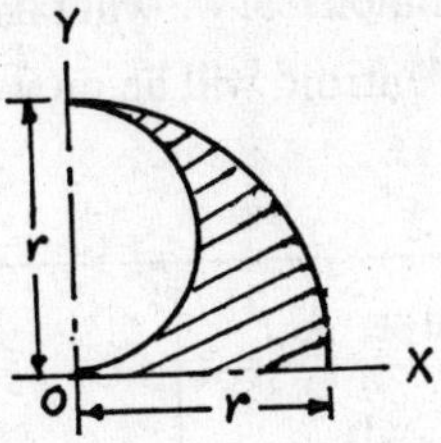

Solution :

$\bar{x}$ and $\bar{y}$ of elements w.r.t. origin 'O'.

Element	Area A	$\bar{x}$	$\bar{y}$	$A\bar{x}$	$A\bar{y}$
Blank semicircle	$-\frac{\pi r^2}{8}$	$\frac{2r}{3\pi}$	$\left(\frac{r}{2}\right)$	$\frac{-r^3}{12}$	$-\frac{\pi r^3}{16}$
Quareter circle	$\frac{\pi r^2}{4}$	$\frac{4r}{3\pi}$	$\frac{4r}{3\pi}$	$\frac{r^3}{3}$	$\frac{r^3}{3}$
Total	$\frac{\pi r^2}{8}$			$0.25\ r^3$	$0.13698\ r^3$

$$\bar{x} = \frac{0.25\ r^3}{\frac{\pi}{8} r^2} = 0.636\ r$$

$$\bar{y} = \frac{0.13698\ r^3}{\frac{\pi}{8} r^2} = 0.3488\ r$$

Problem 10. : Determine the dimension b that will locate the centroidal axis at 4 inches above the base of the section shown in fig.

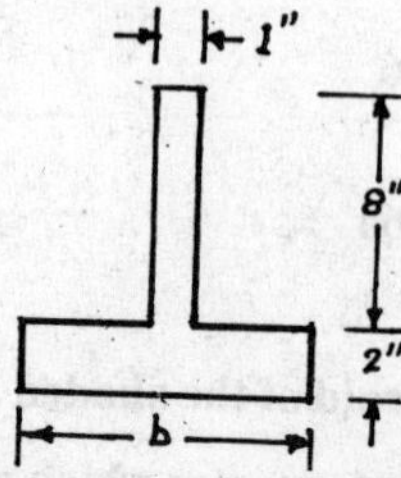

Solution : As figure is symmetrical we will take centre line as y axis and x axis as bottom line of figure. Centroid will lie on y axis.

$\therefore \ \bar{x} = 0$

Element	Area A	$\bar{y}$	$A\bar{y}$
Horizontal	2. b	1	2 b
Vertical	8	6	48
Total	8 + 2b		48 + 2b

We know $\bar{y} = 4$

$$\therefore \ 4 = \frac{48 + 2b}{8 + 2b}$$

$$\therefore \ 32 + 8b = 48 + 2b$$

$$\therefore \ b = 2.67 \text{ inches}$$

Problem 11. : The surface of a plate of uniform thickness is given by the shaded area in fig. If this plate is suspended from a hinge at A, what angle will line AB make with the vertical ?

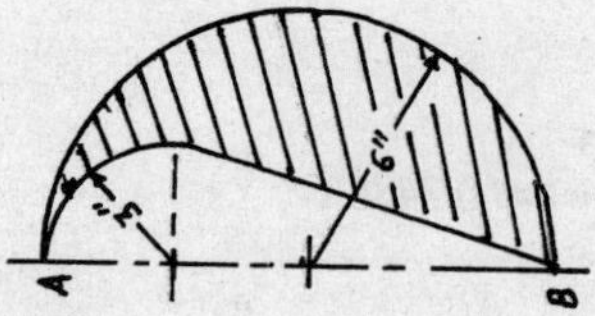

Solution : $\angle\theta = \tan^{-1}\left(\frac{3}{9}\right) = 18.435^{\circ}$

Taking B as origin.

Element	Area A	$\bar{x}$	$\bar{y}$	A $\bar{x}$	A $\bar{y}$
Quarter circle radius = 3	$-\frac{9\pi}{4}$	$\frac{4\times 3}{3\pi}$	$9+\frac{4\times 3}{3\pi}$	− 9	− 72.62
Semicircle radius = 6	$\frac{36\pi}{2}$	$\frac{4\times 6}{3\pi}$	6	144	108 π
Triangle	$-\frac{27}{2}$	1	6	$-\frac{27}{2}$	− 81
Total	35.96			121.5	185.67

$$\therefore \quad \bar{x} = \frac{121.5}{35.96} = 3.378$$

$$\bar{y} = \frac{185.67}{35.96} = 5.163 \text{ from B}$$

$$= 6.8367 \text{ from A}$$

$$\varphi = \tan^{-1}\left(\frac{3.378}{6.8367}\right) = 26.3^o$$

where φ is the angle between vertical AB and centroid. Same will be the angle made by AB with vertical when plate is suspended at A.

Problem 12. : With respect to the given axes, locate the centroid of the shaded area shown in fig.

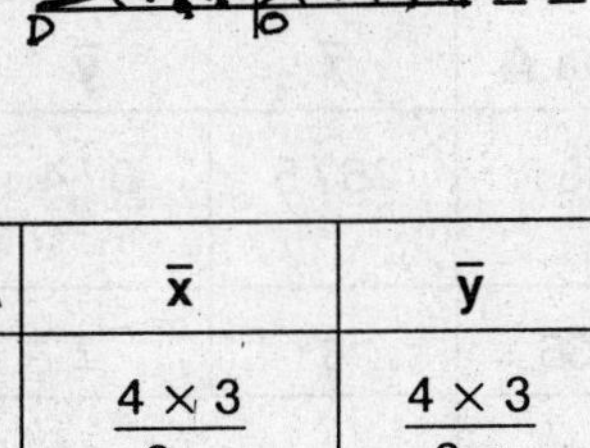

Solution :

Element	Area A	$\bar{x}$	$\bar{y}$	A $\bar{x}$	A $\bar{y}$
ABO quarter circle	$\frac{9\pi}{4}$	$\frac{4\times 3}{3\pi}$	$\frac{4\times 3}{3\pi}$	9	9

AOCD Sq.	9	− 1.5	1.5	− 13.5	13.5
ACD quarter circle	$-\frac{9\pi}{4}$	$-3\left(3-\frac{4\times3}{3\pi}\right)$	$3-\frac{4\times3}{3\pi}$	12.206	− 12.206
Total	9			7.71	10.294

$$\therefore \quad \bar{x} = \frac{7.71}{9} = 0.856$$

$$\bar{y} = \frac{10.294}{9} = 1.143$$

Problem 13. : Locate the centroid of the shaded area enclosed by the curve $y^2 = \alpha x$ and the straight line shown in fig.

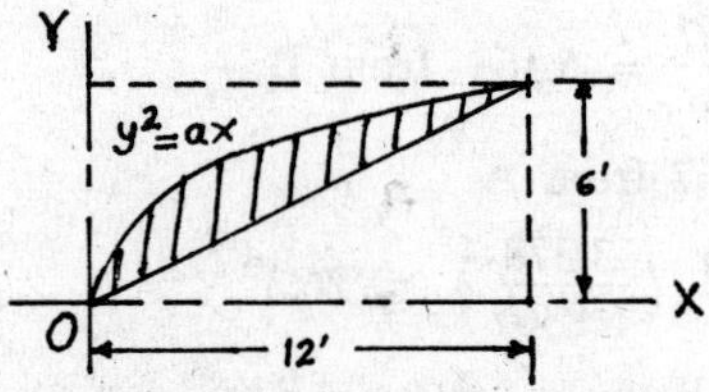

Solution : As we have seen in 1st problem centroid for $y^2 = ax$ is given by $\bar{x} = \frac{3}{5}x$, $\bar{y} = \frac{3}{8}y$ and area is $\frac{2}{3}xy$.

$\therefore$ For the curve covered area

$$\bar{x} = \frac{3}{5} \times 12 = \frac{36}{5}, \quad \bar{y} = \frac{3}{8} \times 6 = \frac{9}{4}$$

$$\text{and area} = \frac{2}{3} \times 12 \times 6 = 48$$

Centroid of shaded area

Element	Area A	$\bar{x}$	$\bar{y}$	A $\bar{x}$	A $\bar{y}$
Curve covered area	48	36 / 5	9 / 4	345.6	108
Triangle	− 36	8	2	− 288	− 72
Total	12			57.6	36

$$\bar{x} = \frac{57.6}{12} = 4.8$$

$$\bar{y} = \frac{36}{12} = 3$$

Problem 14. : Locate the centroid of the cross section of the culvert shown in fig.

Hint : Integration is unnecessary if the area is subdivided into elements to be found in table.

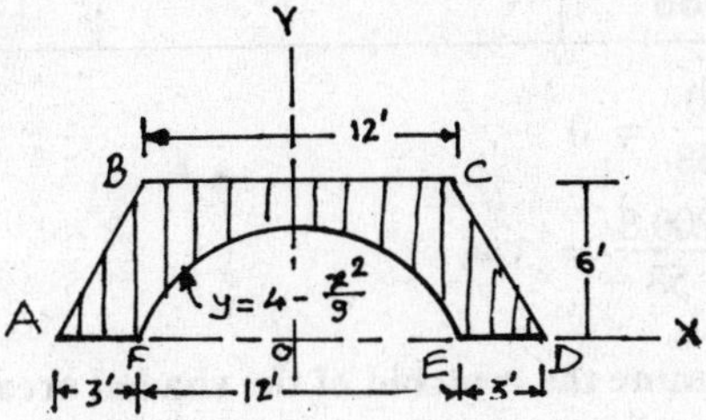

Solution : First finding the centroid of unshaded area under curve

$$y = 4 - \frac{x^2}{9}$$

$$dA = \left(4 - \frac{x^2}{9}\right) dx$$

$$A = \int_{-6}^{6} 4 - \frac{x^2}{9}\, dx = \left[4x - \frac{x^3}{27}\right]_{-6}^{6} = 32$$

$$\bar{y} = \frac{\int_{-6}^{6} \frac{1}{2} y^2\, dx}{32} = \frac{1}{64} \int_{-6}^{6} \left(16 - \frac{8}{9}x^2 + \frac{x^4}{81}\right) dx$$

$$\bar{y} = \frac{1}{64} \left[16x - \frac{8}{27}x^3 + \frac{x^5}{81 \times 5}\right]_{-6}^{6}$$

$$= \frac{1}{64} \left[51.2 + 51.2\right] = 1.6$$

Finding centroid of shaded area

Element	Area A	$\bar{x}$	$\bar{y}$	A $\bar{x}$	A $\bar{y}$
Δ ABF	9	– 7	2	– 63	18
Square BCEF	72	0	3	0	216
Δ CDE	9	7	2	63	18
Curve FGE	– 32	0	1.6	0	– 51.2
Total	58			0	200.8

$$\therefore \quad \bar{x} = \frac{0}{58} = 0$$

$$\bar{y} = \frac{200.8}{58} = 3.46$$

Problem 15. : Determine the centroid of the shaded area shown in fig. with respect to the given X–Y axes.

Hint : Project the known location of the centroid of the semicircle upon X–Y directions.

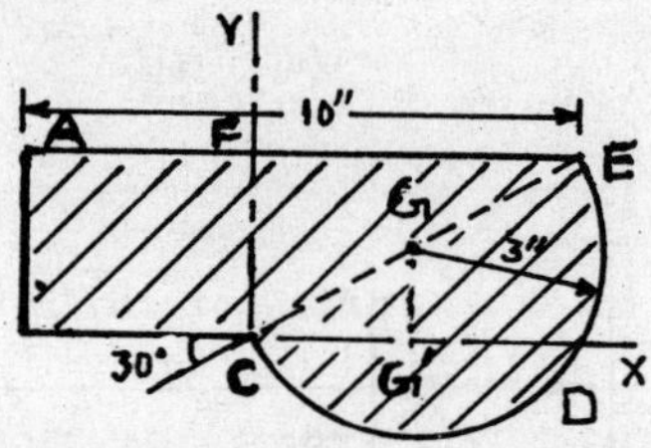

Solution : First finding location of centroid of semicircle with respect to x and y axis.

Centroid of semicircle will be along GD and $\frac{4 \times 3}{3\pi} = \frac{4}{\pi}$ away from G

Now $CG = 3 \quad \therefore \quad CG' = 3 \cos 30^{o} = 2.598$

$GP = \frac{4}{\pi} \quad \therefore \quad G'P' = \frac{4}{\pi} \sin 30^{o} = \frac{2}{\pi} = 0.637$

$\therefore \quad CP' = 2.598 + 0.637 = 3.2346$

Distances of centroid along x axis.

Now $GG' = 3 \sin 30 = 1.5$

$GG'' = \frac{4}{\pi} \cos 30 = 1.1026$

$\therefore$ PP′ = G′G″ = 0.3973

Distance of centroid along y axis.

Distance AF = 10 − 6 cos 30 = 4.8 ; CF = 6 sin 30 = 3

Now finding centroid of shaded area

Element	Area A	$\bar{x}$	$\bar{y}$	A $\bar{x}$	A $\bar{y}$
Square ABCF	14.41	− 2.4	1.5	− 34.7281	21.615
Δ CFE	7.8	1.732	2	13.5096	15.6
Semicircle CDE	$\frac{\pi \times 9}{2}$	3.2346	0.3973	45.73	5.6167
Total	36.347			24.5118	42.8317

$$\bar{x} = \frac{24.5118}{36.347} = 0.674$$

$$\bar{y} = \frac{42.8317}{36.347} = 1.178$$

Chapter Six

MOMENT OF INERTIA

The mathematical definiion of moment of inertia, $I = \int \rho^2 \, dA$, indicates that an area is divided into small parts such as dA, and each area is multiplied by the square of its moment arm about the reference axis. Thus, as shown in fig. if the coordinates of the differential area dA are (x, y), the moment of inertia about the X axis is the summatioon of the product of each area dA by the square of its moment arm y.

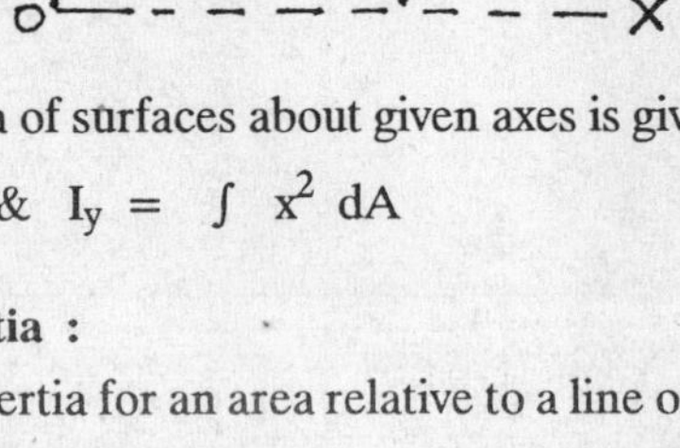

Moment of Inertia of surfaces about given axes is given by

$I_x = \int y^2 \, dA$ & $I_y = \int x^2 \, dA$

Polar Moment of Inertia :

The moment of inertia for an area relative to a line or axis perpendiicular to the plane of the area is called the polar moment of inertia and is denoted by the symbol J. From the fig. the moment of inertia of an area in the XY plane with respect to Z axis is $I = \int \rho^2 \, dA$ and thus we finally obtain,

$$J_z = \int r^2\, dA = \int (x^2 + y^2)\, dA$$
$$= \int x^2\, dA + \int y^2\, dA$$

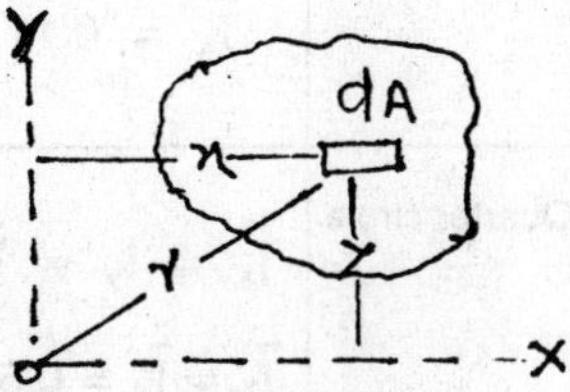

Radius of Gyration :

It is the uniform distance from the reference axis at which the entire area is supposed to be distributed. It is usually designated by k. Moment of Inertia in terms of k is given by

$$I = Ak^2$$

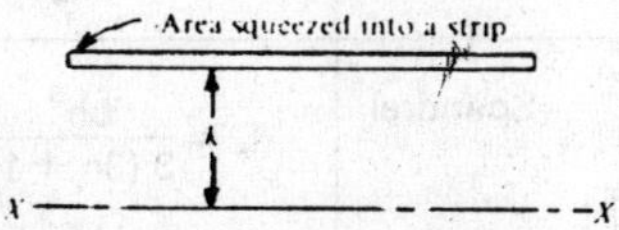

Moment of Inertia for Geometrical Shapes :

Shape	Moment of Inertia	Radius of Gyration
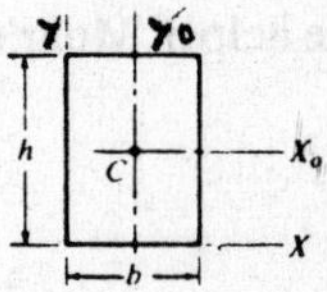 Rectangle	$\bar{I}_x = \dfrac{bh^3}{12}$ $I_x = \dfrac{bh^3}{12}$	$\bar{k}_x = \dfrac{h}{\sqrt{12}}$ $k_x = \dfrac{h}{\sqrt{3}}$
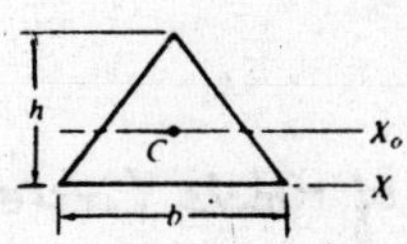 Any triangle	$\bar{I}_x = \dfrac{bh^3}{36}$ $I_x = \dfrac{bh^3}{36}$	$\bar{k}_x = \dfrac{h}{\sqrt{18}}$ $k_x = \dfrac{h}{\sqrt{6}}$
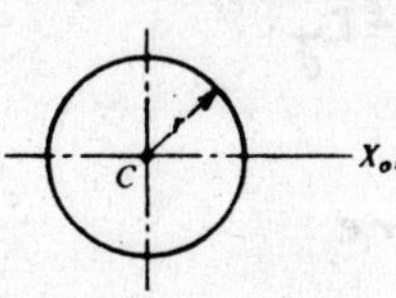 Circle	$\bar{I}_x = \dfrac{\pi r^4}{4}$ $\bar{J} = \dfrac{\pi r^4}{2}$	$\bar{k}_x = \dfrac{r}{2}$ $\bar{k}_z = \dfrac{r}{\sqrt{2}}$

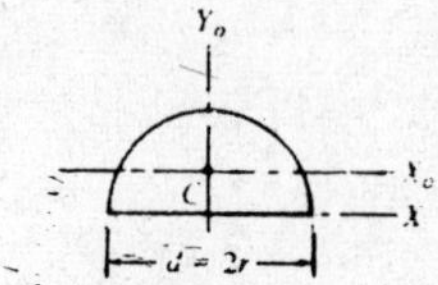 Semicircle	$I_x = \bar{I}_y = \dfrac{\pi r^4}{8}$ $\bar{I}_x = 0.11\, r^4$	$k_x = \bar{k}_y = \dfrac{r}{2}$ $\bar{k}_x = 0.264\, r$
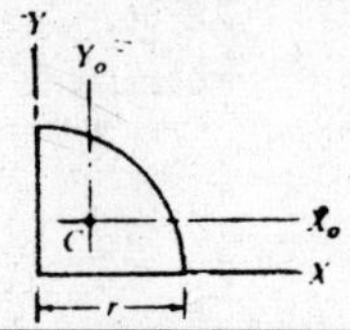 Quarter circle	$I_x = I_y = \dfrac{\pi r^4}{16}$ $\bar{I}_x = \bar{I}_y = 0.55\, r^4$	$k_x = k_y = \dfrac{r}{2}$ $\bar{k}_x = \bar{k}_y = 0.264r$
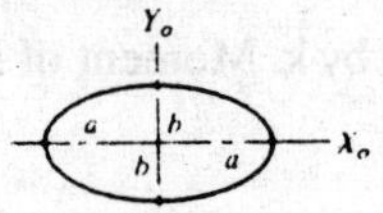 Ellipse	$\bar{I}_x = \dfrac{\pi ab^3}{4}$ $\bar{I}_y = \dfrac{\pi ba^3}{4}$	$\bar{k}_x = \dfrac{b}{2}$ $\bar{k}_y = \dfrac{a}{2}$
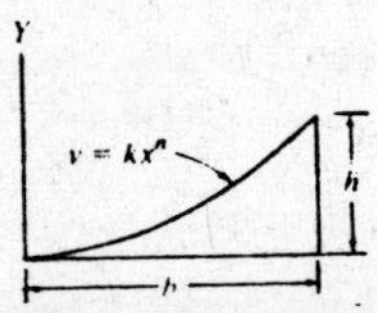 Spandrel	$I_x = \dfrac{bh^3}{3\,(3n+1)}$ $I_y = \dfrac{hb^3}{n+3}$	

Mohr's Circle for Moment Inertia :

Due to simplicity and flexibility in finding M.I. with respect to inclined axes, maximum and minimum M.I. principal axes with the help of Mohr's circle, it is very popular.

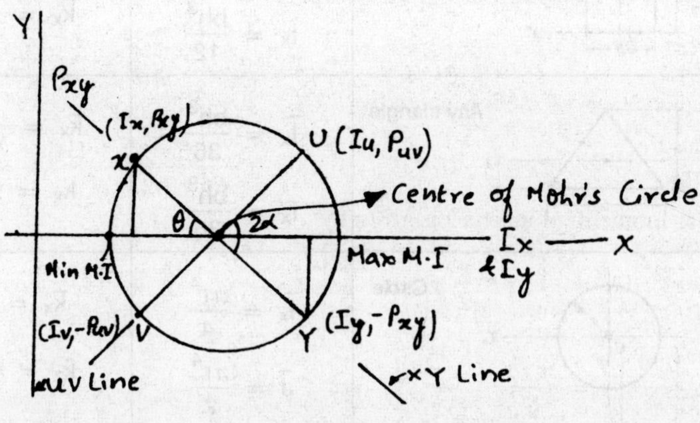

Steps in drawing Mohr's circle are explained below

(1) Draw two perpendicular axes X and Y. Take Y axis to represent P_{xy} and X axis to represent M.I. i.e. I_x, I_y

(2) Plot co–ordinates (I_x, P_{xy}) & $(I_y, -P_{xy})$.

(3) Join these two points by a straight line

(4) Point of intersection of this line and X axis (M.I. axis) is centre of Mohr's circle. Assuming the above line as diameter draw a circle (Mohr's circle) with the centre just found.

(5) To find M.I. and P_{xy} about axis U and V which are inclined at counter clockwise angle α to given X and Y axes then on Mohr's circle the UV line is laid at counterclockwise angle 2α from the line joining (I_x, P_{xy}) & $(I_y, -P_{xy})$ i.e. XY line. Find the co–ordinates at ends of UV line on Mohr's circle.

These co–ordinates will be (I_u, P_{uv}) & $(I_v, -P_{uv})$. Thus we get I_u, I_v and P_{uv}.

(6) Maximum and Minimum M.I. is where $P_{xy} = 0$. So Minimum M.I. is point on M.I. axis on right side and Maximum M.I. is point on M.I. axis on left side. Find the angle between horizontal M.I. axis and XY line. Say it is θ

Then, $\frac{\theta}{2}$ is the rotation of XY to get the principal axes.

Problem 1.: Determine the moment of inertia of the quarter circle shown in fig. with respect to the given axes.

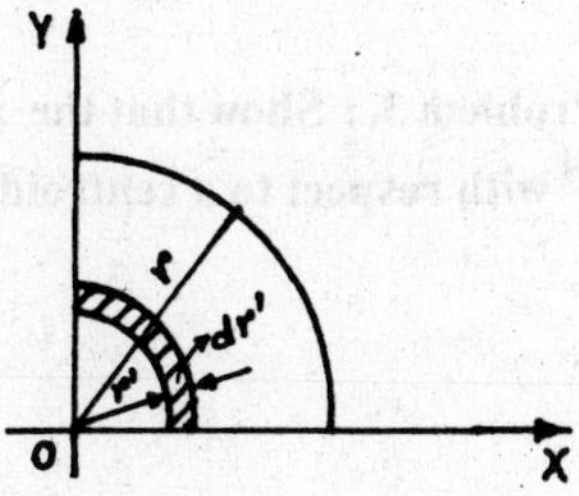

Solution : Area of element is given by

$$dA = \frac{2\pi r'}{4} \cdot dr' = \frac{\pi}{2} r' \cdot dr'$$

Now polar moment of inertia J is given by

$$J = \int_0^r r'^2 \, dA = \int_0^r r'^2 \frac{\pi}{2} r' \, dr$$

$$= \frac{\pi}{2} \left[\frac{r'^4}{4} \right]_0^r = \frac{\pi r^4}{8}$$

Now we know $J = I_x + I_y$

Due to symmetry of shaded area, $I_x + I_y$

$$\therefore \quad J = 2 I_x = \frac{\pi r^4}{8}$$

$$\therefore \quad I_x = \frac{\pi r^4}{16} \text{ and } I_y = \frac{\pi r^4}{16}$$

Problem 2. : Determine the moment of inertia of the semicircle shown in fig. with respect to the given axes.

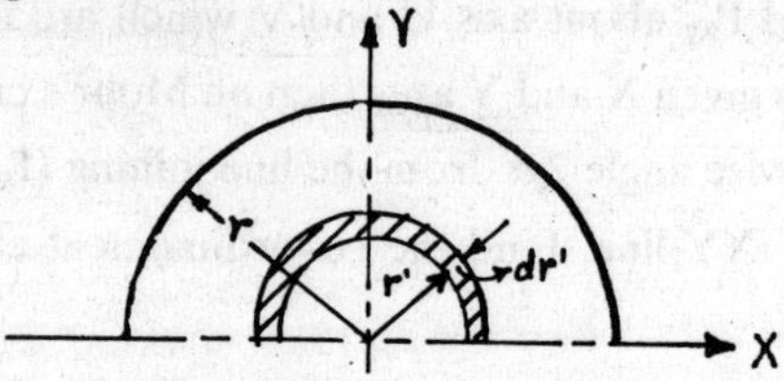

Solution : Area of element $= \pi r' \cdot dr = dA$

$$J = \int_0^r r'^2 \, dA = \int_0^r r'^2 \, \pi r' \, dr = \frac{\pi r^4}{4}$$

Also $J = I_x + I_y = 2I_x$ ..... ($I_x = I_y$ due to symmetry)

$$\therefore \quad 2I_x = \frac{\pi r^4}{4}$$

$$\therefore \quad I_x = \overline{I}_y = \frac{\pi r^4}{8}$$

Problem 3. : Show that the moment of inertia of a semicircle of radius is 0.11 r^4 with respect to a centroidal axis parallel to the diameter.

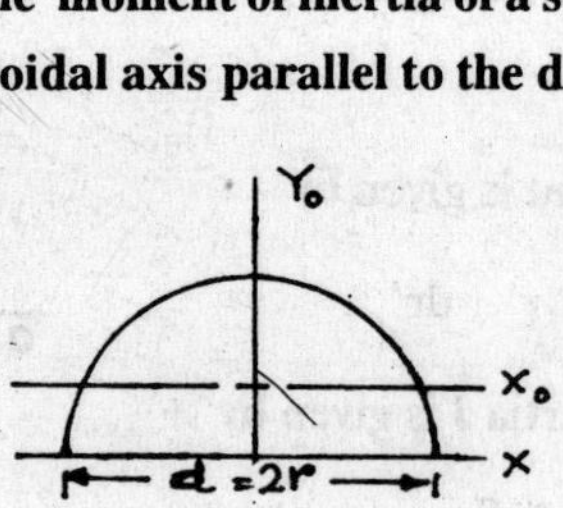

Solution : As we know M.I. of semicircle about its diameter $= \frac{\pi}{8} r^4$

Centroid of semicircle is $\frac{4r}{3\pi}$ above dia.

$$I_x = \bar{I}_x + Ad^2$$

$$\therefore \quad \bar{I}_x = I_x - Ad^2$$

$$= \frac{\pi}{8} r^4 - \frac{\pi r^2}{2} \cdot \left(\frac{4r}{3\pi}\right)^2 = 0.11\, r^4$$

Problem 4.: Determine the moment of inertia for the quarter circle shown in fig. with respect to centroidal X axis.

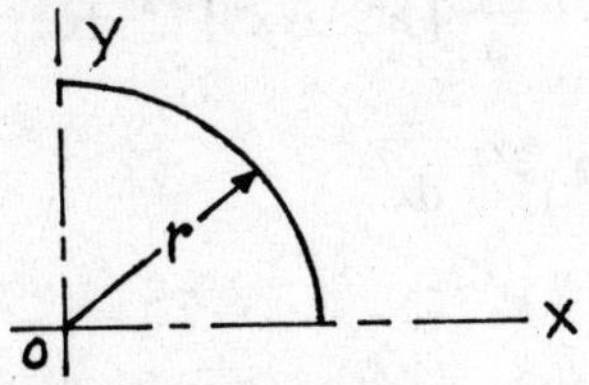

Solution : M.I. of quarter circle about x axis

$$I_x = \frac{\pi r^4}{16}$$

$$d = \text{distance of centroid from x axis} = \frac{4r}{3\pi}$$

$$\bar{I}_x = I_x - Ad^2$$

$$= \frac{\pi r^4}{16} - \frac{\pi r^2}{4} \cdot \left(\frac{4r}{3\pi}\right)^2 = 0.055\, r^4$$

Problem 5.: Determine the moment of inertia with respect to the X axis for the area enclosed by the ellipse whose equation is

$\frac{x^2}{a^2} + \frac{y^2}{b^2} = 1$. Also determine the radius of gyration.

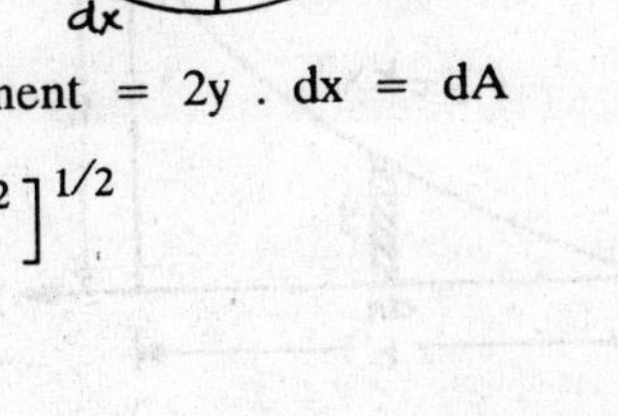

Solution : Area of element $= 2y \cdot dx = dA$

$$y = \left[\left(1 - \frac{x^2}{a^2}\right) \cdot b^2\right]^{1/2}$$

$$\therefore \quad dA = 2\left[b^2\left(1-\frac{x^2}{a^2}\right)\right]^{1/2} dx$$

$$= \frac{2b}{a}\,[a^2 - x^2]^{1/2}\, dx$$

$$\bar{I}_x = \int_{-a}^{a} y^2\, dA$$

$$= \int_{-a}^{a} \frac{b^2}{a^2}(a^2 - x^2) \cdot \frac{2b}{a}[a^2 - x^2]^{1/2}\, dx$$

$$= \frac{2b^3}{a^3}\int_{-a}^{a} \left(a^2 - x^2\right)^{3/2} dx$$

Substituting $x = a \sin\theta$,

$$\therefore \quad dx = a\cos\theta\, d\theta$$

at $x = a,\ \theta = \frac{\pi}{2}$ and $x = -a,\ \theta = \frac{-\pi}{2}$

$$\therefore \quad I_x = \frac{2b^3}{a^3}\int_{-\pi/2}^{\pi/2} a^3\cos^3\theta \cdot a\cos\theta\, d\theta$$

$$= 2ab^3 \int_{-\pi/2}^{+\pi/2} \cos^4\theta\, d\theta = 2ab^3 \times \frac{\pi}{8} = \frac{\pi\, ab^3}{4}$$

Radius of gyration $I_x = A\, k_x^2$

$$\therefore \quad \frac{\pi\, ab^3}{4} = \pi\, ab \cdot k_x^2$$

$$\therefore \quad k_x = \frac{b}{2}$$

Problem 6. : Determine the moments of inertia with respect to the X and Y axes for the shaded area under the spandrel $y = kx^n$ shown in fig.

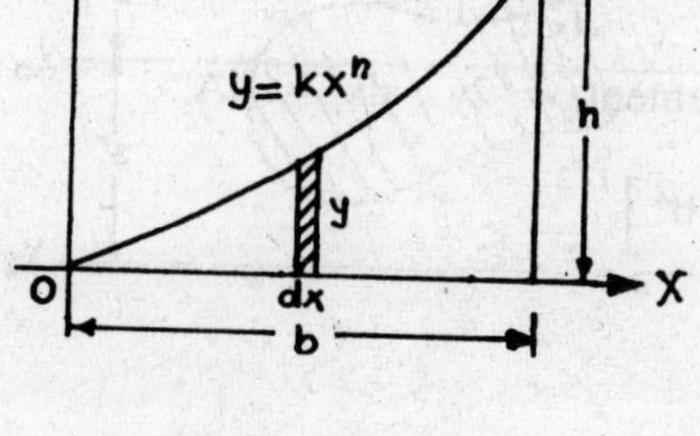

Solution : Equation of curve is $y = kx^n$

Since (b, h) is point on curve so $h = k\,b^n \quad \therefore \quad k = \dfrac{h}{b^n}$

$\therefore$ Equatioin of curve is $y = \dfrac{h}{b^n} x^n$

Area of element $dA = y\,dx$

$$= \frac{h}{b^n} x^n\,dx$$

Since all portions of vertical element are not at same distance from x axis we will have to take $dI_x = \dfrac{1}{3} y^2\,dA$

$$\therefore \quad I_x = \int_0^b \frac{1}{3} y^2\,dA = \int_0^b \frac{1}{3}\frac{h^2}{b^{2n}} x^{2n} \cdot \frac{h}{b^n} x^n\,dx$$

$$= \frac{h^3}{3b^{3n}} \int_0^b x^{3n}\,dx = \frac{h^3}{3b^{3n}} \left[\frac{x^{3n+1}}{3n+1}\right]_0^b$$

$$= \frac{bh^3}{3\,(3n+1)}$$

$$I_y = \int_0^b x^2\,dA = \int_0^b x^2 \cdot \frac{h}{b^n} x^n \cdot dx$$

$$= \frac{h}{b^n}\left[\frac{x^{n+3}}{n+3}\right]_0^b = \frac{h\,b^3}{n+3}$$

Problem 7. : The shaded section shown in fig. has an area of 5 in.^{2} and a moment of inertia of 30 in.^{4} about axis X_1. Find the moment of inertia about axis X_2.

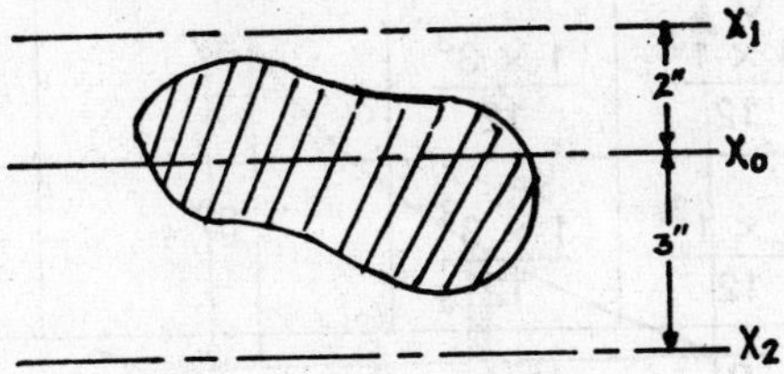

Solution : $I_{x1} = I_{x0} + Ad^2$

$30 = I_{x0} + 5 \times (2)^2$

$\therefore \quad I_{x0} = 10$

$I_{x2} = I_{x0} + Ad^2$

$= 10 + 5 \times (3)^2 = 55 \text{ inch}^4$

Problem 8. : Blocks of wood are glued togehter to form the shaded section shown in fig. Find the moment of inertia of the shaded section about is horizontal and vertical centroidal axes.

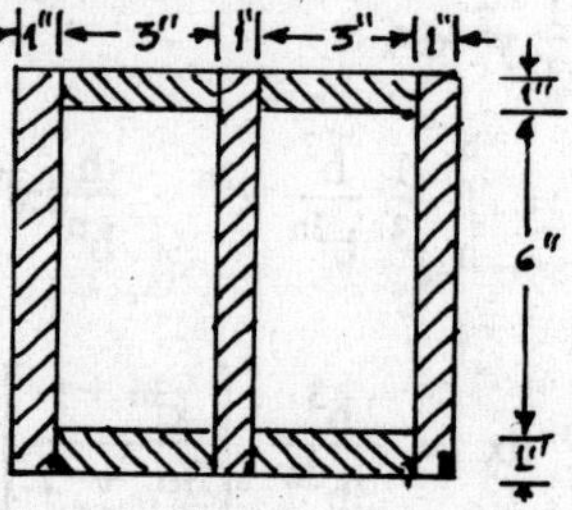

Solution :

Element	$\bar{I}_x$	$\bar{I}_y$	I_x	I_y
(1)	$\frac{1 \times 8^3}{12}$	$\frac{8 \times 1^3}{12}$	42.67	$\frac{8}{12} + 8 \times 4^2 = 128.67$
(2)	$\frac{3 \times 1^3}{12}$	$\frac{1 \times 3^3}{12}$	$\frac{3}{12} + 3 \times 3.5^2 = 37$	$\frac{27}{12} + 3 \times 2^2 = 14.25$
(3)	$\frac{3 \times 1^3}{12}$	$\frac{1 \times 3^3}{12}$	37	14.25
(4)	$\frac{1 \times 8^3}{12}$	$\frac{8 \times 1^3}{12}$	42.67	0.67
(5)	$\frac{3 \times 1^3}{12}$	$\frac{1 \times 3^3}{12}$	37	14.25
(6)	$\frac{3 \times 1^3}{12}$	$\frac{1 \times 3^3}{12}$	37	14.25

(7)	$\frac{1 \times 8^3}{12}$	$\frac{8 \times 1^3}{12}$	42.67	128.67
Total			276	315.01

$$\therefore \quad I_x = 276 \text{ in}^4$$

$$I_y = 315 \text{ in}^4$$

Problem 9. : Determine the moment of inertia of the T–section shown in fig. about its centroidal X_o axis.

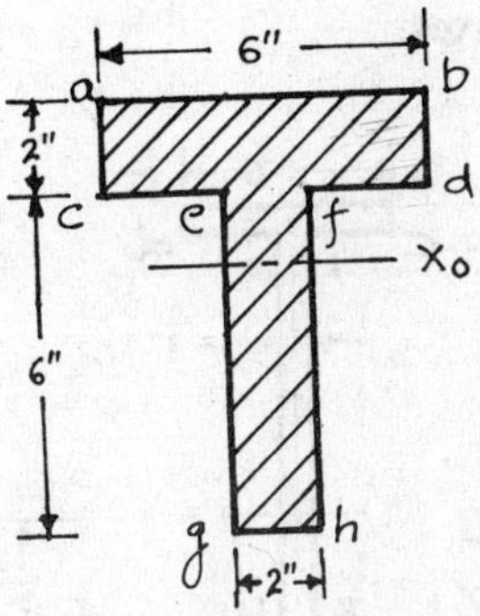

Solution : Let us first find centroid of T section. Taking reference as bottom of T

Element	Area A	$\bar{y}$	$A\bar{y}$
area abcd	12	6 + 1	84
area efgh	12	3	36
Total	24		120

$$\therefore \quad \bar{y} = \frac{120}{24} = 5 \text{ inches from bottom}$$

M.I. of T section about centroidal axis.

Element	$\bar{I}_x$	Area A	d	I_{x_0}
area abcd	$\frac{6 \times 2^3}{12}$	12	2	$4 + 12 \times 2^2 = 52$
area efgh	$\frac{2 \times 6^3}{12}$	12	2	$36 + 12 \times 2^2 = 84$
Total				136

$\therefore \quad I_{x0} = 136 \text{ in}^4$

Problem 10. : Determine the moment of inertia of the area shown in fig. with respect to its centroidal axes.

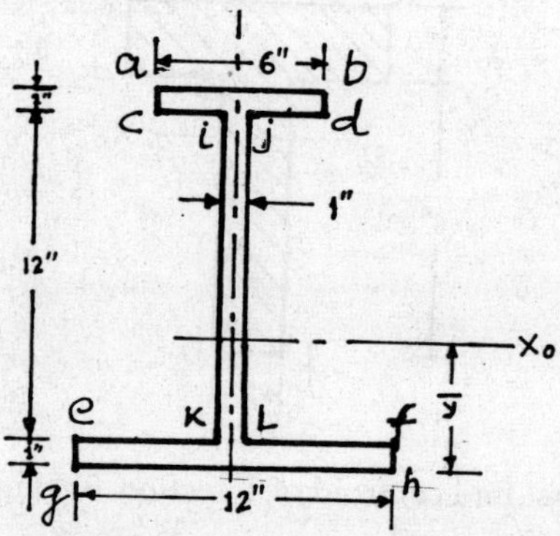

Solution : Let us first find centroid of I section. Taking reference – bottom of I

Element	Area A	$\bar{y}$	A $\bar{y}$
area abcd	6	13.5	81
area efgh	12	7	84
area ijkl	12	0.5	6
Total	30		171

$\therefore \; \bar{y} = \dfrac{171}{30} = 5.7$ inches from bottom

M.I. of I section about centroid axis

Element	$\bar{I}_x$	Area A	d	I_{xo}	$\bar{I}_y = I_{yo}$
area abcd	$\dfrac{6 \times 1^3}{12}$	6	13.5 – 5.7	$\frac{1}{2} + 6 \times (7.8)^2$	$\dfrac{1 \times 6^3}{12} = 18$
area efgh	$\dfrac{1 \times 12^3}{12}$	12	7 – 5.7	$144 + 12 \times (1.3)^2$	$\dfrac{12 \times 1^3}{12} = 1$
area ijkl	$\dfrac{12 \times 1^3}{12}$	12	5.2	$1 + 12\,(5.2)^2$	$\dfrac{1 \times 12^3}{12} = 144$
Total				855.3	163

$\therefore \; I_{xo} = 855.3 \text{ in}^4$

$I_{yo} = 163 \text{ in}^4$

Problem 11. : Find the moment of inertia about the indicated Y axis for the shaded area shown in fig.

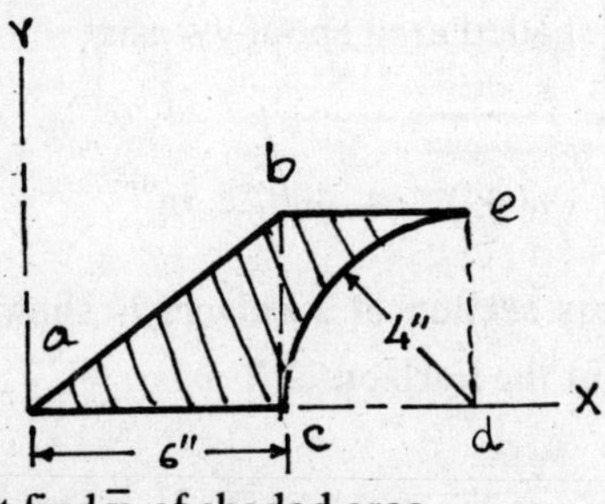

Solution : Let us first find $\bar{x}$ of shaded area

Element	Area A	$\bar{x}$	A $\bar{x}$
Δ abc	12	4	48
area bcde	16	8	128

quarter circle	$-\dfrac{\pi \times 4^2}{4}$	$6+4-\dfrac{4 \times 4}{3\pi}$	-104.33
Total	15.433		71.67

$$\therefore \quad \bar{x} = \frac{71.67}{15.433} = 4.644$$

M.I. of shaded area about centroidal y axis.

Element	$\bar{I}_y$	Area A	d	I_{yo}
Δ abc	$\dfrac{4 \times 6^3}{36}$	12	0.644	$24 + 12\,(0.644)^2$
area bcde	$\dfrac{4 \times 4^3}{12}$	16	3.356	$21.33 + 16\,(3.356)^2$
quarter circle	0.055×4^4	-4π	8.3 – 4.644	$14.08 + (-4\pi)\,(3.658)^2$
Total	15.43			76.43

$$\therefore \quad I_{yo} = 76.43$$

$\therefore$ M.I. of shaded area about y – axis

$$I_y = I_{yo} + Ad^2$$

$$= 76.43 + (15.43)\,(4.644)^2 = 409.28 \text{ in}^4$$

Problem 12. : The cross section of a culvert is shown in fig. Compute the moment of inertia about the horizontal X axis.

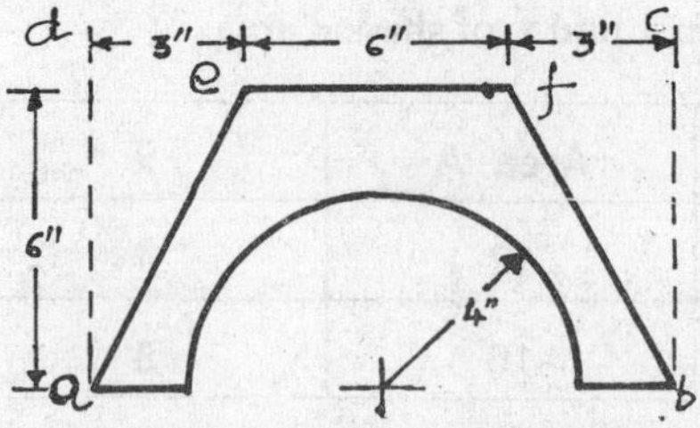

Solution :

Element	$\bar{I}_x$	Area A	d	I_x
area abcd	$\frac{12 \times 6^3}{12}$	72	1	$216 + 72 \times 1^2$
Δ aed	$-\frac{3 \times 6^3}{36}$	+ 9	0	− 18
Δ bcf	$-\frac{3 \times 6^3}{36}$	+ 9	0	− 18
Semicircle	-0.11×4^4	$\frac{\pi \times 4^2}{2}$	4 − 1.697	$[28.16 + 25.13\ (2.303)^2] \times -1$
Total				90.6

$\therefore \quad I_x = 90.6 \text{ in}^4$

Problem 13. : Find the moment of inertia about the axis Y_1 of the shaded area under the second-degree curve shown in fig.

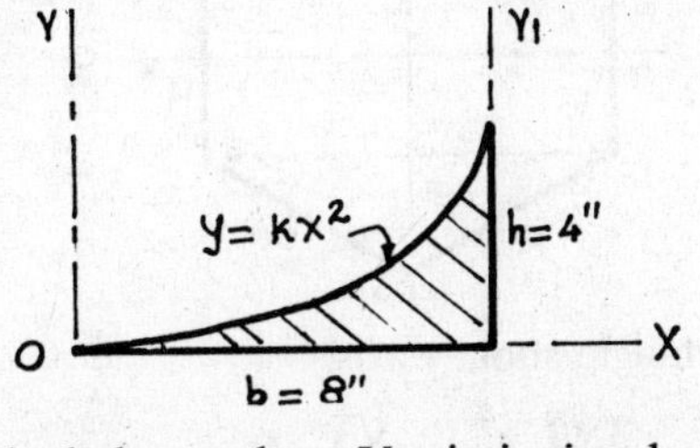

Solution : M.I. of shaded area about Y axis is givey by $\frac{hb^3}{n + 3}$ .... here

$h = 4,\ b = 8,\ n = 2$

$$I_y = \frac{4 \times 8^3}{5} = 409.6 \text{ inches}^2$$

Area under the curve $y = kx^n$ is given by $\frac{1}{n + 1}\ b \cdot h$ and $\bar{x}$ for the same

are is given by $\frac{n + 1}{n + 2} \cdot b$

For $y = kx^2$

$$\bar{x} = \frac{3}{4} \times 8 = 6 \text{ inch}$$

and area $A = \frac{1}{3} \times 8 \times 4 = \frac{32}{3}$

Now $I_y = I_{yo} + Ad^2$

$\therefore \quad I_{yo} = I_y - Ad^2$

$$= 409.6 - \frac{32}{3} \times (6)^2 = 25.6 \text{ in}^4$$

$I_{y1} = I_{yo} + Ad^2$

$$= 25.6 + \left(\frac{32}{3}\right) (8 - 6)^2 = 68.27 \text{ in}^4$$

Problem 14. : Compute the moment of inertia with respect to an axis passing through two opposite apexes of a regular hexagon of side a.

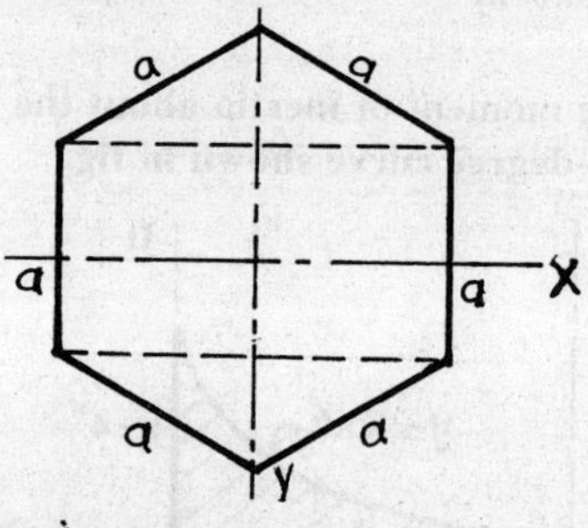

Solution : M.I. of rectangle about y axis.

$$= \frac{a \times (a\sqrt{3})^3}{12} = \frac{3\sqrt{3}\ a^4}{12} = \frac{\sqrt{3}}{4} a^4$$

M.I. of 4 triangles w.r.t. y axis

$$= 4 \times \frac{\left(\frac{a}{2} \times \left(\frac{a\sqrt{3}}{2}\right)^3\right)}{12} = \frac{3\sqrt{3}}{3 \times 2 \times 8} a^4 = \frac{\sqrt{3}}{16} a^4$$

$\therefore$ M.I. of hexagon about y axis

$$= \frac{\sqrt{3}}{4} a^4 + \frac{\sqrt{3}}{16} a^4 = \frac{5\sqrt{3}}{16} a^4$$

Problem 15.: Compute the moment of inertia of the 10–in. by 15–in rectangle shown in fig. about the X axis to which it is inclined at an angle $\theta = \sin^{-1}\frac{4}{5}$.

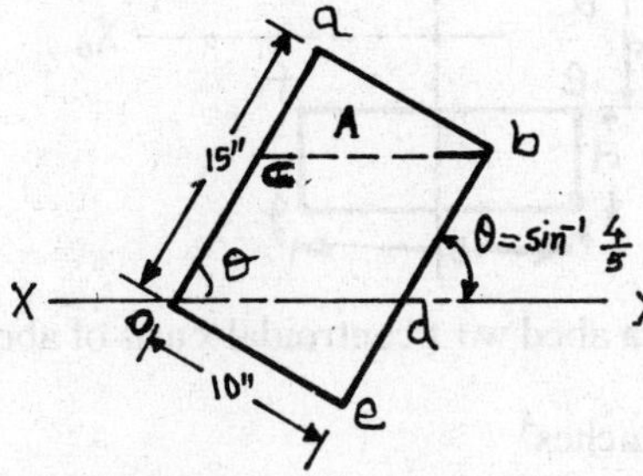

Solution :

From fig. $od = cb = \dfrac{oe}{\sin\theta} = \dfrac{10}{0.8} = 12.5$ in

$ef = ag = oe \cos\theta = 10 \cdot \cos(53.12) = 6$ in

$ae = \left[10^2 + 15^2\right]^{1/2} = 18$ in

$gf = ae - ag - ef = 18 - 6 - 6 = 6$ in

M.I. of Δ oed w.r.t. x axis $= \dfrac{od \cdot (fe)^3}{12} = \dfrac{12.5\,(6)^3}{12} = 225$ inches^4

M.I. of ▱ cbod w.r.t x axis $= \dfrac{od\ (fg)^3}{3} = \dfrac{12.5\ (6)^3}{3} = 900$ inches^4

M.I. of Δ abc w.r.t. cedntroidal axis

$$= \frac{cb\ (ag)^3}{36} = \frac{12.5\,(6)^3}{36} = 75 \text{ inches}^4$$

M.I. of Δ abc w.r.t x axis $= 75 + \dfrac{1}{2}(12.5)(6)(8)^2 = 2475$ inches^4

Total M.I. w.r.t x axis $= 225 + 900 + 2475 = 3600$ inches^4

Problem 16.: Determine the distance d at which the two 2–inches by 6–inches rectangles shown in fig. should be spaced so that $I_x = I_y$.

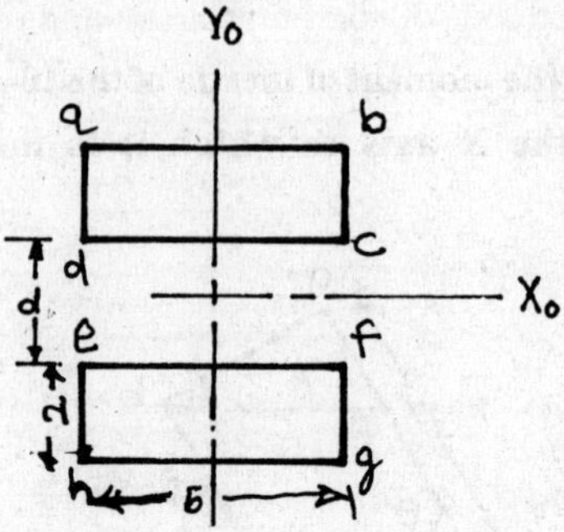

Solution : M.I. of area abcd w.r.t centroidal x axis of abcd

$$= \frac{6 \times 2^3}{12} = 4 \text{ inches}^4$$

M.I. of area abcd w.r.t centroidal y axis of abcd

$$= \frac{2 \times 6^3}{12} = 36 \text{ inches}^4$$

Similarly $I_{x(efgh)} = 4$

and $I_{y(efgh)} = 36$

$I_{y0} = I_{y(abcd)} + I_{y(efgh)} = 36 + 36 = 72$

$$I_{x0} = I_{x(abcd)} + A \cdot \left(\frac{d}{2}\right)^2 + I_{x(efgh)} + A \left(\frac{d}{2}\right)^2$$

$$= 4 + 12\left(1 + \frac{d}{2}\right)^2 + 4 + 12\left(1 + \frac{d}{2}\right)^2$$

For I_{y0} to be equal to I_{x0}

$$72 = 8 + 24\left(1 + \frac{d}{2}\right)^2$$

$\therefore \quad d = 1.266$ inches

Problem 17.: Find the product of inertia with respect to the specified XY axes for the shaded area of fig.

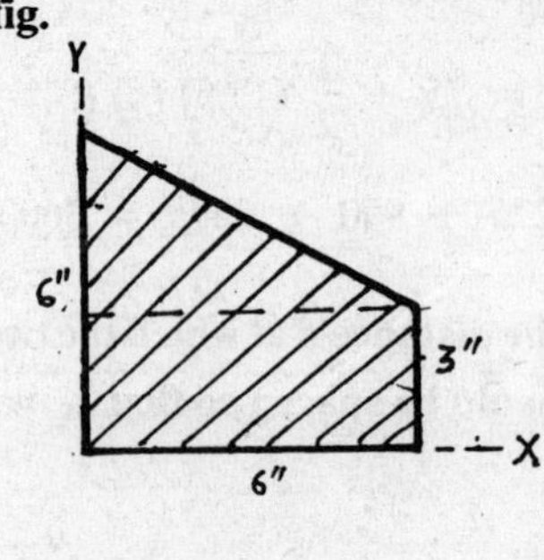

Solution : Shaded portion is divided in triangle and rectangle .

As we know for triangle

$$\bar{P}\,xy = -\frac{b^2 h^2}{72}$$

$$\therefore \quad \bar{P}xy \text{ for triangle} = -\frac{3^2 \cdot 6^2}{72} = -4.5 \text{ in}^4$$

$$\text{For triangle } Pxy_1 = \bar{P}xy + A\bar{x} \cdot \bar{y}$$

$$= -4.5 + \frac{1}{2} \times 3 \times 6 \times 2 \times (3 + 1) = 67.5 \text{ in}^4$$

$$\text{For rectangle } Pxy_2 = A \cdot \bar{x} \cdot \bar{y}$$

$$= (6 \times 3) \cdot (1.5)(3) = 81 \text{ in}^4$$

$$\therefore \quad Pxy = Pxy_1 + Pxy_2$$

$$= 67.5 + 81 = 148.5 \text{ in}^4$$

Problem 18. : Compute the product of inertia of the triangular area shown in fig. with respect to the given X and Y axes.

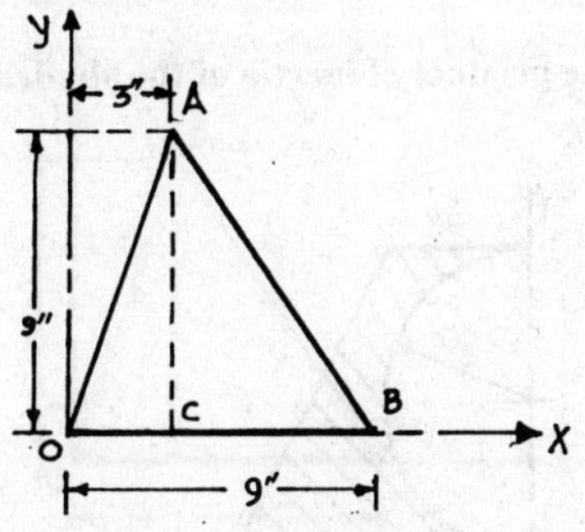

$$\textbf{Solution : } \bar{P}xy_{(OAC)} = \frac{3^2 \cdot 9^2}{72} = 10.125 \text{ in}^4$$

For Δ OAC $-$ $\bar{x} = 2$ in & $\bar{y} = 3$ in

$$\bar{P}_{xy\,(ABC)} = \frac{-6^2\,9^2}{72} = -40.5 \text{ in.}^4$$

For Δ ABC $\rightarrow$ $\bar{x} = 5$ in., $\bar{y} = 3$ in.

$$\therefore \quad \text{Centroid of } \Delta\,OAB = \bar{x} = \frac{2 \times \frac{27}{2} + 5 \times 27}{81/2} = 4 \text{ in.}$$

$$\bar{y} = \frac{3 \times \frac{27}{2} + 3 \times 27}{81/2} = 3 \text{ in.}$$

$$P_{xy(OAC)} = \bar{P}_{xy(OAC)} + (A\,\bar{x}\,\bar{y})_{(OAC)}$$

$$= 10.125 + \left(\frac{27}{2}\right) \times 2 \times 3 = 91.125 \text{ in}^4$$

$$P_{xy(ABC)} = \overline{P_{xy(ABC)}} = (A\,\bar{x}\,\bar{y})_{(ABC)}$$

$$= -40.5 + 27 \times 5 \times 3 = 364.5 \text{ in}^4$$

$$P_{xy} = 91.125 + 364.5 = 455.625 \text{ in}^4$$

Now Pxy w.r.t centroidal axis

$$P_{xy} = \overline{P_{xy}} + A\,\bar{x}\,\bar{y}$$

$$\therefore \quad \overline{P_{xy}} = P_{xy} - A\,\bar{x}\,\bar{y}$$

$$= 455.625 - \left(27 + \frac{27}{2}\right)(4)\,(3)$$

$$= -30.375 \text{ in}^4$$

Problem 19.: Find the product of inertia of the shaded area in fig. about the specified X and Y axis.

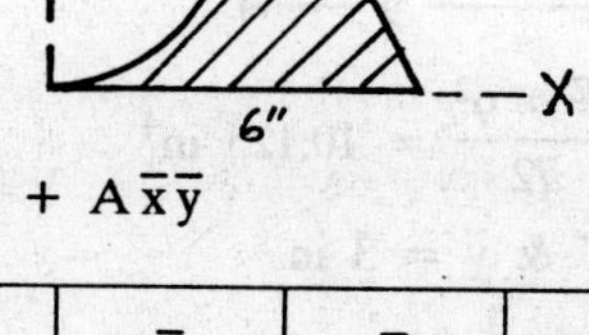

Solution : $P_{xy} = \overline{P_{xy}} + A\,\bar{x}\,\bar{y}$

Element	$\overline{P_{xy}}$	$\bar{x}$	$\bar{y}$	Area	Pxy
Semicircle	0	$\frac{4 \times 3}{3\pi}$	3	$\frac{9\pi}{2}$	$-\frac{9\pi}{2} \times \frac{4}{\pi} \times 3$
Rectangle	0	1.5	3	18	$18 \times 3 \times 1.5$

Triangle	$-\frac{3^2 \times 6^2}{72}$	4	2	9	$-4.5+9\times4\times2$
Total					94.5 in^4

$\therefore$ Pxy of shaded area $= 94.5\ in^4$

Problem 20. : Determine the product of inertia of the quarter circular area as shown in fig. with respect to the given X and Y axes.

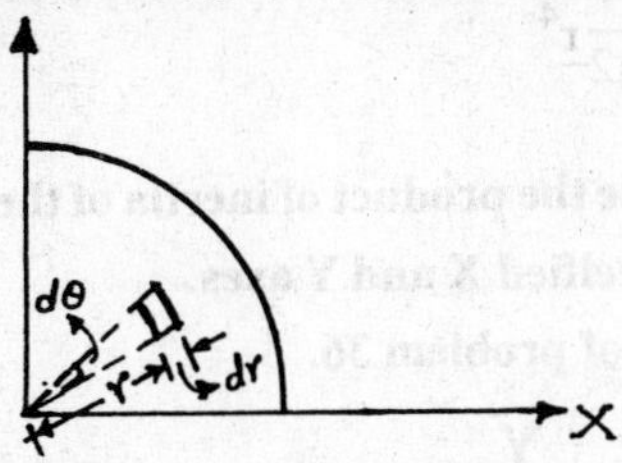

Solution : Area of element of $dA = r \cdot d\theta \cdot dr$

$x = r\cos\theta \ ; \ y = r\sin\theta$

$$Pxy = \int_0^r \int_0^{\pi/2} r\cos\theta \cdot r\sin\theta \cdot r\,d\theta \cdot dr$$

$$= \int_0^r -\frac{r^3}{2}\left[\frac{\cos 2\theta}{2}\right]_0^{\pi/2} dr$$

$$= \int_0^r \frac{r^3}{2} \times \frac{2}{2}\,dr$$

$$\therefore \quad Pxy = \frac{r^4}{8}$$

Problem 21.: Use the result of the previous problem to determine the product of inertia of the shaded area shown in fig. with respect to the given X and Y axes.

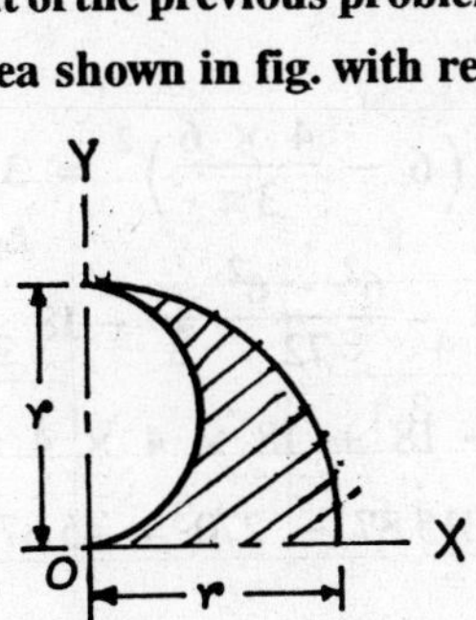

Solution : As we have seen Pxy for quarter circle $= \frac{r^4}{8}$

Pxy for semicircle

$$= \overline{Pxy} + A \cdot \bar{x}\,\bar{y}$$

$$= 0 + \frac{\pi}{2}\left(\frac{r}{2}\right)^2 \cdot \frac{2r}{3\pi} \cdot \frac{r}{2} = \frac{r^4}{24}$$

$\therefore$ Pxy of shaded area

$$= \frac{r^4}{8} - \frac{r^4}{24} = \frac{r^4}{12}$$

Problem 22.: Compute the product of inertia of the shaded area shown in fig. with respect to the specified X and Y axes.

Hint : Use the result of problem 36.

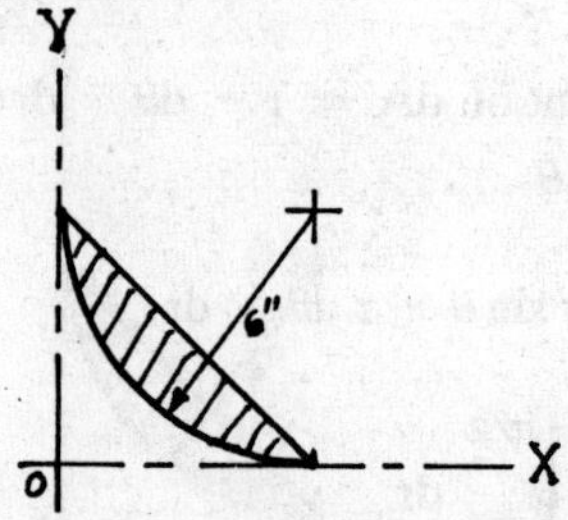

Solution : $Px_1\,y_1$ for quarter circle $= \frac{6^4}{8} = 162 \text{ in}^4$

$\overline{P_{xy}}$ for quarter circle

$\Rightarrow Px_1\,y_1 = \overline{Px_1y_1} + A\,\bar{x} \cdot \bar{y}$

$$\therefore \quad \overline{Px_1\,y_1} = 162 - \frac{\pi}{4}(6)^2 \cdot \left(\frac{4 \times 6}{3\pi}\right)^2 = -21.34$$

Pxy for quarter circle,

$$= -21.34 + \frac{\pi}{4}(6)^2\left(6 - \frac{4 \times 6}{3\pi}\right)^2 = 315.87 \text{ in}^4$$

$\overline{Px_1\,y_1}$ for triangle ABC $= -\frac{6^2 \cdot 6^2}{72} = -18$

Pxy for triangle ABC $= -18 + 18 \times 4 \times 4 = 270 \text{ in}^4$

$\therefore$ Pxy of shaded area $= 315.87 - 270 = 45.87 \text{ in}^4$

Problem 23. : For a certain area it is known that $I_x = I_y = 120$ in.^{4} and $P_{xy} = -60$ in.^{4}. Determine the moments of inertia and product of inertia with respect to U and V axes which are rotated 60o clockwise to the X and Y axes respectively.

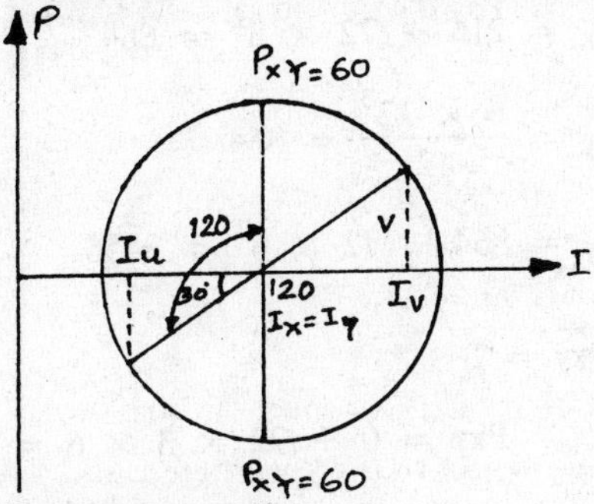

Solution :

$I_u = 120 - 60 \cos 30 = 68.04 \text{ in}^4$

$I_v = 120 + 60 \cos 30 = 171.96 \text{ in}^4$

$P_{uv} = 60 \sin 30 = 30 \text{ in}^4$

Problem 24. : A certain area has the following properties $I_x = 40$ in^4, $I_y = 100$ in.^{4}, $P_{xy} = 40$ in.^{4} Determine the maximum and minimum moments of inertia and also the angle that the axis of maximum inertia makes with the X axis. Illustrate by a diagram.

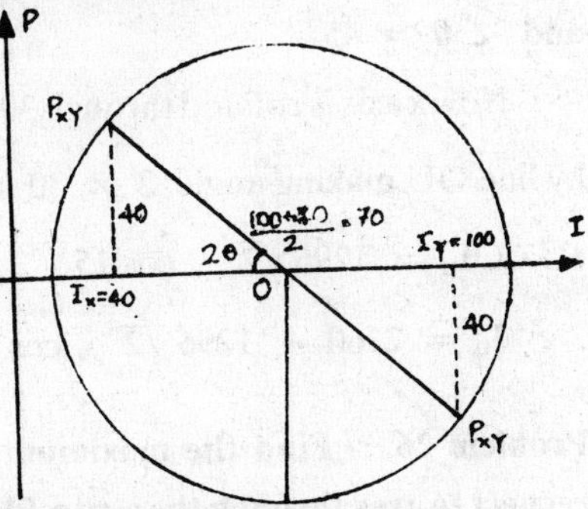

Solution : Radius of Mohr's circle

$$= OP_{xy} = \left[(40)^2 + (40-70)^2\right]^{1/2} = 50.$$

$\therefore$ Max $I = 70 + 50 = 120 \text{ in}^4$

Min. $I = 70 - 50 = 20 \text{ in}^4$

$\sin 2\theta = \dfrac{40}{50} \quad \therefore \quad 2\theta = 53.13^\circ$

Angle that the axis of maximum inertia makes with x axis

$$= \frac{180 - 53.13}{2} = 63.4^\circ$$

Problem 25. : For the 6–in. rectangular area shown in fig. Find the moment of inertia about the indicated U axis.

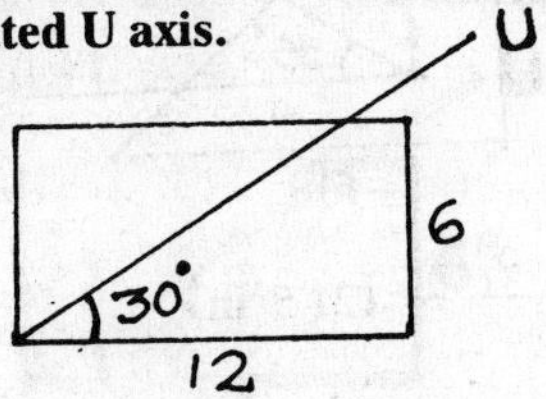

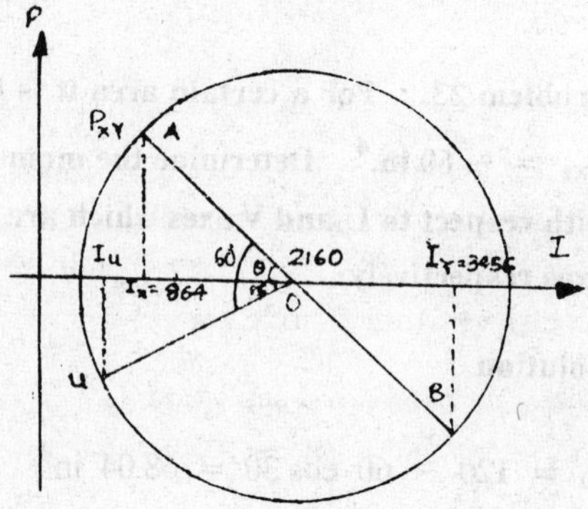

Solution :

$$\bar{I}_x = \frac{12 \times 6^3}{12} = 216$$

$$I_x = 216 + 72 \times 3^2 = 864$$

$$\bar{I}_y = \frac{6 \times 12^3}{12} = 864$$

$$I_y = 864 + 72 \times 6^2 = 3456$$

$$\overline{Pxy} = 0$$

$$\therefore \quad Pxy = 0 + 72 \times 3 \times 6 = 1296$$

Centre of Mohr's circle $= \dfrac{864 + 3456}{2} = 2160$

Distance between Ix and O $= 2160 - 864 = 1296$

$\therefore$ Radius of Mohr's circle $= 1296 \cdot \sqrt{2}$

and $\angle \theta = 45^o$

Now x axis is rotated through 30° same is represented here in Mohr's circle by line OU making angle $2 \times 30 = 60^o$ with AB

$\therefore \quad OI_u = 1296\sqrt{2} \cdot \cos 15$

$\therefore \quad I_u = 2160 - 1296\sqrt{2} \cdot \cos 15 = 390 \text{ in}^4$

Problem 26. : Find the maximum and minimum moments of inertia with respect to axes through the centroid of the 9–in by 6–in. triangular area shown in fig. Also compute the moment of inertia with respect to the given centroidal U axis.

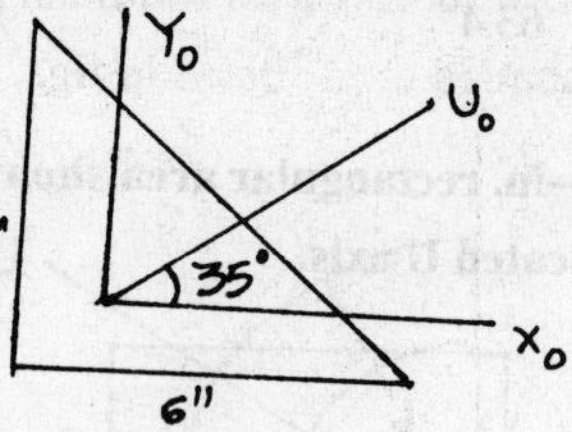

Solution : $\quad I_{xo} = \dfrac{6 \times 9^3}{36} = 121.5 \text{ in}^4$

$$I_{y0} = \frac{9 \times 6^3}{36} = 54 \text{ in}^4$$

$$P_{x0\,y0} = -\frac{6^2 \cdot 9^2}{72} = -40.5 \text{ in}^4$$

$$\text{Centre of Mohr's circle} = \frac{121.5 + 54}{2} = 87.75$$

Distance between I sub y and 'O'

$$= 87.75 - 54 = 33.75$$

$\therefore$ Radius of Mohr's circle

$$= \left[(40.5)^2 + (33.75)^2\right]^{1/2} = 52.72$$

$$\theta = \sin^{-1}\left(\frac{40.5}{52.72}\right) = 50.2^\circ$$

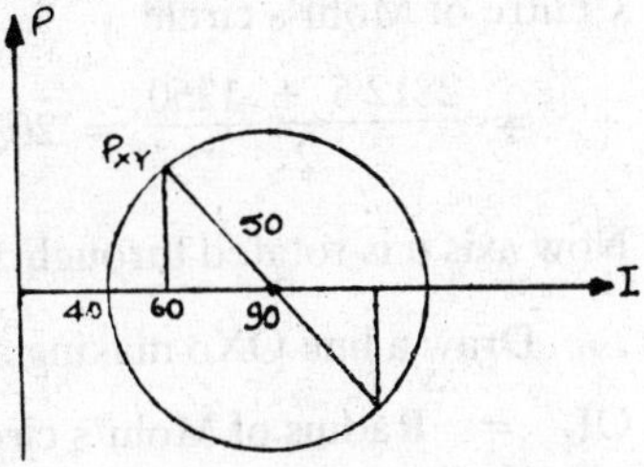

Now X axis is rotated through 35°. This is represented in Mohr's circle by line OU making an angle of 70° with AB.

$$OI_u = [52.72 \cos(19.8)] = 49.6$$

$$\therefore \quad I_u = 87.75 + 49.6 = 137.35 \text{ in}^4$$

Problem 27. : For a certain area, it is known that $I_z = 120$ in.^{4} ; $I_y = 60$ in.^{4}, and min. U = 40 in.^{4} Find P_{xy}.

Solution : $I_x = 120 \text{ in}^4$, $I_y = 60 \text{ in}^4$, $I_{min} = 40 \text{ in}^4$

$$\therefore \quad \text{Centre of Mohr's circle} = \frac{120 + 60}{2} = 90$$

$$\text{Radius of Mohr's circle} = 90 - 40 = 50$$

$$\therefore \quad Pxy = \left[(50)^2 - (30)^2\right]^{1/2} = 40 \text{ in}^4$$

Problem 28. : Find the maximum and minimum of inertia with respect to axes through point O of the shaded area shown in fig.

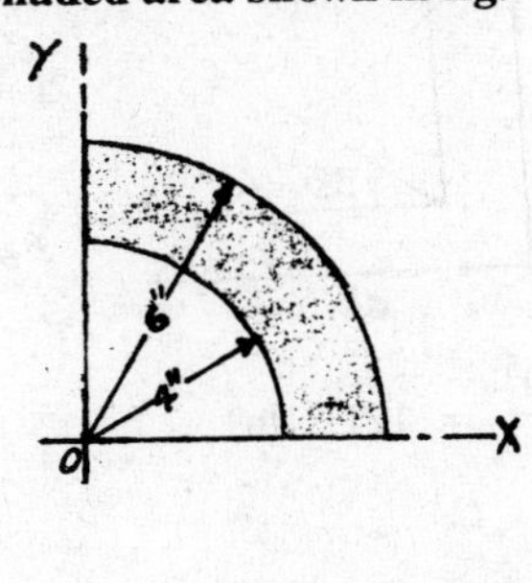

Solution : $Iv = \dfrac{10 \times 15^3}{12} = 2812.5 \text{ in}^4$

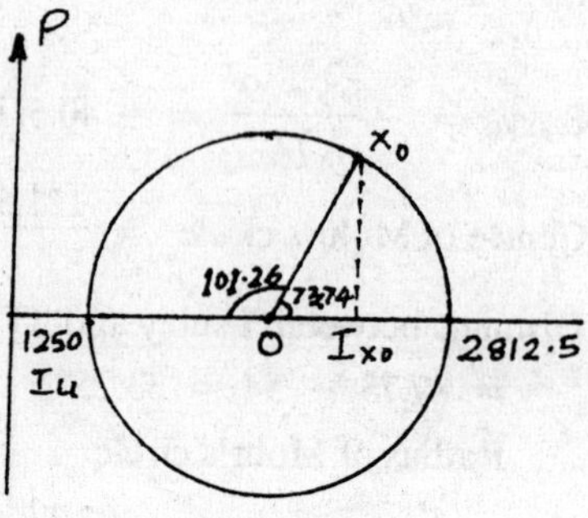

$Iu = \dfrac{15 \times 10^3}{12} = 1250 \text{ in}^4$

$Puv = 0$

Centre of Mohr's circle

$$= \frac{2812.5 + 1250}{2} = 2031.25$$

Now axis u is rotated through $\theta = 53.13^\circ$

$\therefore$ Draw a line OXo making an angle of $2 \times 53.13^\circ = 106.26^\circ$ with OI_u

OI_{x_o} = Radius of Mohr's circle

$= 2031.25 - 1250 = 781.25$

$\therefore \; OI_{x_o} = 781.25 \times \cos(73.74) = 218.75$

$\therefore \; I_{x_o} = 2031.25 + 218.75 = 2250 \text{ in}^4$

$I_x = I_{xo} + Ad^2$ (where d = 3 i.e. distance between x and x_o axis, by applying simple geometry).

$= 2250 + (10 \times 15) \times (3)^2 = 3600 \text{ in}^4$

Chapter Seven

VIRTUAL WORK

This is another approach to solve equilibrium problems based on relationship between applied forces and possible small motion of the system. Here "possible" indicate that the motion does not in fact take place or infinitesimally small motion is considered for the sake of analysis. Such motion is called virtual motion.

Here we consider three cases of virtual work. Principle of a virtual work for

(a) a particle, (b) a rigid body, and (iii) a system of ideally connected rigid bodies. Principle is same for all above cases.

Principle of Virtual work :

" If system of forces acting on a body or a system (or a particle) be in equilibrium and if system be imaginated to undergo a small displacement consistent with the geometrical conditions then the algebraic sum of the virtual works done by the forces of the system is zero.

i.e. Resultant Virtual Work done $\delta U = F \cdot \delta r = 0$

where F is applied force and δr is virtual displacement.

The virtual work of a force is equal to sum of the virtual work of its components.

Problem 1. : For the lazy tongs mechanism shown in fig. determine the relation between P and R for any position angle θ. How is this relation changed if the weights of the links is considered ? Each link weighs 2W lb.

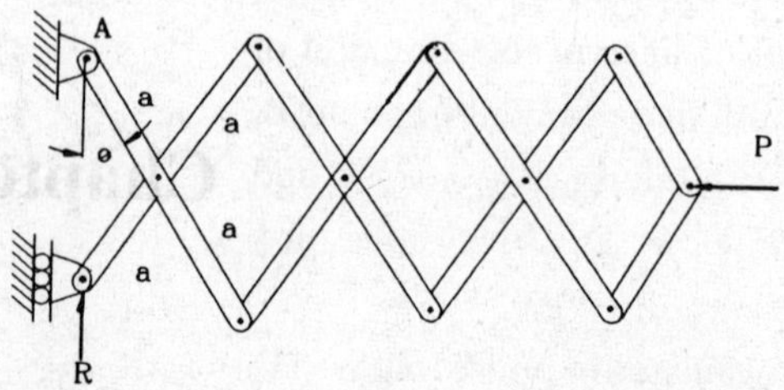

Solution : Here the motion to be considered is of point B and of roller hinge R. We are not considering the motion of links as they are weightless and will not add to work, what-so-ever be the displacement of the links.

Taking 'O' as origin.

Distance AO $= y_A = 2a \cos\theta$ (In terms of θ and a)

and distance x of B from AO (interms of θ & a)

$X_B = 7a \sin\theta$.

Now virtual displacement of B due to application of force P is

$\delta x_B = 7a \cos\theta + \delta\theta$

and virtual displacement of 'O' due to reaction R is

$\delta y_A = -2a \sin\theta \,.\, \delta\theta$

Applying virtual work equation.

$$P \cdot \delta x_B + R \cdot \delta y_A = 0$$

$$\therefore \; P \cdot (7a \cos\theta \cdot \delta\theta) + R(-2a \cdot \sin\theta \cdot \delta\theta) = 0$$

$$\therefore \; R = \frac{7P}{2 \tan\theta}$$

Now, if the weights of links is considered then virtual work due to these weights is zero since theme is no displacement along the direction of weights of links. The above relation will be still true.

Problem 2. : Solve the above problem if the mechanism is rotated 90° counterclockwise so that it is above the supports. Find the relation between force P and reaction R for the following. Weight of each link is 2W.

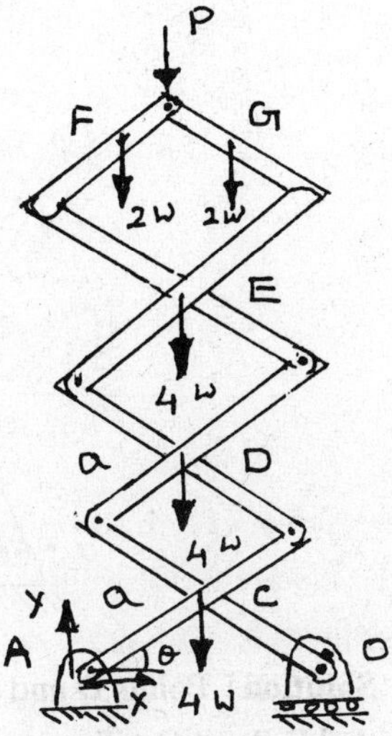

Solution : Here the mechanism being vertical, there is displacement of links in the direction of weight of link. So we will find positions of points B, F, G, E, D, C and O with reference to point A and angle θ. Points B, F, G, E, D, C will move in y direction and point O in x direction.

With 'A' as origin, coordinates of points will be.

$$y_B = 7\,a \sin\theta \;;\quad y_F = \frac{13}{2}\,a \sin\theta$$

$$y_G = \frac{13}{2}\,a \sin\theta \;;\quad y_E = 5\,a \sin\theta$$

$$y_D = 3\,a \sin\theta \;;\quad y_C = a \sin\theta$$

$$X_o = 2\,a \cos\theta$$

Now virtual displacement of all the points due to application of force P at B is as below.

$$\delta\, y_B = 7\,a \cos\theta \cdot \delta\theta \;;\quad \delta\, y_F = 6.5\,a \cos\theta \cdot \delta\theta$$

$$\delta\, y_E = 5\,a \cos\theta \cdot \delta\theta \;;\quad \delta\, y_D = 3\,a \cos\theta \cdot \delta\theta$$

$$\delta\, y_C = a \cos\theta \cdot \delta\theta \;;\quad \delta\, X_o = -2\,a \sin\theta \cdot \delta\theta$$

Applying virtual work equation.

$$P \cdot \delta\, y_B + 2w \cdot \delta\, y_F + 2w \cdot \delta\, y_G + 4w\, \delta\, y_E$$
$$+ 4w \cdot \delta\, y_D + 4w \cdot \delta\, y_C + R\, \delta\, x_A = 0$$

$$\therefore\; P \cdot (7a \cos\theta \cdot \delta\theta) + 4w\,(6.5\,a \cos\theta \cdot \delta\theta + 5\,a \cos\theta \cdot \delta\theta$$
$$+ 3\,a \cos\theta \cdot \delta\theta + a \cos\theta \cdot \delta\theta) + R \cdot (-2a \sin\theta \cdot \delta\theta) = 0$$

$$\therefore\; 7P \cos\theta + 4w \cos\theta\,(6.5 + 5 + 3 + 1) - 2R \sin\theta = 0$$

$$\therefore\; 2R \sin\theta = 7P \cos\theta + 62\,w \cos\theta$$

$$\therefore\; R = 3.5\,P \cot\theta + 31\,w \cot\theta.$$

Problem 3. : Use the method of virtual work to determine the horizontal component of the reaction at C for the frame shown in fig.

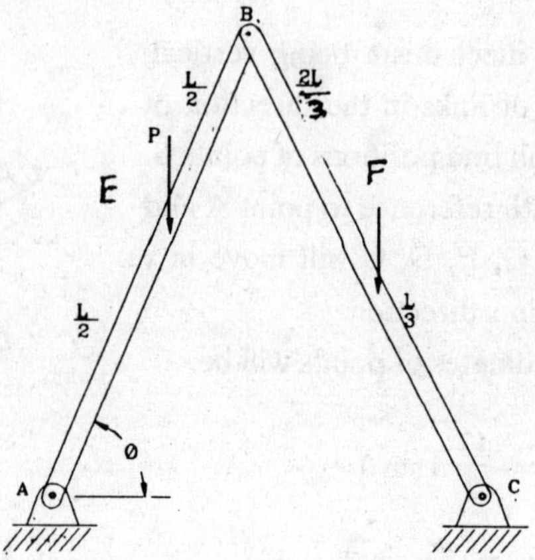

Solution : Points D and E will have some vertical displacement due to force P and F. Point C will move horizontally.

Taking A as origin, positions of D, E and C are given as .

$$y_D = \frac{L}{3}\sin\theta\ ,\ y_E = \frac{L}{2}\sin\theta\ ,\ x_C = 2L\cos\theta .$$

Virtual displacement of D, E and C is given by

$$\delta y_D = \frac{L}{3}\cos\theta\cdot\delta\theta\ ,\ \delta y_E = \frac{L}{2}\cos\theta\cdot\delta\theta\ ,$$

$$\delta X_C = -2L\sin\theta\ \delta\theta$$

Applying virtual work equation

$$P\cdot\delta Y_E + F\cdot\delta y_D + C_x\cdot\delta X_C = 0$$

$$P\cdot\frac{L}{2}\cos\theta\cdot\delta\theta + F\cdot\frac{L}{3}\cos\theta\cdot\delta\theta + C_x(-2L\sin\theta)\,\delta\theta = 0$$

$$\therefore\ C_x = \left(\frac{P}{4}+\frac{F}{6}\right)\cot\theta .$$

Problem 4. : Solve the previous problem if the force F acts horizontally to the right instead of vertically.

Solution : Point E will move vertically, point D will move horizontally, point C will move horizontally.

Positions of E, D and C depending upon direction of motion is given by

(with respect to origin A)

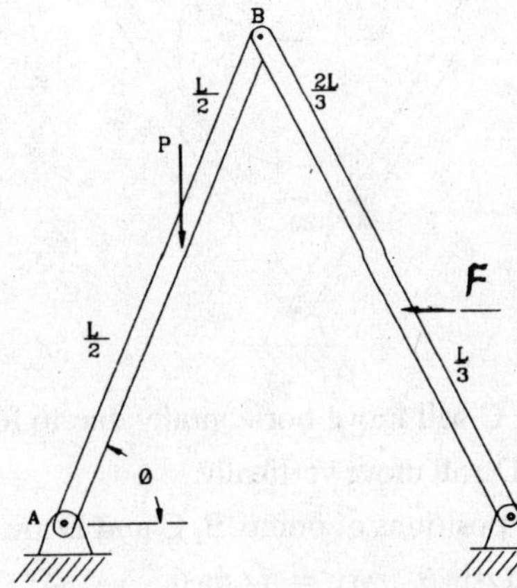

$$y_E = \frac{L}{2}\sin\theta\ ; \qquad X_D = L\cos + \frac{2L}{3}\cos\theta = \frac{5}{3}L\cos\theta$$

$$X_C = 2L\cos\theta$$

Virtual displacement of E, D and C with respect to A is given by

$$\delta y_E = \frac{L}{2}\cos\theta\,\delta\theta$$

$$\delta y_D = -\frac{5}{3}L\sin\theta\,\delta\theta$$

$$\delta X_C = -2L\sin\theta\,\delta\theta$$

Virtual work equation

$$P\delta y_E + F\delta y_D + C\delta X_C = 0$$

$$P\cdot\frac{L}{2}\cos\theta\,\delta\theta + F\left(-\frac{5}{3}L\sin\theta\,\delta\theta\right) + C_x\,(-2L\sin\theta\,\delta\theta) = 0$$

$$C_x = \frac{P}{4}\cot\theta - \frac{5}{6}F$$

Problem 5. : Find the force P acting on each of the two links shown in fig. that will support the cylinder of weight W. Neglect friction and the weights of the links.

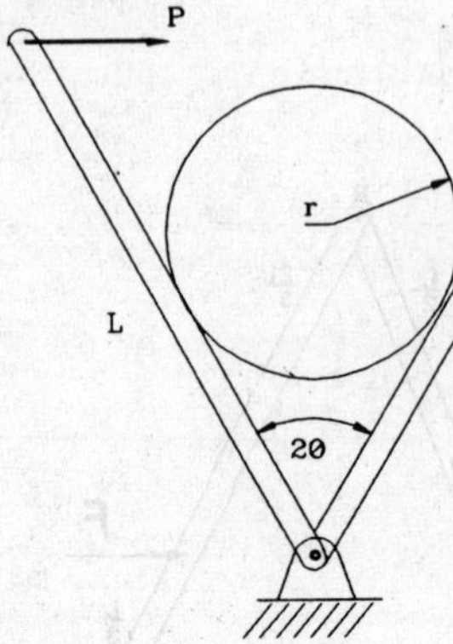

Solution : Points B and C will move horizontally due to forces P applied at ends B and C and point D will move vertically.

Taking point A as origin, positions of points B, C and D are given as

$X_B = L \sin\theta$, $X_C = L \sin\theta$, $y_D = r/\sin\theta$

Virtual displacement of B, C and D with respect of A is given by

$$\delta x_B = L\cos\theta \cdot \delta\theta,\ \delta x_C = L\cos\theta \cdot \delta\theta,\ \delta y_D = \frac{-r\cos\theta}{\sin^2\theta}\,\delta\theta$$

Applying virtual work equation.

$$P \cdot \delta x_B + P \cdot \delta x_C + W \cdot \delta y_D = 0$$

$$P \cdot l\cos\theta \cdot \delta\theta + P \cdot L\cos\theta \cdot \delta\theta + W\left(\frac{-r\cos\theta}{\sin^2\theta}\right)\delta\theta = 0$$

$$\therefore\ P = \frac{W \cdot r}{2L\sin^2\theta}$$

Problem 6. : In fig. the link BD is pinned at C to a collar which is free to slide along the fixed vertical shaft. Find the force P supplied by a hydraulic cylinder on the collar to maintain W for any value of the angle θ

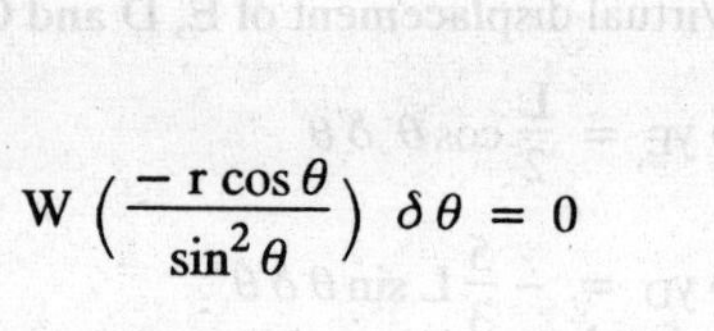

Solution : Both points C and D will move vertically

Taking point A as origin, positions of point C and D are given as

$y_C = 2a\sin\theta$, $y_D = 2a\sin\theta + b\sin\theta$

$= (2a + b)\sin\theta$

Virtual displacement of points C and D is given by

$\delta y_C = 2a\cos\theta \cdot \delta\theta$, $\delta y_D = (2a + b)\cos\theta \cdot \delta\theta$

Applying virtual work equation

$W \cdot \delta y_D - P \cdot \delta y_C = 0$

$P \cdot 2a\cos\theta \cdot \delta\theta - W \cdot (2a + b)\cos\theta \cdot \delta\theta = 0$

$P = W\left(1 + \dfrac{b}{2a}\right)$

Problem 7. : As shown in fig. a horizontal force P is applied to the end D of three identical links smoothly pinned together. If each link is of weight W, determine equilibrium position as defined by the angles θ , φ , and α

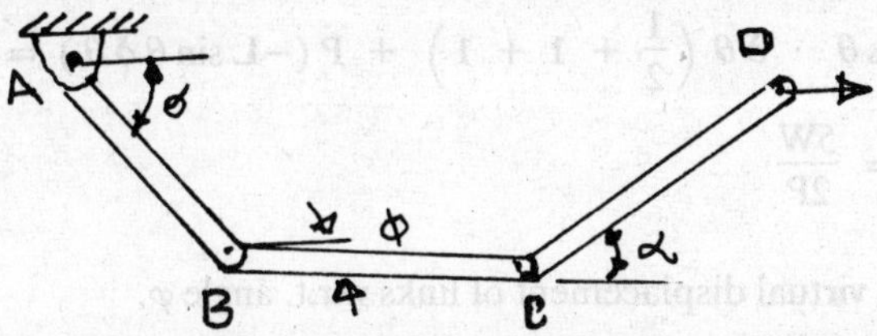

Solution : For points E, F and G only vertical motion is to be considered while for point D horizontal motion is to be considered.

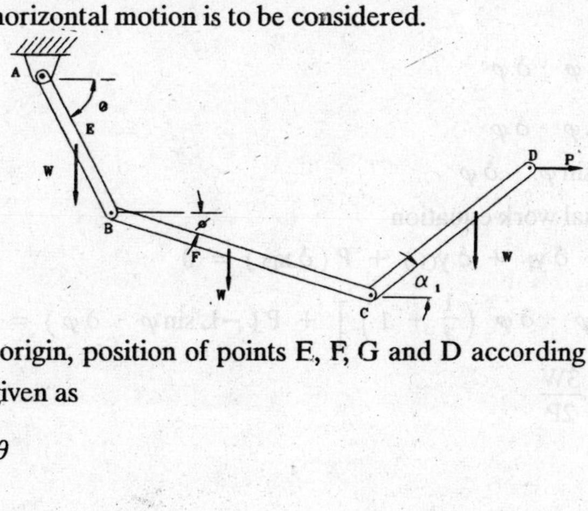

Taking A as origin, position of points E, F, G and D according to motion of direction is given as

$y_E = \dfrac{L}{2}\sin\theta$

$$y_F = L\sin\theta + \frac{L}{2}\sin\phi$$

$$y_G = L\sin\theta + L\sin\phi - \frac{L}{2}\sin\alpha$$

$$x_D = L(\cos\theta + \cos\phi + \cos\alpha)$$

considering virtual displacement of links with respect to angle θ (differentiating w.r.t. θ)

$$\delta y_E = \frac{L}{2}\cos\theta \cdot \delta\theta$$

$$\delta y_F = L\cos\theta \cdot \delta\theta$$

$$\delta y_G = L\cos\theta \cdot \delta\theta$$

$$\delta x_D = -L\sin\theta \cdot \delta\theta$$

Applying virtual work equation

$$w \cdot (\delta y_E + \delta y_F + \delta y_G) + P(\delta x_D) = 0$$

$$W \cdot (L\cos\theta \cdot \delta\theta \left(\frac{1}{2} + 1 + 1\right) + P(-L\sin\theta\,\delta\theta) = 0$$

$$\therefore \quad \tan\theta = \frac{5W}{2P}$$

Considering virtual displacement of links w.r.t. angle φ. (differentiating w.r.t. ϕ)

$$\delta y_E = 0$$

$$\delta y_F = \frac{L}{2}\cos\varphi \cdot \delta\varphi$$

$$\delta y_G = L\cos\varphi \cdot \delta\varphi$$

$$\delta x_D = -L\sin\varphi \cdot \delta\varphi$$

Applying virtual work equation

$$W \cdot (\delta y_E + \delta y_F + \delta y_G) + P(\delta x_D) = 0$$

$$W \cdot \left[L\cos\varphi \cdot \delta\varphi \left(\frac{1}{2} + 1\right)\right] + P(-L\sin\varphi \cdot \delta\varphi) = 0$$

$$\therefore \quad \tan\varphi = \frac{3W}{2P}$$

Virtual displacement of links w.r.t. angle α is (differential w.r.t. α)

$\delta y_E = 0$

$\delta y_F = 0$

$$\delta y_G = -\frac{L}{2}\cos\alpha \cdot \delta\alpha$$

$\delta x_D = -L\sin\alpha \cdot \delta\alpha$

Applying virtual work equation

$$W \cdot (\delta y_E + \delta y_F + \delta y_G) + P\delta x_D = 0$$

$$W \cdot \left(-\frac{L}{2}\cos\alpha \cdot \delta\alpha\right) + P(-L\sin\alpha \cdot \delta\alpha) = 0$$

$$\therefore \tan\alpha = \frac{-W}{2P}$$

Problem 8.: Solve the above problem if, in addition to the weights of the links, load of 2W each are suspended from B to C.

Solution : In addition to motions of points E, F, G and D we will have to consider motions of points B and C since the weights are attached at those points.

Positions of those points

$y_B = L\sin\theta$

$y_C = L\sin\theta + L\cos\phi$

differentiating with respect to θ.

$\delta y_B = L\cos\theta \cdot \delta\theta$, $\delta y_C = L\cos\theta \cdot \delta\theta$

Applying virtual work equation

$$W(\delta y_E + \delta y_F + \delta y_G) + 2W(\delta y_B + \delta y_C) + P(\delta x_D) = 0$$

$$W\left[L\cos\theta \cdot \delta\theta\left(\frac{1}{2} + (1+1)\right)\right] + 2W(2L\cos\theta\,\delta\theta) + P(-L\sin\theta\,\delta\theta) = 0$$

$$\tan\theta = \frac{13W}{2P}$$

Now differentiating w.r.t. φ

$\delta y_B = 0$

$\delta y_C = L\cos\phi\ \delta\phi$

Applying virtual work equation

$$W\left[L\cos\phi \cdot \delta\phi\left(\frac{1}{2}+1\right)\right] + 2W\,(L\cos\phi\,\delta\phi)$$

$$+\ P(-L\sin\phi\,\delta\phi) = 0$$

$\therefore\ \tan\phi' = 7W/2P$

Now differentiating w.r.t. α

$\delta y_B = 0$

$\delta y_C = 0$

$\therefore\ \tan\alpha = -\dfrac{W}{2P}$ (as in earlier problem).

Chapter Eight

KINEMATICS & KINETICS OF PARTICLES

Dynamics :

Dynamics can be divided into two main branches :

(a) Kinematics and (b) Kinetics.

In kinematics motion of particle is studied without considering the forces that produce or change this motion.

In kinetics motion of particle is studied with the unbalanced force system that produce or change this motion.

Translation and Rotation :

If a body moves in such a way that all its particle move in parallel planes, then the body is said to have motion of translation or plane motion. The translation is called rectilinear, if the path of motion is straight ; and curvilinear, if the path of motion is curved. In translation all the particles of a body move in same or parallel paths and therefore have same value of displacement, velocity and acceleration.

If a body rotates about its geometric axis, then body is said to have motion of rotation. For example, motion of pulley, shafts, flywheel. The combined motion of rotation and translation may be assumed to be motion of pure rotation about some centre. In combined motion, this centre about which rotation is assumed to take place, goes on changing from one instant to another. This centre is known as instantaneous centre (I.C.).

Equations of Rectilinear Motion : Uniform acceleration

Under uniform acceleration "a"	Under force of gravity
(1) $v = u \pm at$	$v = u \pm gt$
(2) $s = ut \pm \frac{1}{2}at^2$	$h = ut \pm \frac{1}{2}gt^2$
(3) $v^2 = u^2 \pm 2as$	$v^2 = u^2 \pm 2gh$

where 'u' is initial velocity, 'v' is final velocity,, 't' is the time & 'S/h' is horizontal / vertical distance travelled.

Note : Negative sign must be used when there is retardation.

Rectilinear Motion :

When the motion is under variable acceleration the ground equations can not be written. For such cases, displacement, velcoity and acceleration of body is expressed in the simple form of mathematical equations and various other conditions of its motion can be obtained by elementry calculus. Simple variable acceleration problems can also be solved using motion diagrams, which is discussed later in the chapter.

Curvilinear Motion :

Curvilinear motion in a plane is motion along a plane curve (path). For example, motion of train rounds a curve on a level track. The velocity and acceleration of a point on such a curve can be expressed in
(a) Rectangular components ; or (b) Tangential and normal components.

Rectangular Components :

Let a body starts moving along a curved path with a velocity v and acceleration a. The velocity at any point is tangent to the path of motion of the particle. At any point P let its velocity is v. Then its horizontal and vertical components will be v_x ($= v\cos\alpha$) and v_y ($= v\sin\alpha$), where α is the angle the tangent at P makes with horizontal. Components of acceleration in two perpendicular direction x and y will be

$a_x = a\cos\alpha$ and $a_y = a\sin\alpha$, respectively.

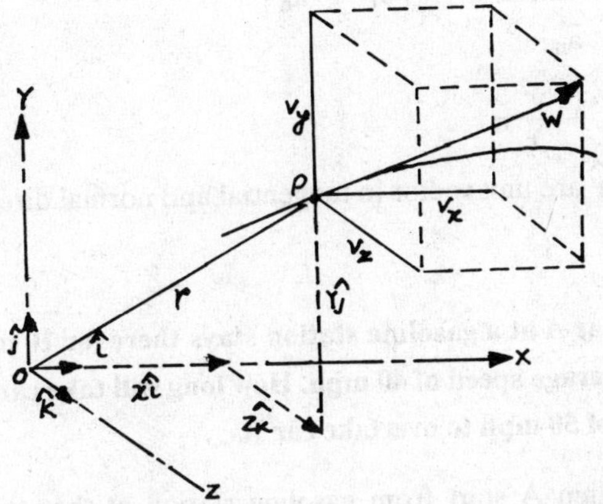

If $\hat{i}$, $\hat{j}$ and $\hat{k}$ are the unit vectors in the direction x, y and z respectively. Then position vector of any particle is given by

$$s = x\hat{i} + y\hat{j} + z\hat{k}.$$

$$\text{Velocity } \vec{V} = \frac{ds}{dt} = \frac{dx}{dt}\hat{i} + \frac{dy}{dt}\hat{j} + \frac{dz}{dt}\hat{k}$$

$$= \dot{x}\hat{i} + \dot{y}\hat{j} + \dot{z}\hat{k}$$

$$= v_x\hat{i} + v_y\hat{j} + v_z\hat{k}$$

Similarly,

$$\vec{a} = \frac{dv}{dt} = \frac{d^2s}{dt^2} = \frac{d^2x}{dt^2}\hat{i} + \frac{d^2y}{dt^2}\hat{j} + \frac{d^2z}{dt^2}\hat{k}$$

$$= \dot{x}\hat{i} + \dot{y}\hat{j} + \dot{z}\hat{k}$$

$$= a_x\hat{i} + a_y\hat{j} + a_z\hat{k}$$

Normal and Tangential Component :

Consider again a body in curvilinear motion. Velocity at any point is tangent to the path of motion of the particle. There its velocity v will be tangent velocity v_t ($= v$). Component of velocity in the direction normal to this will be v_n ($= 0$).

Tangent acceleration $a_t = \frac{dv}{dt}$;

and the normal accleration $a_n = \frac{v^2}{r}$

Total acceleration $a = [a_t^2 + a_n^2]^{1/2}$

$\vec{a} = \vec{a_t} + \vec{a_n}$

$= \frac{dv}{dt}\hat{t} + \frac{v^2}{r}\hat{n}$

where $\hat{t}$ and $\hat{n}$ are unit vector in tangential and normal direction.

Problem 1. : Car A at a gasoline station stays there for 10 min after a car B passes at an average speed of 40 mph. How long will take car A moving at an average speed of 50 mph to overtake car B.

Solution : When A start from gasoline station at that time the distance travelled by B,

$= \frac{40 \times 10}{60} = \frac{20}{3} = 13.33$ miles

Let A overtake B after time 't' min.

$\therefore$ Distance travelled by A in t min. $= \frac{50}{60} \times t$ miles ..... (1)

Also total distance travelled by B

$= \frac{40}{60} \times + 13.33$ ..... (2)

Both the distances travelled are same, comparing equatiion (1) and (2), we get

$\therefore \frac{5t}{6} = \frac{4t}{6} + 13.33$

$t = 40$ min.

Problem 2. : A ship being launched slides down the ways with a constant acceleration. She takes 4 sec. to slide the first foot. How long will she take to slide down the ways if their length is 900 ft.

Solution : Ship takes 4 sec for first feet.

Now, $s = ut + \frac{1}{2} at^2$

$$1 = 0 \times t + \frac{1}{2} \times a \times 4^2 \quad (\because \ u = 0)$$

$$a = \frac{1}{8} \text{ feet/sec}^2$$

So for , $s = 900$ ft & $u = 0$

$$s = ut + \frac{1}{2} at^2$$

$$900 = 0 + \frac{1}{2} \times \frac{1}{8} \times t^2$$

$$t = 120 \text{ sec.}$$

Problem 3. : A ball is thrown vertically into the air at 120 ft / sec. After 3 sec. another ball is thrown vertically. What initial velocity must the second ball have to pass the first ball at 100 ft from the ground.

Solution : For first ball :

$$s = 100 \text{ ft}, \ u = 120 \text{ fps}, \ a = -32.2 \text{ ft/sec}^2$$

$$\therefore \ 100 = 120 \times t + \frac{1}{2} (-32.2) \times t^2$$

$$\Rightarrow 16.1 t^2 - 120 t + 100 = 0$$

$$\therefore \ t = 6.5 \text{ sec or } 0.96 \text{ sec.}$$

For second ball : $s = 100$, $t = 6.5 - 3 = 3.5$ sec.

$$s = ut + \frac{1}{2} at^2$$

$$\therefore \ 100 = u \times 3.5 + \frac{1}{2} \times (-32.2) \times 3.5^2$$

$$u = 84.9 \text{ ft./sec}$$

Problem 4. : A stone is dropped down a well and 5 sec. later the sound of the splash is heard. If the velocity of sound is 1120 fps. What is the depth of the well.

Solution : For Stone :

$$u = 0, \ a = g = 32.2 \text{ ft/sec}^2, \ s = h, \ t = t \text{ sec.}$$

$\therefore \quad h = 0 + \frac{1}{2} (32.2) \times t^2$

$h = 16.1\,t^2$ ...... (1)

For Sound :

$u = 1120$ ft/sec. , $a = 0$ (Because mass is less)

$s = h \; ; \; t = (5 - t)$

$h = 1120 \times (5 - t)$

$h = 1120\,(5 - t)$ ...... (2)

From equation (1) and (2)

$16.1\,t^2 = 1120\,(5 - t)$

$16.1\,t^2 = -\,1120\,t + 5600$

$16.1\,t^2 + 1120\,t - 5600 = 0$

$t = 4.64$ sec.

Putting value of t in equation (1)

$h = 16.1 \times (4.64)^2$

$h = 353.3$ ft.

Problem 5. : A stone is dropped into a well with no initial velocity and 4.5 sec. later the splash was heard. Then a second stone is thrown downward into the well with an initial velocity v_o and the splash is heard in 4.0 sec. If the velocity of sound is constant at 1120 fps, determine the initial velocity of the second stone.

Solution : For First stone : $u = 0 \; ; \; s = h \; ; \; a = 32.2$ fps^2

$\therefore \quad h = 0 - \frac{1}{2} (-\,32.2)\,t^2$

$h = 16.1\,t^2$ .... (1)

For Sound : $u = 1120$ fps ; $a = 0$; $s = h$;

$t_1 = (4.5 - t)$ for 1st stone ; $t_2 = (4 - t)$ for 2nd stone

$\therefore \quad h = 1120 \times (4.5 - t)$ .... (2)

From eq^n. (1) & (2) $\Rightarrow$ $16.1\,t^2 = 5040 - 1120\,t$

$16.1\,t^2 + 1120\,t - 5040 = 0$

$t = 4.2$ sec.

$\therefore \quad h = 16.1 \times (4.2)^2 = 284$ ft

For Second stage stone : $a = 32.2$; $s = h$

$$\therefore \quad h = ut - \frac{1}{2}(-32.2)\, t^2$$

$$h = ut + 16.1\, t^2 \qquad \ldots\ldots (3)$$

For Sound : $h = 1120\,(4 - t) + \frac{1}{2} \times 0 \times t^2$

$$\therefore \quad h = 4480 - 1120\, t \qquad \ldots\ldots (4)$$

$$\Rightarrow 284 = 4480 - 1120\, t$$

$\therefore \quad t = 3.74$ sec.

Putting this value of t in (3)

$$284 = u \times 3.74 + 16.1 \times (3.74)^2$$

$$u = 15.72 \text{ ft/ sec.}$$

Problem 6. : A train moving with constant acceleration travels 24 ft during the 10th sec of its motion and 18 ft during the 12th sec of its motion. Find its initial velocity.

Solution : $S_{10} = u\,(10) + \frac{1}{2}\, a\,(10)^2$; $S_9 = u\,(9) + \frac{1}{2}\, a\,(9)^2$

$\therefore$ The distance travelled between 9 and 10th sec.

$$= S_{10} - S_9 = 24 = u + \frac{1}{2}\, 19a \qquad \ldots\ldots (1)$$

Similarly the distance travelled between 11th and 12th sec.

$$S_{12} - S_{11} = 18 = u + \frac{1}{2}\, 23\, a \qquad \ldots\ldots (2)$$

Solving (1) and (2)

$$u = 52.5 \text{ fps.}$$

Problem 7. : An automobile starting from rest speeds up to 40 fps with a constant acceleration of 4 fps^2. run at this speed for a time and finally comes to rest with a deceleration of 5 fps^2 ? If the total distance travelled is 1200 ft find the total time required.

Solution : I Phase : $u = 0$; $v = 40$ fps , $a = 4$ fps^2

$v = u + at$

$\Rightarrow 40 = 0 + 4 \times t$

$\therefore\ t = 10$ sec.

$$s = ut + \frac{1}{2} at^2 = 0 + t + \frac{1}{2} \times 4 \times t^2$$

$\therefore\ s_1 = 200$ ft.

II phase : $u = 40$ fps , $a = 0$

$$s = ut + \frac{1}{2} at^2$$

$\therefore\ s_2 = 40\, t_2$ .... (1)

III Phase : $a = -5$ fps^2 , $u = 40$ fps , $v = 0$

$v = u + at$

$\Rightarrow 0 = 40 + (-5 \times t)$

$\therefore\ t = 8$ sec.

$v^2 = u^2 + 2\,as$

$\Rightarrow 0 = 40^2 + 2 \times (-5) \times s_3$

$\therefore\ s_3 = 1600/10 = 160$ ft

Now $s_1 + s_2 + s_3 = 1200$

$200 + s_2 + 160 = 1200$

$\therefore\ s_2 = 840$ ft

Putting this in equation (1)

$$t_2 = \frac{s_2}{40} = \frac{840}{40} = 21 \text{ sec.}$$

$\therefore$ Total time required $= t_1 + t_2 + t_3$

$= 10 + 8 + 21 = 39$ sec.

Problem 8. : An auto A is moving at 20 fps and accelerating 5 fps^2 to overtake an auto B which is 384 ft ahead. If auto B is moving at 60 fps and deccelerated at 3 fps^2, how soon will A pass B.

Solution : Auto A : $u = 20$ fps , $a = 5 fps^2$

$S_A = ut + \frac{1}{2} at^2$

$= 20 \times t + \frac{1}{2} \times 5 \times t^2$

$= 20t + 2.5\,t^2$ ...... (1)

Auto B : $u = 60$ fps , $a = -3\ fps^2$

$S_B = 60 \times t + \frac{1}{2}(-3)\ t^2$

$= 60t - 1.5\,t^2$ ......(2)

Also $S_A = 384 + S_B$

Putting this in equation (1)

$S_B = 20t + 2.5\,t^2 - 384$

$\Rightarrow 60t - 1.5t^2 = 20t + 2.5t^2 - 384$ (from eq^n. 2)

$4t^2 - 40t - 384 = 0$

$t = 16$ sec.

Problem 9. : A balloon rises from the ground with a constant acceleration of 3 fps^2. Five second later a stone is thrown vertically up from the lauching site. What must be the minimum velocity of the stone for it to just touch the balloon. Note the balloon and the stone have the same velocity at contact.

Solution : For balloon : $u = 0$, $a = 3\ fps^2$, $S = h$,

$V_B = u + at = 0 + 3 \times t$

$V_B = 3t$ ......(1)

$s = ut + \frac{1}{2} at^2 = 0 + \frac{1}{2} \times 3t^2$

$s = 1.5\,t^2$ ...... (2)

For Stone : $a = -g = -32.2\ fps^2$, $s = h$, $t_s = (t - 5)$

$v_s = u + at = u - 32.2\,(t - 5)$

$v_s = u - 32.2\,t + 161$

As $v_s = V_B$

$\therefore \quad 3t = u - 32.2\,t + 161$

$u = 35.2\,t - 161 \qquad \ldots\ldots (3)$

$$s = ut + \frac{1}{2}\,at^2$$

$$s = u\,(t - 5) + \frac{1}{2}(-32.2)\,(t - 5)^2 \qquad \ldots\ldots (4)$$

equating (2) and (4)

$$1.5\,t^2 = u\,(t - 5) - 16.1\,(t - 5)^2$$

$$1.5\,t^2 = (35.2\,t - 161)\,(t - 5) - 16.1\,(t - 5)^2$$

Solving for 't' and substituting it in equation (3) we get,

$u = 66.4$ fps

Problem 10. : The rectilinear motion of a particle is governed by the equation $s = r \sin \omega t$ where r & ω are constants. Show that the acceleration is $a = -\omega^2 s$.

Solution : $\quad s = r \sin \omega t$

$$\therefore \quad \frac{ds}{dt} = \omega\, r \cos \omega t$$

$$\therefore \quad \frac{dv}{dt} = -\omega^2\, r \sin \omega t$$

$$a = -\omega^2\, r \sin \omega t$$

$$a = -\omega^2\, s$$

Problem 11. : The motion of a particle along a straight line is defined by $s = \frac{1}{3}\,t^3 - 36\,t$.

(a) Find the average acceleration during the fourth second.

(b) When the particle reverses its direction, what is its acceleration ?

Solution : Points to remember :

(1) When particle reaches at maximum velocity or the stage of constnt velocit then acceleration a = 0

(2) When particles reverse the direction, velocity = 0

(3) When displacement is maximum, velocity = 0

(4) When particle returns to its original position then displacement is zero

$$s = \frac{1}{3}t^3 - 36\,t$$

$$V = \frac{ds}{dt} = \frac{3t^2}{3} - 36 = t^2 - 36$$

$$a = \frac{dv}{dt} = 2t$$

for $t = 3$ sec, $a_3 = 6$ fps^2

for $t = 4$ sec, $a_4 = 8$ fps^2

so average acceleration $= \frac{a_3 + a_4}{2} = \frac{6+8}{2} = 7$ fps^2

(B) When particle reverses its direction $v = 0$

$v = t^2 - 36$

$t^2 - 36 = 0$

$\therefore\ t = 6$ sec.

So $a = 2t = 12$ fps^2

Problem 12. : The velocity of a particle moving along the X axis is defined by $v = kx^3 - 4x^2 + 6x$ where v is in fps, x is in ft and k is a constant. If $k = 1$ compute the value of the acceleration when $x = 2$ ft .

Solution : $v = kx^3 - 4x^2 + 6x$

Now $a = \frac{dv}{dt} = \frac{dv}{dx} \cdot \frac{dx}{dt}$

$$a = v\frac{dv}{dx}$$

$\therefore\ a = (kx^3 - 4x^2 + 6x)\,(3kx^2 - 8x + 6)$

If $k = 1$ & $x = 2$ft

$\therefore\ a = [8 - 16 + 12]\,[12 - 16 + 6]$

$a = 8$ fps^2

Problem 13. : In the previous problem find the smallest value of k that will make the acceleration equal to 16 fps^2 at $x = 3$ ft.

Solution :

$$a = (kx^3 - 4x^2 + 6x)(3kx^2 - 8x + 6)$$

(From previous problem)

$$\Rightarrow 16 = [k(27) - 36 + 18][27k - 24 + 6]$$

$$\Rightarrow 16 = (27k - 18)^2$$

$$\Rightarrow 27k - 18 = \pm 4$$

$$k_{min} = 0.52$$

Problem 14. : The motion of a particle in rectilinear motion is defined by $a = 6\sqrt{v}$ where a is in fps^2 and v is in fps. When t = 2 sec, v = 36 fps and s = 30 ft. Determine the value of s at t = 3 sec.

Solution : $a = \dfrac{dv}{dt} = 6\sqrt{v}$

$$\frac{dv}{\sqrt{v}} = 6\,dt$$

Integrating both sides $\int v^{-1/2}\,dv = \int 6\,dt$

$$2\sqrt{v} = 6t + c_1$$

Initially t = 2 , v = 36 fps,

$$\therefore 2 \times 6 = 12 + c_1$$

$$c_1 = 0$$

So $2 \times \sqrt{v} = 6t$

$$v = 9t^2$$

$$\frac{ds}{dt} = 9t^2$$

$$ds = 9t^2\,dt$$

Integrating both sides $\int ds = \int 9t^2\,dt$

$$s = 3t^3 + c_2$$

Now s = 30 when t = 2 sec,

$$\therefore 30 = 3 \times 8 + c_2$$

$$c_2 = 6$$

so $s = 3t^3 + 6$

Now s at t = 3 sec. = 87 ft.

Problem 15. : The rectilinear motion of a particle is governed by $a = -8s^{-2}$ where a is in fps^2 and s is in feet. When t = 1 sec., S = 4 and v = 2fps. Determine the acceleration of the particle at t = 2 sec.

Solution : $a = \frac{dv}{dt} = -8s^{-2}$ .... (1)

$\Rightarrow v \cdot \frac{dv}{ds} = -8s^{-2}$

$v\,dv = -8s^{-2}\,ds$

Integrating on both sides

$\frac{v^2}{2} = 8s^{-1} + c_1$

when v = 2, s = 4

$\Rightarrow 2 = 2 + c_1 \;,\; c_1 = 0$

$\therefore \frac{v^2}{2} = 8s^{-1} + 0$

or $v = \frac{ds}{dt} = 4s^{-1/2}$

$s^{1/2}\,ds = 4\,dt$

Integrating both sides $\int s^{1/2}\,ds = \int 4\,dt$

$\frac{2}{3}s^{3/2} = 4t + c_2$

when s = 4 , t = 1

$c_2 = 4/3$

Now at t = 2 sec, from above equation.

$\frac{2}{3}s^{3/2} = 8 + \frac{4}{3}$

$\therefore s = 14^{2/3}$

Putting value of s in equation (1)

$a = -8s^{-2} = -8\,[14^{2/3}]^{-2}$

$= -8\,[14]^{-4/3}$

$\therefore a = -0.237\ f/sec^2$

Problem 16. : The rectilinear motion of particle is governed by $a = 12t - 6t^2$. It starts from rest when $t = 0$. Determine its velocity when it returns to its starting position.

Solution : When the particle returns to its original position at that time $s = 0$, so we have to find v and t at this position.

$$a = \frac{dv}{dt} = 12t - 6t^2$$

$$dv = (12t - 6t^2)\, dt$$

Integrating both sides, we get

$$v = 6t^2 - 2t^3$$

Now $v = \dfrac{ds}{dt} = 6t^2 - 2t^3$

Integrating both sides $\int ds = \int (6t^2 - 2t^3)\, dt$

$$s = 2t^3 - \frac{1}{2}t^4 \qquad \ldots\ldots (1)$$

When the particle comes to its origional position then $s = 0$

So putting $s = 0$ in equation (1), we get

$$t = 4 \text{ sec.}$$

Using this value of t

$$v = 6t^2 - 2t^3 = -32 \text{ fps.}$$

MOTION CURVES

The method is especially useful when motion has distinct phases each requiring its associated set of equations to determine displacement, velocity and acceleration.

Motion Diagrams :

Motion diagrams are the curves which shows the variation of displacement, velcoity and acceleration with time. These are constructed in accordance with the following laws.

First law of motion diagrams :

It is used to determine the shape of a diagram.

The slope of the curve at any point in any diagram	=	The length of the ordinate at the corresponding point in the next higher diagram.

Second law of motion diagrams :

It is used to determine values of ordinates representing displacement, velocity or acceleration.

The difference in length of any two ordinate in any diagram.	=	The area between corresponding ordinates in the next higher diagram.

Consider the a-t curve as shown in figure and assume that the velocity v_1 and displacement s_1, are known at the time t_1. The velocity v_2 at any other time t_2 is found by using the relationship,

$$\int_{v_1}^{v_2} dv = \int_{t_1}^{t_2} a\,dt$$

$$v_2 - v_1 = \int_{t_1}^{t_2} a\,dt = (\text{area})_{a-t}$$

Similarly,

$$\int_{s_1}^{s_2} ds = \int_{v_1}^{v_2} v\,dt$$

$$s_2 - s_1 = (\text{area})_{v-t}$$

It is to be remembered that diagrams must be drawn on sheet from top to bottom in the sequence acceleration–time (a–t), velocity–time (v–t) and displacement–time (s–t).

Degree of Curve :

Exponent with which ordinate changes with abscissa is called degree of curve.

Diagram	Equation	Degree	Area	Centroid
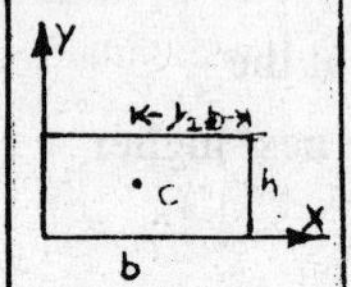	$y = kx^o$	zero	bh	b / 2
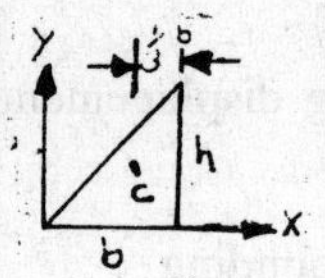	$y = kx$	first	bh / 2	b / 3
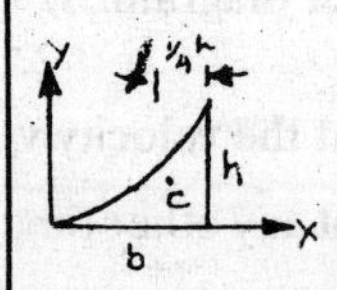	$y = kx^2$	second	bh / 3	b / 4
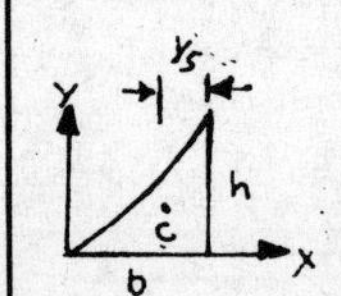	$y = kx^3$	third	bh / 4	b / 5
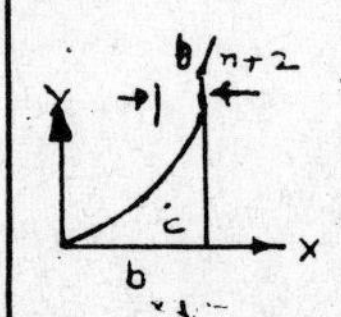	$y = kx^n$	n^{th} degree	$\frac{bh}{n+1}$	$\frac{b}{n+2}$

If the acceleration is represented by zero degree curve then velocity $\left(= \int_t a\,dt\right)$ will be the first degree curve and displacement $\left(= \int_t \int_t a\,dt\,dt\right)$ will be the second degree curve. Thus v–t curve is of one higher degree than a–t curve and s–t curve is two degree higher curve than a–t curve.

Method of transform to higher degree curve :

(i) If the zero degree curve is positive then first degree curve is increasing and if the zero degree curve is negative then the first degree curve is decreasing as shown in fig. (i).

(ii) If the first degree curve is increasing (+ve or –ve) the second degree curve is concave type and its increasing or decreasing direction depends upon positive or negativeness of first degree curve as shown in fig. (ii).

(iii) If the first degree curve is decreasing (+ve or –ve) the second degree curve is convex type and its increasing or decreasing direction depends upon positive and negativeness of first degree curve as shown in fig. (iii)

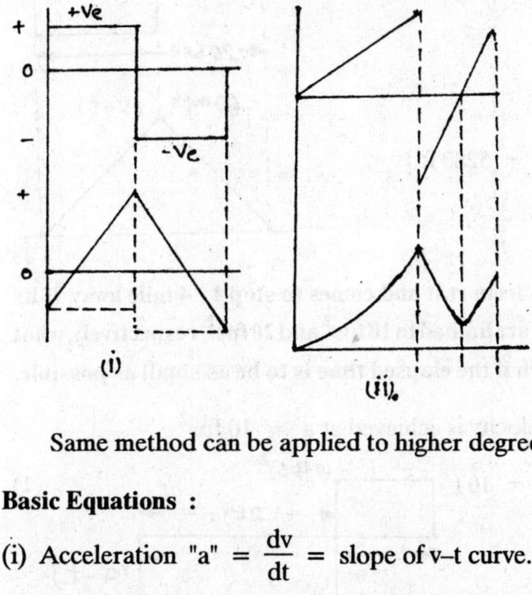

Same method can be applied to higher degree curves also.

Basic Equations :

(i) Acceleration "a" $= \dfrac{dv}{dt} =$ slope of v–t curve.

(ii) Change in velocity between any two points 1 and 2

$v_2 - v_1 = \Delta v =$ Area of a–t curve between 1 and 2.

(iii) Velocity "v" $= \dfrac{ds}{dt} =$ slope of s–t curve.

(iv) Change in displacement between any two points 1 and 2

$= s_2 - s_1 = \Delta s =$ Area of v–t curve.

$= v_1 (\Delta t) + (\text{Area})_{a-t} \ (t_2)$

Where $(\text{Area})_{a-t}$ = Area bounded by a–t curve during corresponding change in velocity as displacement ; v_1 = velocity at the start of time interval ; and t_2 = moment arm to the centroid of area bounded from outer right edge.

Problem 17. : An auto is accelerated form rest to a top speed of 60 mph and then immediately decelerated to a stop. If the total elapsed time is 20 sec, determine the distance covered. The acceleration and deceleration are both constant but not necessarily of the same magnitude.

Solution : S = area (v – t) curve

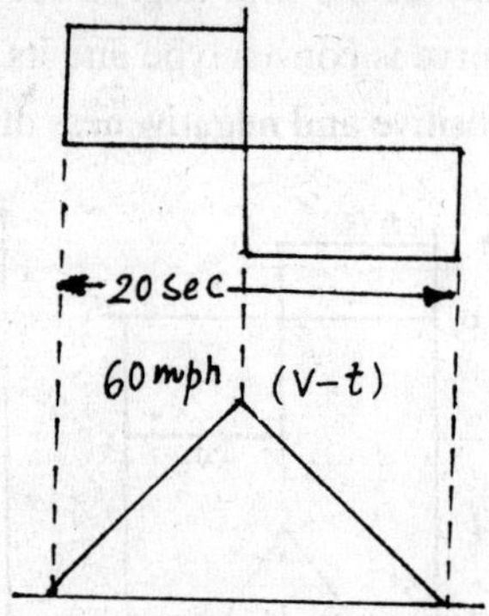

$$s = \frac{1}{2} \times 20 \times \frac{60}{3600}$$

$$= \frac{20}{120} \text{ mps}$$

$$= \frac{20}{120} \times 5280 \; [\text{ 1 mile } = 5280 \text{ ft }]$$

$$= 880 \text{ ft .}$$

Problem 18. : An auto starts from rest and comes to stop 1 / 4 mile away. If its acceleration and deceleration are limited to 10 fps^2 and 20 fps^2 respectively, what maximum speed does it reach if the elapsed time is to be as small as possible.

Solution : The maximum velocity is achieved at a = 10 fps^2

$$\Delta vu = \text{Area } (a - t) = 10\,t \quad \ldots. (1)$$

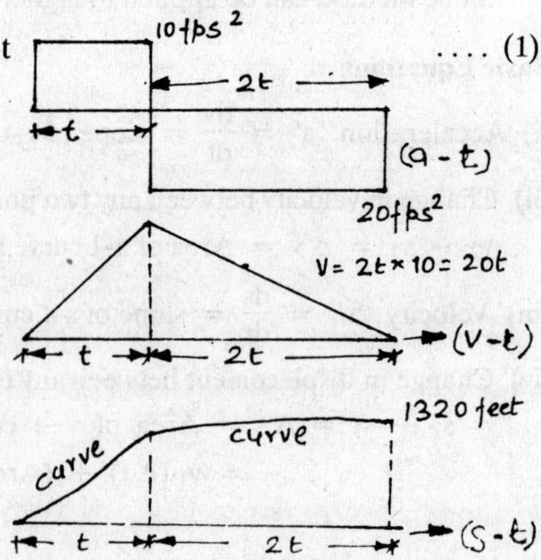

Also total distance covered

s = area v – t

$$= \frac{1}{4} \text{ mile } = 1320 \text{ ft}$$

$$1320 = \frac{1}{2} \times 3t \times v$$

$$1320 \times 2 = 3t \times 10\,t \quad \text{from (1)}$$

$\therefore \quad t = 9.38$ sec.

Max. speed $= 9.38 \times 10$ fps

$= 93.8$ fps.

Problem 19. : A car and a truck are both moving at 45 mph along a highway. To pass the truck safely, the car must move 200 ft relative to the truck. Determine the minimum time to do this if the car can accelerate at 6 fps^2 but must not exceed a speed of 60 mph.

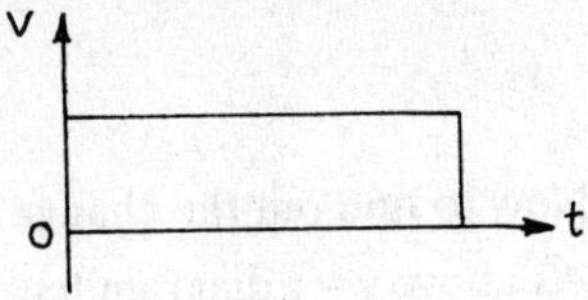

Solution : $V_{truck} = \dfrac{45 \times 5280}{3600} = 66$ fps

$V_{car} = \dfrac{60 \times 5280}{3600} = 88$ fps

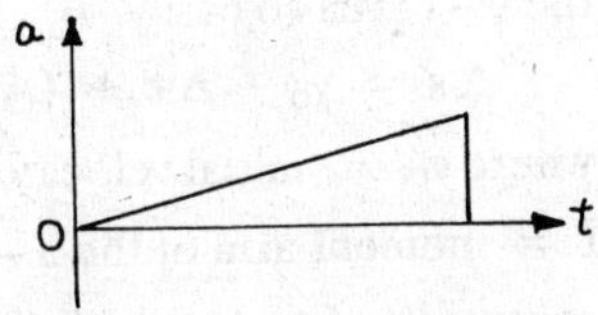

Suppose car takes time t to pass the truck safely.

s = Area (v – t) curve.

$200 = \dfrac{1}{2} \times t \times 22$

$t = 200 \,/\, 11$ sec.

But Δv = area (a – t)

$\Delta v = 6t = \dfrac{200}{11} \times 6 = 110$

Hence relative velocity is 110. But V_{car} should not exceed 88 fps. so taking $V_{car} = 88$fps,

$\Delta v = 110 - 88 = 22$

Also $\Delta v = 6t = 22$.

$\therefore \quad t = 3.67$ sec

Problem 20. : The acceleration of an object moving along a straight path decreases uniformly from 10 f / sec^2 to zero in 12 sec at which time its velocity is 6 ft / sec. Find its initial velocity and the change in position during the 12 sec interval.

Solution : Δv = Area of (a – t) curve

$$= \frac{1}{2} \times 12 \times 10 = 60$$

$$\Delta v = v_1 - v_o = 60$$

When $v_1 = 6$ ft/sec.

$$6 - v_0 = 60,$$

$$v_0 = -54 \text{ fps}$$

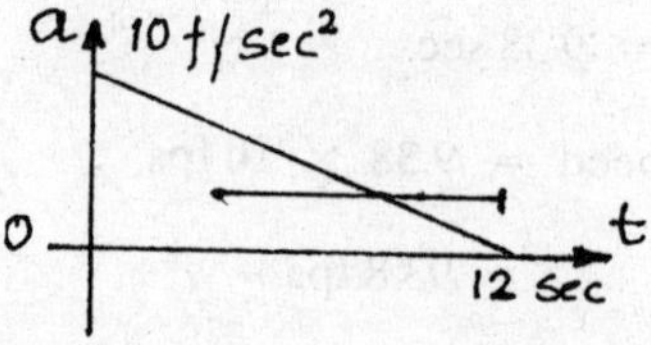

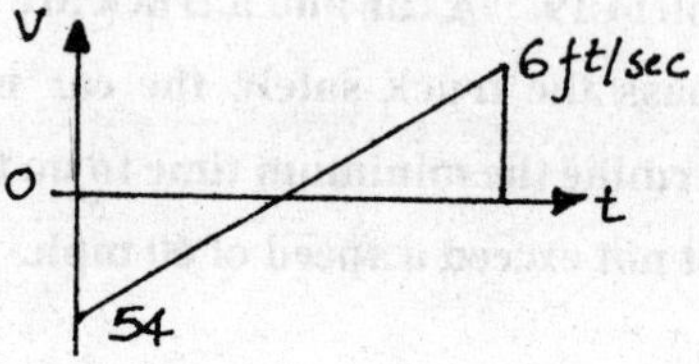

Now to find out the change is position during the 12 sec interval.

As above v – t diagram have positive and negative area so we cannot find out the v – t area so easily.

$$\Delta s = v_o \cdot \Delta t + (\text{Area under a–t diagram}) \times \bar{t}$$

where v_o = initial velocity, Δt = time interval,

$\bar{t}$ = moment arm of the a – t curve and is measured as the distance between centroid and end of the time interval.

$$\Delta s = -54 \times 12 + \frac{1}{2} \times 12 \times 10 \times 8$$

$$\Delta s = -648 + 480 = 168 \text{ ft.}$$

Problem 21. : The velocity of a particle changes from – 8 fps to 58 fps during 12 sec interval during which its acceleration increases uniformly with time from an initial value of 3 fps^2. Determine the displacement covered in the 12 sec interval.

Solution :

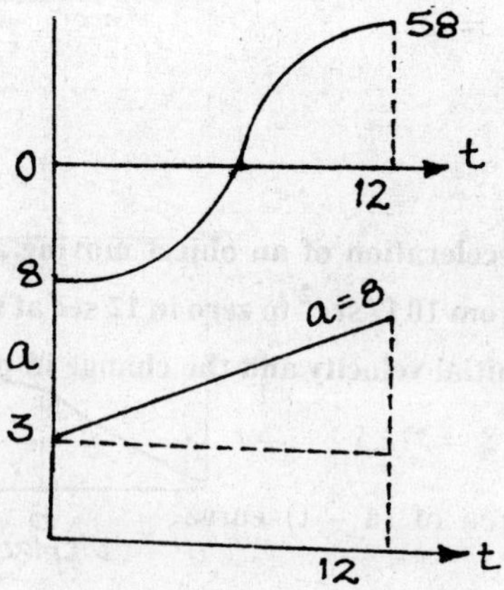

$\Delta v =$ Area under $a - t$ curve

$$v_1 - v_0 = 12 \times 3 + \frac{1}{2} \times 12 \times h$$

$$58 - (-8) = 36 + 6h \quad \therefore \quad h = \frac{30}{6} = 5 \text{ ft}$$

$\therefore \quad a = 5 + 3 = 8$

By the $v - t$ diagram we cannot easily find out Δs, so use formula

$$\Delta s = v_o \Delta t + \text{Area under } (a - t) \text{ diagram } \times \bar{t}$$

$$= -8 \times 12 + \left[12 \times 3 \times t_1 + \frac{1}{2} \times 12 \times 5 \times t_2\right]$$

$$= -96 + \left[12 + 3 \times 6 + \frac{1}{2} \times 12 \times 5 \times 4\right]$$

$\therefore \quad \Delta s = 240$ feet.

Problem 22. : A particle is moving with the acceleration values varying with time as shown below.

Interval of Time (sec)	**0 - 5**	**5 - 10**	**10 - 15**
Acceleration (m/sec^2)	**+ 1.0**	**+ 2.0**	**– 1.5**

Determine :

(i) The velocity and (ii) Displacement of the particle at time t = 5 sec, t = 10 sec, and t = 15 sec. Plot the v-t and s-t diagrams.

Solution :

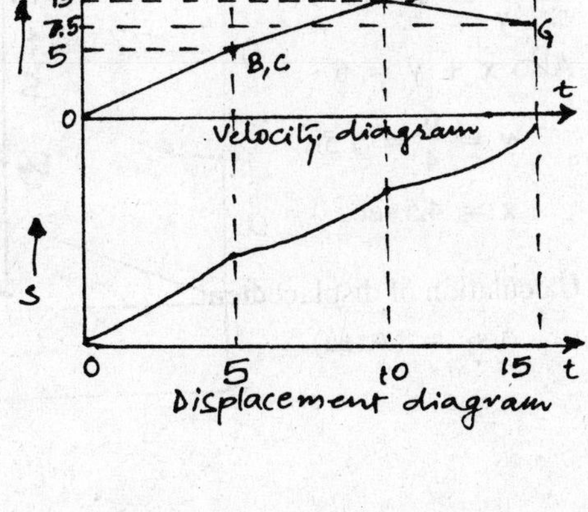

(i) $v_B = (\text{Area})_{a-t} = OAB5$

$$= 1.0 \times 5 = 5 \text{ m/s}$$

$$v_D = v_B + (\text{Area})_{a-t}$$

$$= 5 + 5 \times 2 = 15 \text{ m/s}$$

$$v_G = (\text{Area})_{a-t}$$

$$= 1.5 \times 5 = 7.5 \text{ m/s}$$

(ii) $s_B = (\text{Area})_{v-t}$

$$= \frac{1}{2} \times 5 \times 5 = 12.5$$

$$s_F = s_B + 5 \times 5 + \frac{1}{2} + 5 \times 10$$

$$= 12.5 + 25 + 25 = 62.5 \text{ m}$$

$$s_G = s_F + 7.5 \times 5 + \frac{1}{2} \times 5 \times 7.5$$

$$= 62.5 + 37.5 + 18.75 = 118.75 \text{ m}.$$

Problem 23.: The curved portions of the v-t curve shown in figure are second degree parabolas with horizontal slope at t = 0 and t = 12 sec. Sketch that a-t and s-t curves if s_0 is zero.

Solution : Calculation of acceleration

(i) $\Delta v_1 = v_A - v_0 = (\text{Area})_{a-t}$

$$18 = \frac{1}{2} \times 6 \times OA\,(a_A)$$

$$\therefore \; a_A = \frac{18 \times 2}{6} = 6 \text{ m/s}^2$$

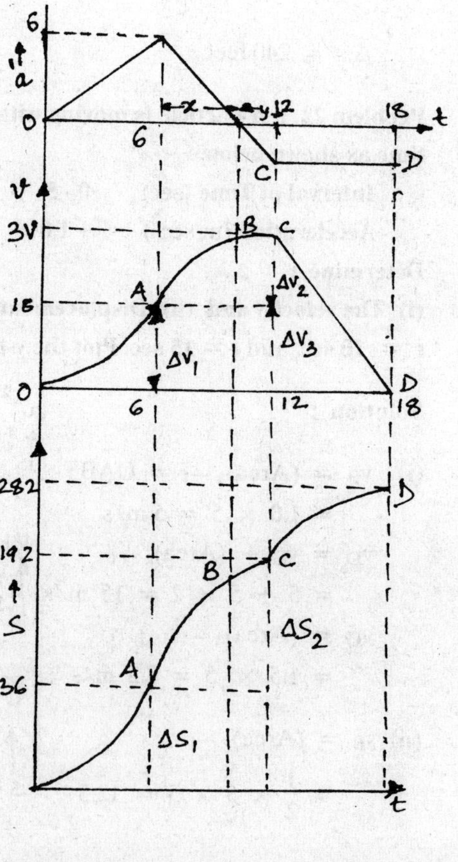

Similarly

$$\Delta v_2 = \frac{1}{2} \times (12 - 6) \times (\Delta a)$$

$$(30 - 18) = \frac{1}{2} \times 6 \times (6 + a_c)$$

$$6 + a_c = 4$$

$$a_c = -2 \text{m/s}^2$$

$$\text{Now } \frac{6}{2} = \frac{x}{y}$$

$$\frac{x}{y} = 3$$

Also $x + y = 6$

$$y = \frac{6}{4} = 1.5$$

$$x = 4.5 \text{ sec.}$$

Calculation of displacement

$\Delta s_1 = (\text{Area})_{v-t}$

$$s_A - s_0 = \frac{1}{3} \times 6 \times 18 = 36$$

$$s_A = 36 \text{ m}$$

Here $s_0 = 0$

$$\Delta s_2 = (\text{Area})_{v-t}$$

$$= 18 \times 6 + \left[12 \times 6 - \frac{1}{3} \times 6 \times 12\right]$$

$$= 108 + 72 - 24 = 156$$

$$s_C - s_B = 156$$

$$\therefore \quad s_C = 192 \text{ m}$$

$$s_D - s_C = (\text{Area})_{v-t}$$

$$= \frac{1}{2} \times (18 - 12) \times 30 = 90$$

$$\therefore \quad s_D = 90 + 192 = 282 \text{ m} \quad \text{at } t = 18 \text{ sec.}$$

Problem 24.: The a-t curve for a partcile having rectilinear motion is shown in fig. at t = 0, the velocity is 244 cm/s and the particle is 1829 cm/s to the left of the origin of displacement. Draw the v-t and s-t curves, specifying values of v and s at t = 4 sec, 7 sec and 13 sec.

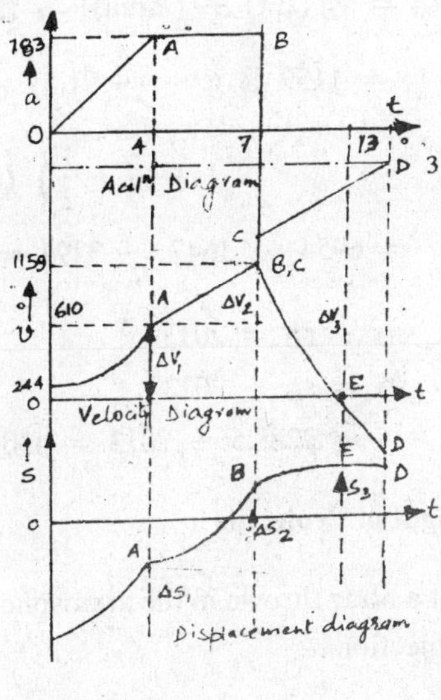

Solution : (i) Calculation of velocity

$$v_A - V_o = \frac{1}{2} \times 4 \times 183$$

$$\therefore \quad v_A = 244 + 366 = 610 \text{ cm/s}$$

$$v_B - v_A = (7 - 4) \times 183$$

$$\therefore \quad v_B = 610 + 549 = 1159 \text{ cm/s}$$

$$v_D - v_B = -\frac{1}{2}(366 - 91.5)(13 - 7)$$

$$- 91.5 \times 6$$

$$\therefore \quad v_D = v_B - 823 - 549$$

$$= 1159 - 1372.5 = -213.5 \text{ cm/s}$$

Velocity at point E is zero.

$$v_E - v_B = -\frac{1}{2}(366 - 91.59)\, t - 91.5\, t$$

$$= -137.25\,t - 91.5t$$

$$0 - 1159 = -228.75\,t$$

$$\therefore \quad t = 5.066 \text{ sec.}$$

$$t_e = t_B + t$$

$$= 7 + 5.066 = 12.066 \text{ sec.}$$

Similarly calculate

$$(\Delta s) = (\text{Area})_{a-t}$$

$$\Delta s_1 = 244 \times 4 + \frac{1}{3} \times 4 \times 366 = 1464 \text{ cm}$$

$$s_A - s_D = 1464$$

$$s_A = s_D + 1464$$

$$= -1829 + 1464 = -365 \text{ cms.}$$

$$\Delta s_2 = 610 \times (7 - 4) + \frac{1}{2} \times (7 - 4) \times 549$$

$$= 1830 + 823.5$$

$$s_B - s_A = 2653.5 \text{ cm}$$

$$s_B + 365 = 2653.5$$

$$s_B = 2288.5 \text{ cm}$$

$$\Delta s_3 = v_B (\Delta t) + (\text{Area})_{a-t}\,(t_2)$$

$$= 1159 \times 6 + (-91.5) \times 6 \times \left(\frac{1}{2} \times 6\right)$$

$$+ \left(-\frac{1}{2}\right) (-366 - 91.5) \times 6 \times \frac{2}{3} \times 6$$

$$= 6954 - 1647 - 3294 = 2013 \text{ cm}$$

$$s_D - s_B = 2013$$

$$s_D = s_B + 2013$$

$$= 2288.5 + 2013 = 4301.5 \text{ cm.}$$

Flight of Projectile :

Let a body thrown in the atmosphere with a initial velocity of v_o at an angle of projection α.

Horizontal component of velocity $= v_{ox} = v_o \cos \alpha$

Vertical component of velocity $= v_{oy} = v_o \sin \alpha$

Horizontal displacement

$$X = v_{ox} \cdot t + \frac{1}{2} a_x t^2$$

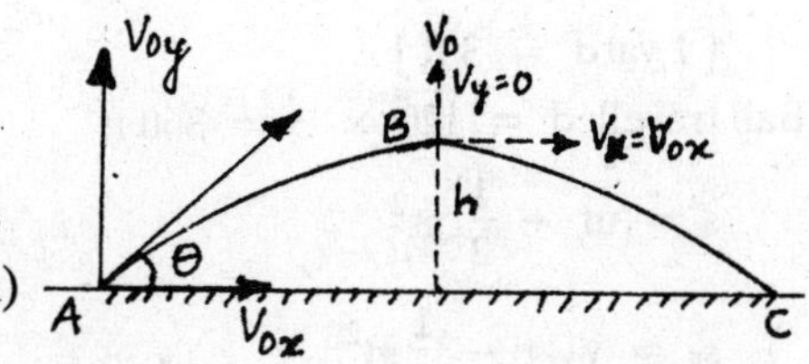

As $a_x = 0$, $X = v_o \cos \alpha \cdot t$

The horizontal range of projectile

$R = v_o \cos \alpha\, T,$ ...... (1)

where T is time of flight.

Now the vertical displacement

$$Y = v_{oy} t + \frac{1}{2} a_y t^2$$

when $Y = 0$, $t = T$ and $a_y = -g$

$$0 = v_{oy} T - \frac{1}{2} g T^2$$

$$\therefore \quad T = \frac{2 v_{oy}}{g}$$

$$T = \frac{2 v_o \sin \alpha}{g}$$

Substituting value of T in equation (1)

$$R = v_o \cos \alpha \times \frac{2 v_o \sin \alpha}{g}$$

$$R = \frac{v_o^2 \sin 2\alpha}{g}$$

Now to find maximum height 'H' of projectile.

Consider , $v_y^2 = v_{oy}^2 - 2g Y \quad (\because \; a_y = -g)$

At maximum height $v_y = 0$, $Y = h_{max}$

$$0 = v_o^2 \sin^2 \alpha - 2g\, h_{max}$$

$$\therefore \quad h_{max} = \frac{v_o^2 \sin^2 \alpha}{2g}$$

Problem 25. : A golf ball is hit from an elevated tee to a green, the distance horizontally is 120 yards. If the initial velocity of the ball is 100 fps at 53.1° to the horizontal, how high is the tee above the green.

Solution :

$v_{ox} = 100 \cos 53.10 = 60$ f/sec.

$v_{oy} = 100 \sin 53.10 = 80$ f/sec.

[1 yard = 3 ft]

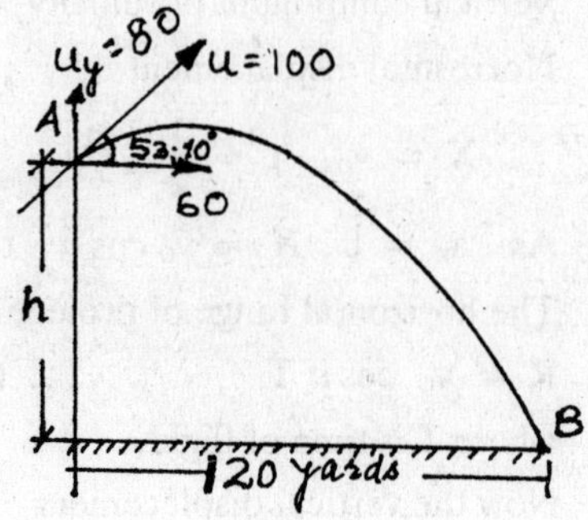

Ball travelled = $120 \times 3 = 360$ ft

$$s = ut + \frac{1}{2} at^2$$

$$s_y = v_{oy} t - \frac{1}{2} gt^2$$

$$s_y = 80\,t - \frac{1}{2} \times (32.2) \times t^2 \quad \ldots\ldots (i)$$

$$s_x = v_{ox} t + \frac{1}{2} a_x t^2 \;;\; a_x = 0$$

$$360 = 60\,t$$

$t = 6$ sec. putting in equation (i)

$$s_y = 80 \times 6 - \frac{1}{2} \times 32.2 \times 6^2$$

$= -100$ ft (below reference axes)

Problem 26. : A golf ball is driven with the initial velocity of 161 fps upward at 30° to the horizontal from an elevated tee 64.4 ft above a level fairway. What horizontal distance will the ball travel before hitting the fairway.

Solution :

$v_{ox} = 161 \cos 30 = 139.43$ fps

$v_{oy} = 161 \sin 30 = 80.5$ fps

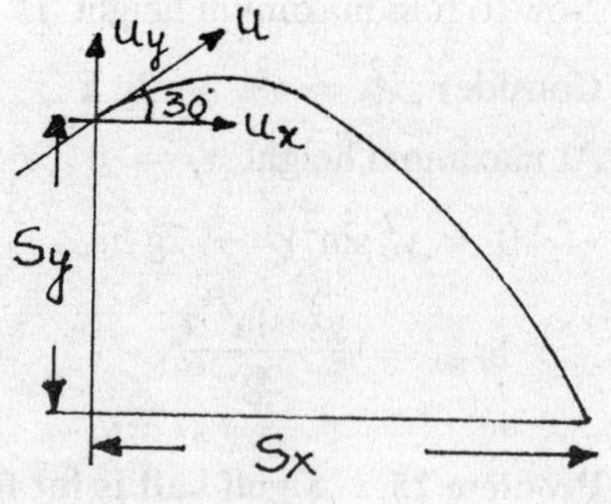

$$s = ut + \frac{1}{2} at^2$$

$$s_x = v_{ox} \cdot t$$

$$s_y = v_{oy} \times t - \frac{1}{2} gt^2$$

$$64.4 = 161 \sin 30 \times t - \frac{1}{2} \times 32.2 \times t^2$$

$$64.4 = 80.5\,t - 16.1\,t^2$$

$$t^2 - 5t - 4 = 0$$

$$t^2 - 5t + 6.25 = 4 + 6.25$$

or $(t - 2.5)^2 = 10.25$

$\therefore\ t = 5.7$ sec.

$s_x = 139.43 \times 5.7 = 794.78$ fts $= 265$ yards [3 feet = 1 yard]

Problem 27. : A stunt man is to drive an auto across the water–filled gap shown in fig. Determine the auto's minimum take–off velocity and the angle θ of the landing ramp.

Solution :

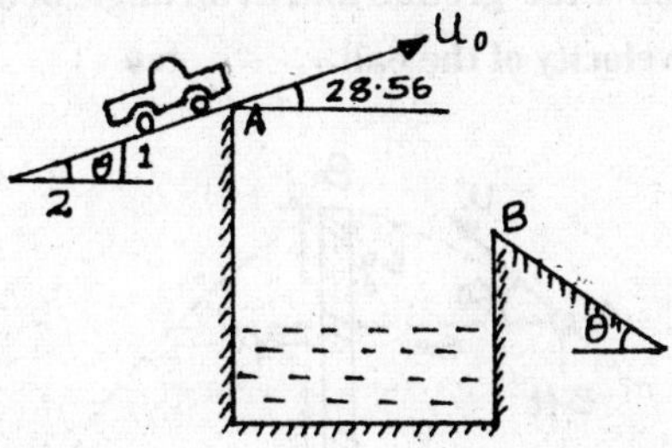

$$\alpha = \tan^{-1}\frac{1}{2} = 28.56^{o}$$

Let v_o = initial velocity of auto

$$s_x = v_{ox} \times t$$

$$40 = v_o \cos 28.56 \times t$$

$$t = \frac{45.54}{v_o} \qquad \ldots\ldots \text{(i)}$$

also $s_y = v_{oy}\, t - \frac{1}{2} gt^2$

$$= v_o \sin 28.56 \times t - \frac{1}{2} gt^2$$

$$= v_o \sin 28.5 \times \frac{45.54}{v_o} - \frac{1}{2}\, 32.2 \times \left(\frac{40}{v_o}\right)^2 = -10 \text{ (given)}$$

Solving this $v_o = 32.76$ f/sec.

$v_{ox} = 32.76 \cos 28.56 = 29.3$ f/sec.

$v_{oy} = 32.76 \sin 28.56 = 15.66$ f/sec.

also $v_y^2 = v_{oy}^2 + 2a_y\, s_y$

$$v_y^2 = (15.66)^2 - 2\,(32.2)\,(-10)$$

$$v_y = 29.3 \text{ f/sec.}$$

$$v_x^2 = v_{ox}^2 + 2a_x s_x \quad \text{(eliminate)}$$

$$v_x = v_{ox} = 29.3 \text{ fps.} \quad (\text{as } a_x = 0)$$

$$\theta = \tan^{-1}\left(\frac{v_y}{v_x}\right) = \tan^{-1}\frac{29.3}{29.3} = 45^o$$

Problem 28. : A ball is thrown so that it just clears a 25 ft. wall 100 ft away. If it left the hand 5 ft above the ground and at an angle of 60° to the horizontal, what was the initial velocity of the ball.

Solution :

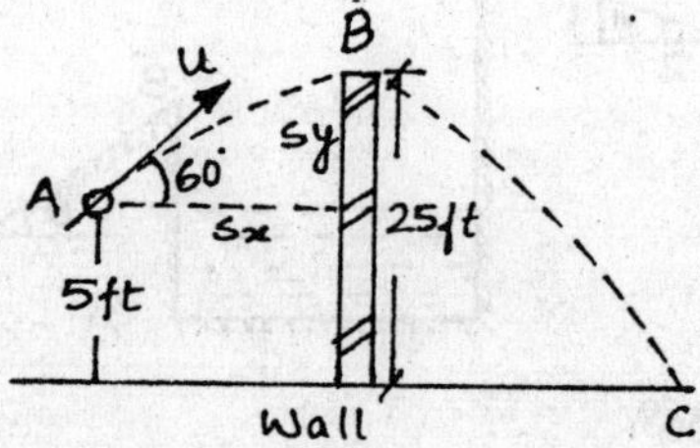

$$s = v_o t + \frac{1}{2} at^2$$

$$s_y = v_{oy} \cdot t - \frac{1}{2} \times 32.2 \times t^2$$

$$s_x = v_{ox}\, t$$

$$100 = v_{ox}\, t$$

$$t = \frac{100}{v_o \cos 60} = \frac{200}{v_o} \qquad \ldots\ldots \text{(i)}$$

$$s_y = v_o \sin 60 \times t - 16.1 \times t^2$$

Putting value of t from (i)

$$20 = v_o \sin 60 \times \frac{100}{v_o \cos 60} - 16.1 \times \left(\frac{100}{v_o \cos 60}\right)^2$$

$$20 = 173.2 - 16.1 \times \frac{10000}{0.25\, v_o^2}$$

$$20 = 173.2 - \frac{644000}{v_o^2}$$

$$20v_o^2 = 173.3\, v_o^2 - 644000$$

$$153.3\, v_o^2 = 644000$$

$$v_o^2 = 4200$$

$$v_o = 64.8 \text{ fps}$$

Problem 29. : In fig. a ball thrown down the incline strikes it at a distance s = 254.5 ft. If the ball rises to a maximum height h = 64.4. ft above the point of release, compute its initial velocity and inclination α .

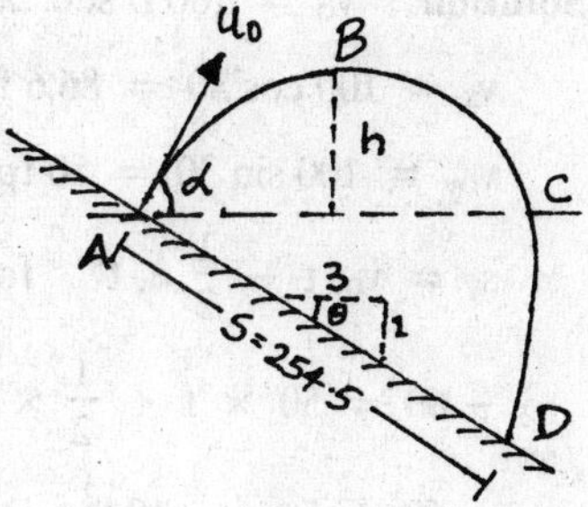

Solution : $h = \dfrac{v_{oy}^2}{2g} = 64.4$ ft.

$$v_{oy} = [\,64.4 \times 2 \times 32.2\,]^{1/2} = 64.4 \text{ f/sec.}$$

$$\theta = \tan^{-1}\frac{1}{3} = 18.43^o$$

$$s_y = 254.5 \sin 18.43 = 80.46 \text{ ft.}$$

$$s_x = \cos 18.43 \times 254.5 = 241.45 \text{ ft.}$$

Also $s_y = v_{oy} \times t - 16.1\, t^2$

$$= v_{oy} \times t - 16.1\, t^2$$

$$64.4 \times t - 16.1\, t^2 = -80.45$$

$$16t^2 - 64.4\, t - 80.45 = 0$$

$$\therefore \quad t = 5 \text{ sec.}$$

As $s_x = v_x \times t$

$$241.45 = v_x \times 5$$

$$v_x = 48.3 \text{ f/sec.}$$

$$\alpha = \tan^{-1}\frac{v_{oy}}{v_x} = \tan^{-1}\frac{64.4}{48.3} = 53.13^o$$

$$v = \left[v_x^2 + v_{oy}^2\right]^{1/2} = \left[(64.4)^2 + (48.3)^2\right]^{1/2}$$

$$v = 80.5 \text{fps.}$$

Problem 30. : A particle has an initial velocity of 100 ft / sec. up to the right at 30° with the horizontal. The components of acceleration are constant at $a_x = -4\ fps^2$ and $a_y = -20\ fps^2$. Compute the horizontal distance covered until the particle reaches a point 60 ft below its original elevation.

Solution : $v_o = 100$ f/sec. at $\theta = 30^o$

$v_x = 100 \cos 30 = 86.6$ fps , $a_x = -4$ ft/sec^2

$v_{oy} = 100 \sin 30 = 50$ fps , $a_y = -20$ ft/sec^2

$s_y = v_{oy} t + \frac{1}{2} a_y t^2$ for 60 ft below

$-60 = 50 \times t + \frac{1}{2} \times -20 \times t^2$

$-60 = 50t - 10t^2$

$10t^2 - 50t - 60 = 0$ $\quad\quad$ $(2t - 6) = 0$

or $t^2 - 5t - 6 = 0$

Solving this we get, $t = 6$ sec.

$s_x = v_x t + \frac{1}{2} a_x t^2$

$= 86.6 \times 6 + \frac{1}{2} \times (-4)(6)^2$

$s_x = 448$ ft

Problem 31. : A particle moves along the path $y = x^2 - 4x + 100$, starting with an initial velocity of $v_o = (4\hat{i} - 16\hat{j})$ fps. If v_x is constant, determine v_y and a_y at $x = 16$ ft.

Solution : $v_o = (4i - 16j)$

$v_{ox} = 4$ f/sec

$v_{oy} = -16$ fps

Now, $y = x^2 - 4x + 100$

$$\frac{dy}{dt} = 2x\frac{dx}{dt} - \frac{4dx}{dt} \quad\quad \ldots\ldots\text{(i)}$$

$$v_y = (2x - 4)\frac{dx}{dt} = (2x - 4)\,v_x$$

$\therefore\ v_y$ (at $x = 16$) $= (2 \times 16 - 4) \times 4 = 112$ fps

Differentiating equation (i) w.r.t. t

$$\frac{d^2y}{dt^2} = 2x\frac{d^2x}{dt^2} + 2\left(\frac{dx}{dt}\right)^2 - \frac{4d^2x}{dt^2}$$

$$a_y = 2x\,a_x + 2(v_x)^2 - 4\,a_x,\ \text{as } a_x = 0$$

$$\therefore\ a_y = 2v_x^2 = 32\ \text{ft/sec}^2$$

Problem 32. : If the velocity of a particle is defined by $v = n\,(2t + 1)\,i + 3j$ fps, and its position vector at t = 1 sec is $r = 4\hat{i} + 3\hat{j}$ ft, determine the path of the particle in terms of its x and y coordinates.

Solution : $v = (2t + 1)\,i + 3j$ f/sec.

at $t = 1$, $x = 4$ ft and $y = 3$ ft

$v_x = 2t + 1$; $v_y = 3$

$$\int dx = \int (2t + 1)\,dt$$

$$x = \frac{2t^2}{2} + t + A$$

$$x = t^2 + t + A$$

$$y = 3t + B$$

at $t = 1$, $x = 4$, so $A = 2$

at $t = 1$, $y = 3$, so $B = 0$

$$\therefore\ x = t^2 + t + 2,\ y = 3t$$

Putting $t = \frac{y}{3}$ in above equation

$$x = \left(\frac{y}{3}\right)^2 + \left(\frac{y}{3}\right) + 2$$

$$x = \frac{y^2}{9} + \frac{y}{3} + 2$$

$$9x = y^2 + 3y + 18$$

Problem 33.: A rocket is released from a jet fighter flying horizontally at 750 mph at an altitude of 8000 ft above its target. The rocket thrust gives it a constant horizontal acceleration of 0.6 g. Determine the angle between the horizontal and the line of sight to the target.

Solution : 750 mph = 1100 fps

$a_x = 0.60\text{ g}$

$v_x = 1100\text{ f/sec.}$

$v_{oy} = 0$

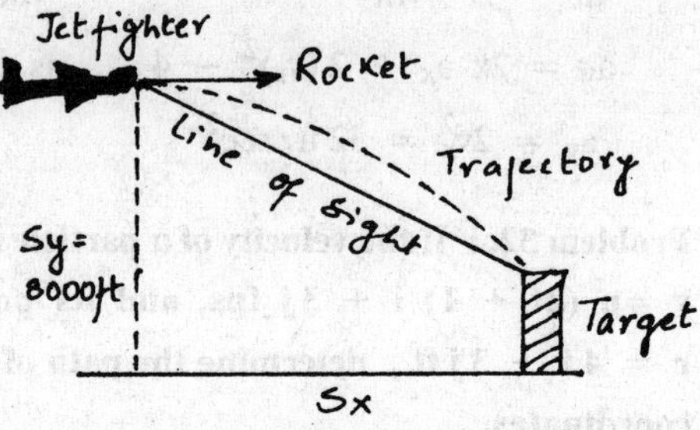

$$s_y = v_{oy}\, t + \frac{1}{2}\, gt^2$$

$$8000 = \frac{1}{2}\,(32.2)\, t^2$$

$$\therefore\ t = 22.3\text{ sec.}$$

$$s_x = v_x\, t + \frac{1}{2}\, at^2$$

$$= 1100 \times 22.3 + \frac{1}{2} \times 0.60 \times 32.2 \times (22.3)^2$$

$$= 29333.82\text{ ft}$$

Hence required angle $= \tan^{-1} \dfrac{s_y}{s_x}$

$$= \tan^{-1} \frac{8000}{29333.82} = 15.25^{\circ}$$

Problem 34. : If the initial velocity of an object is 40 fps, determine the horizontal distance it can cover without rising more than 10 ft.

Solution : $v_o = 40\text{ fps}$; $h = 10\text{ ft}$

$$h = \frac{v_o^2 \sin^2 \alpha}{2g}$$

$$10 = (40)^2\, \frac{\sin^2 \alpha}{2 \times 32.2}$$

$$\sin^2 \alpha = \frac{32.2 \times 2 \times 10}{40 \times 40}$$

$$\alpha = 40^{\circ}$$

$$R = \frac{v_0^2 \sin 2\alpha}{g} = \frac{40^2 \times \sin(2 \times 40)}{32.2}$$

$$= 48.93 \text{ ft.}$$

Problem 35. : The telescoping rod shown in fig. forces the pin P to move along the fixed path $9y = x^2$ where x and y are in inches. At any time t, the x coordinate of P is given by $x = t^2 - 5t$. Determine the y components of velocity and acceleration of P at $x = 6$ in.

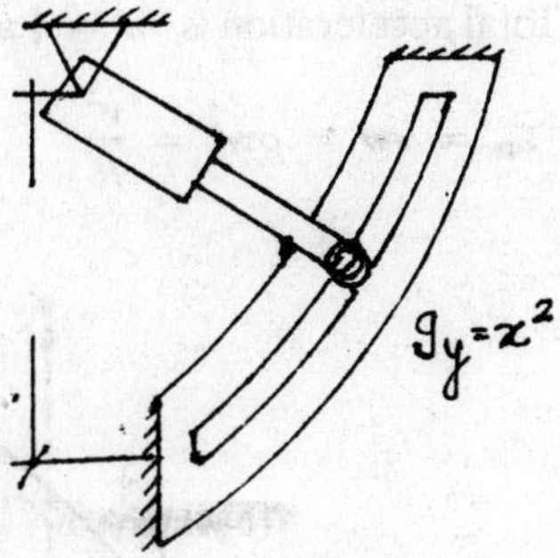

Solution :

Given $9y = x^2$ also $x = t^2 - 5t$

diff. w.r.t. t to both equations

$$\frac{dx}{dt} = 2t - 5 \Rightarrow v_x = 2t - 5 \qquad \text{......(1)}$$

$$9\frac{dy}{dt} = 2x\frac{dx}{dt} \qquad \text{......(2)}$$

Again diff. $v_x = 2t - 5$

$$\frac{dv_x}{dt} = a_x = 2 \qquad \text{......(3)}$$

$$v_y = \frac{2x}{9} v_x \qquad \text{......(4)}$$

As $x = t^2 - 5t$

at $x = 6$, $t^2 - 5t - 6 = 0$,

v_x at $t = 6$ sec from equation (1)

$$v_x = 2 \times 6 - 5 = 7 \text{ fps}$$

$$v_y = \frac{2}{9} \times 6 \times 7 = 9.83 \text{ fps}$$

diff. equation (4) w.r.t. to t

$$a_y = \frac{2}{9}(x \cdot a_x + v_x^2)$$

$$= 13.56 \text{ fps}^2$$

Normal and Tangential Components of Acceleration and Radius of Curvature :

a_t = tangential acceleration in the direction of increasing speed

a_n = normal acceleration directed inward toward the centre of curvature

Total acceleration is $a = [a_n^2 + a_t^2]^{1/2}$

Also $a_n = \nu w = \rho w^2 = \dfrac{\nu^2}{\rho}$

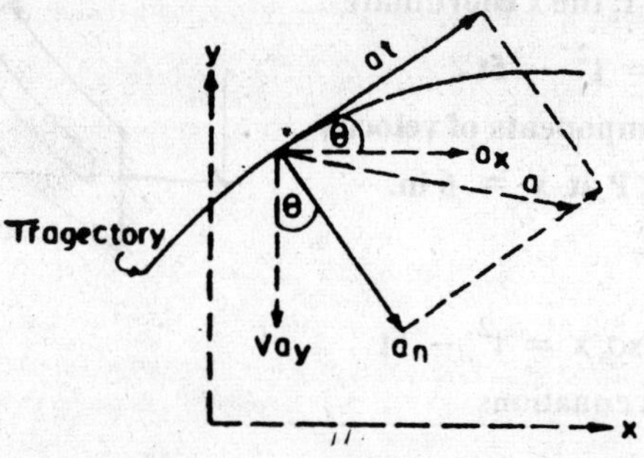

Where ρ = Radius of curvature

w = angular velocity,

and ν = linear velocity.

From Fig.,

$$a_n = a_x \sin\theta + a_y \cos\theta$$

$$a_t = a_x \cos\theta - a_y \sin\theta$$

Now consider $\vec{v} \times \vec{a} = \vec{v} \times (\vec{a_t} + \vec{a_n}) = \vec{v} \times \vec{a_t} + \vec{v} \times \vec{a_n}$

But velocity vector is always along the tangent,

$\therefore \quad \vec{v} \times \vec{a_t} = 0$

$\therefore \quad \vec{v} \times \vec{a} = \vec{v} \times \vec{a_n}$

$\therefore \quad |\vec{v} \times \vec{a}| = |\vec{v} \times \vec{a_n}|$

But $|\vec{v} \times \vec{a_n}| = v \cdot a_n \cdot \sin 90 = v \cdot a_n$

But $a_n = \dfrac{v^2}{\rho}$

$$\therefore \quad |\vec{v} \times \vec{a}_n| = \frac{v^3}{\rho}$$

$$\therefore \quad |\vec{v} \times \vec{a}| = \frac{v^3}{\rho}$$

$$\therefore \quad \rho = \frac{v^3}{|\vec{v} \times \vec{a}|}$$

Problem 36. : A particle moves on a circular path of 40–ft radius so that its arc distance from a fixed point on the path is given by $s = 4t^3 - 10t$ where s is in feet and t is in seconds. Compute the total acceleration at the end of 2 sec.

Solution : $s = 4t^3 - 10t$

Diff. w.r.t. t

$$\frac{ds}{dt} = v = 12t^2 - 10 \qquad \ldots\ldots \text{(i)}$$

at $t = 2$ sec.

$$V = 12 \times 4 - 10 = 38 \text{ fps}$$

again diff. eq. (1) w.r.t. time

$$\frac{dv}{dt} = 24\,t$$

$$a_t = 24\,t$$

$$\text{at } t = 2 \text{ sec.}, \quad a_t = 48 \text{ fps}^2$$

$$a_n = \frac{v^2}{\rho} = \frac{38^2}{40} = 36.1 \text{ fps}^2$$

$$a^2 = a_n^2 + a_t^2$$

$$= [48^2 + 36.1^2]^{1/2} = 60 \text{ fps}^2$$

Problem 37. : A stone is thrown with an initial velocity of 100 fps upward at 60° to the horizontal. Compute the radius of curvature of its path at the position where it is 50 ft horizontally from its initial position.

Solution : $v_o = 100$ fps at $\theta = 60^o$

$$v_{ox} = 100 \cos 60 = 50 \text{ fps}$$

$$v_{oy} = 100 \sin 60 = 86.6 \text{ fps}$$

Now $s_x = v_{ox}\, t + \frac{1}{2} a_x t^2$

as $a_x = 0$

$\therefore \quad 50 = 50 \times t$

or $\quad t = 1$ sec.

Now $v_y = v_{oy}\, t - \frac{1}{2} g t^2$

$$= 86.6 \times 1 - \frac{1}{2} \times 32.2 \times 1 = 54.4 \text{ fps}$$

$$v_x = V_{ox} = 50 \text{ fps}$$

$$v = [v_x^2 + v_y^2]^{1/2}$$

$$= \left[50^2 + 54.4^2\right]^{1/2} = 73.89 \text{ fps}$$

$$\theta = \tan^{-1} \frac{v_y}{v_x} = \tan^{-1} \frac{54.4}{50} = 47.41^\circ$$

Now $a_x = 0$ & $a_y = -32.2 \text{ fps}^2$

$$a_n = a_y \cos\theta = 21.75 \text{ fps}^2$$

$$\rho = \frac{v^2}{a_n} = \frac{73.89^2}{21.75} = 251 \text{ ft.}$$

other way of doing it is

$$\bar{v} = 50\,i + 54.4\,j + 0\,k$$

and $\bar{a} = 0i - 32.2j + 0k$

$$\therefore \quad \bar{v} \times \bar{a} = \begin{vmatrix} i & j & k \\ 50 & 54.4 & 0 \\ 0 & -32.2 & 0 \end{vmatrix} = (-1610)\,k$$

$$\therefore \quad |\bar{v} \times \bar{a}| = 1610$$

$$\rho = \frac{v^3}{|\bar{v} \times \bar{a}|} = \frac{73.89^3}{1610} = 251 \text{ ft.}$$

Problem 38. : A particle has an initial velocity of 60 m/s up to the right at a slope of 0.75. The components of acceleration are constant at $a_x = -3.6\ \text{m/s}^2$ and $a_y = -6\ \text{m/s}^2$. Compute the radius of curvature at the start and at the top of the path.

at $t = 10 \quad a_x = -4.7 \text{ fps}^2$

Differentiating equation (ii)

$$a_y = \frac{dv_y}{dt} = 10 - 4t \quad \text{at } t = 10$$

$$a_y = -30 \text{ fps}^2$$

$$\bar{a} = -4.7i - 30j$$

$$\bar{v} \times \bar{a} = 68.37i \times (-4.7i - 30j)$$

$$\bar{v} \times \bar{a} = 2051.1k \quad \text{as } i \times j = k$$

$$|\bar{v} \times \bar{a}| = 2051.1$$

$$r = \frac{v^3}{|\bar{v} \times \bar{a}|} = \frac{319552.62}{2051.1} = 159.82 \text{ ft.}$$

Problem 41. : Using the data of previous problem, determine the radius of curvature of the path of the particle at t = 12 sec.

Solution : $v_x = 100 - (12)^{3/2} = 58.43$ fps

$$v_y = 100 + 10t - 2t^2$$

$$\bar{v} = 58.43i - 68j$$

$$v = [58.43^2 + 68^2]^{1/2} = 89.65 \text{ fps}$$

$$v_y = -68 \text{ fps}$$

at $t = 12$ sec.

$$a_x = -\frac{3}{2} t^{1/2} = -5.2 \text{ f/sec}^2$$

$$a_y = 10 - 4t = -38 \text{ f/sec}^2$$

$$\therefore \bar{a} = -5.2i - 38j$$

$$a = [a_x^2 + a_y^2]^{1/2} = 38.39 \text{ fps}^2$$

$$\bar{v} \times \bar{a} = \begin{vmatrix} i & j & k \\ 58.43 & -68 & 0 \\ -5.2 & -38 & 0 \end{vmatrix}$$

$$\bar{v} \times \bar{a} = i \begin{vmatrix} -68 & 0 \\ -38 & 0 \end{vmatrix} - j \begin{vmatrix} 58.3 & 0 \\ -5.2 & 0 \end{vmatrix} + k \begin{vmatrix} 58.4.5 & -68 \\ -5.2 & -38 \end{vmatrix}$$

$$= 0 - 0 + k\,(-2220.34 - 353.6) = -2573.94\,k$$

$$|\bar{v} \times \bar{a}| = 2573.94$$

$$r = \frac{v^3}{|\bar{v} \times \bar{a}|} = \frac{89.65^3}{2573.94} = 280 \text{ ft.}$$

Problem 42. : The rectangular components of acceleration for a particle are $a_x = 3t$ and $a_y = 30 - 10t$ where a is in fps^2. If the particle starts from rest at the origin, find the radius of curvature of the path at $t = 2$ sec.

Solution : $a_x = 3t$; $a_y = 30 - 10t$

at $t = 2$ sec.

$a_x = 6 \text{ fps}^2$; $a_y = 10 \text{ fps}^2$

$$\int a_x\, dt = \int 3t\, dt = v_x = \frac{3t^2}{2}$$

$$\int a_y = \int (30 - 10\,t)\, dt = v_y = 30t - \frac{10t^2}{2}$$

at $t = 2$ sec. ; $v_y = 40$ fps , $v_x = 6$ fps

$$\theta = \tan^{-1}\frac{v_y}{v_x} = \tan^{-1}\frac{40}{6} = 81.5^o$$

$$v = [v_x^2 + v_y^2]^{1/2} = 40.45 \text{ fps}$$

$$a_n = a_y \cos\theta - a_x \sin\theta$$

$$= 10 \cos 81.5 - 6 \sin 81.5 = -4.45 \text{ fps}^2$$

$$\rho = \frac{v^2}{a_n} = \frac{40.45^2}{4.456} = 367.2 \text{ ft}$$

Problem 43. : The position vector of a particle is given by

$r = 2t^2\,\hat{i} + 10t\,\hat{j} + \frac{1}{3}t^3\,\hat{k}$ where r is in feet and t is in seconds. Determine the normal and tangential components of acceleration and the principal radius of curvature of the path of the particle at $t = 3$ sec.

Solution : $r = 2t^2\, i + 10t\, j + \frac{1}{3}t^3\, k$

at t = 3 sec.

$x = 2t^2$	$v_x = 4t$	$a_x = 4$	$v_x = 12$	4
$y = 10t$	$v_y = 10$	$a_y = 0$	$v_y = 10$	0
$z = \frac{t^3}{3}$	$v_z = t^2$	$a_z = 2t$	$v_z = 9$	6

$$\bar{v} = 12\hat{i} + 10\hat{j} + 9\hat{k}$$

$$v = 18.927 \text{ fps}$$

$$\bar{a} = 4\hat{i} + 6\hat{k}$$

$$a = 7.21 \text{ fps}^2$$

$$\bar{v} \times \bar{a} = \begin{vmatrix} i & j & k \\ 12 & 10 & 9 \\ 4 & 0 & 6 \end{vmatrix} = 80.59$$

$$r = \frac{v^3}{|\bar{v} \times \bar{a}|} = \frac{(18.027)^3}{80.597} = 72.68 \text{ ft.}$$

$$a_n = \frac{v^2}{r} = 4.47 \text{ fps}^2$$

$$a_t = [a^2 - a_n^2]^{1/2} = 5.66 \text{ fps}^2$$

KINETICS OF PARTCLES :

Consider a block moving right ward on horizontal plane. The FBD of block with different forces acting on H is shown in fig.

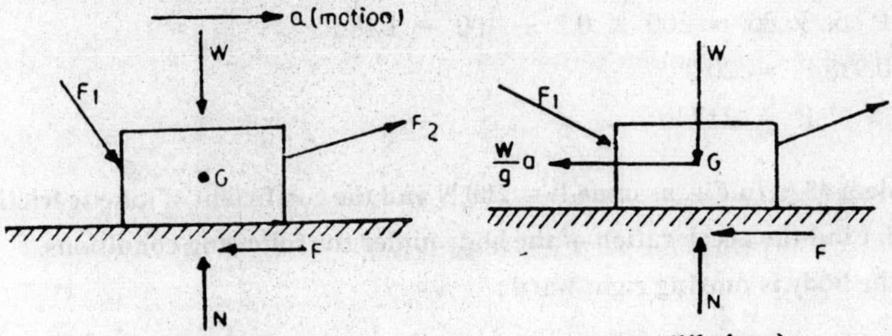

(In motion) (In equilibrium)

If the block is not in motion, to apply easy equation of static equilibrium, a effective of all force $\frac{W}{g}$ a is added through center of gravity in the direction opposite to acceleration as shown in second fig.

The force $\frac{W}{g}$ a is called inertia reaction.

Problem 44. : Determine the force P that will give the body in fig. an acceleration of 0.20 g fps^2. The coefficient of kinetic friction is 0.20.

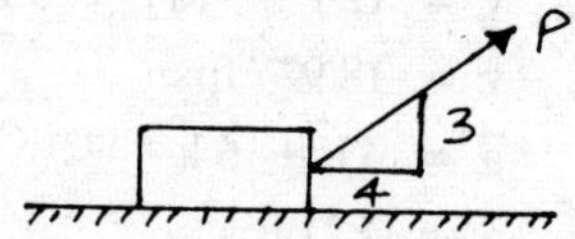

Solution : W = 500 , u = 0.20

$a = 0.20 \text{ g fps}^2$

To keep it in a state of equilibrium, inertia force of 500 × $\frac{a}{9}$ should be their.

$\Sigma F_y = 0$

$-W + P \sin 36.86 + N = 0$

$N = 500 - 0.59 P$ ...... (i)

We know that $F = uN$

$= 0.2 \times (500 - 0.59 P)$

$F = (100 - 1.18 P)$ ..... (ii)

$\Sigma F_x = 0$

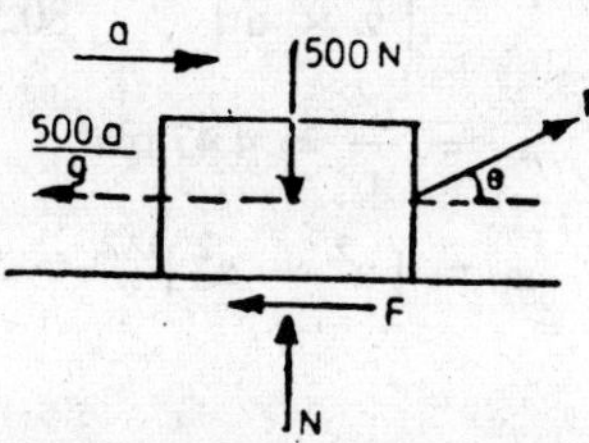

i.e. $P_x - 500 \frac{a}{g} - F = 0$

$P \cos 36.86 - \frac{500 \times 0.20 g}{g} - (100 - 1.18 P) = 0$

$P \cos 36.86 = 500 \times 0.2 + 100 - 1.18 P$

$0.918 P = 200$

$P = 217 \text{ lb}$

Problem 45. : In Fig. assume P = 200 N and the coefficient of kinetic friction is 0.5. Find the acceleration of the body under the following conditions :

(a) the body is moving right ward ;

(b) the body is moving left wart and

(c) the body starts from rest.

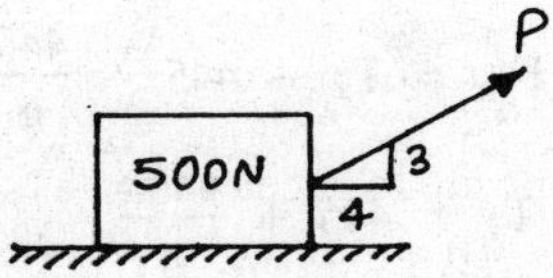

Solution :

(a) From the freebody diagram,

when it is moving right ward,

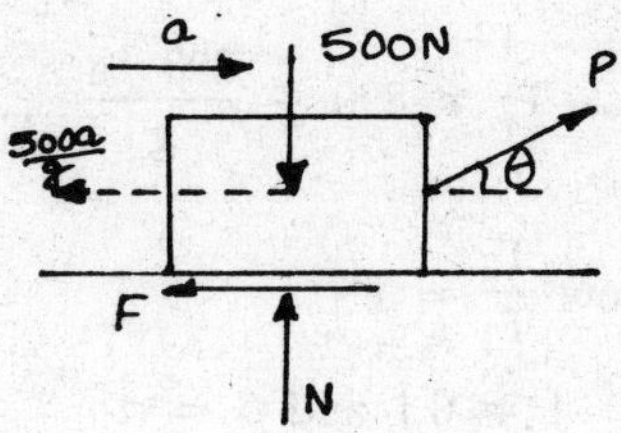

$$\theta = \tan^{-1}\frac{3}{4} = 36.87^o$$

$$\Sigma F_y = 0$$

$$N + 200 \sin 36.87 - 500 = 0$$

$$N = 500 - 200 \times 0.6 = 380 \text{ N}$$

Frictional force $F = 0.5 \times 380 = 190$ N

$$\Sigma F_x = 200 \cos 36.87 - 190 - \frac{500 \times a}{9.8} = 0$$

$$\frac{500 \times a}{9.8} = 200 \times 0.8 - 190 = -30$$

$$\therefore \quad a = -0.588 \text{ m/s}^2$$

$$= 0.588 \text{ (left word)}$$

(b) From FBD of the block moving left ward

$$\Sigma F_x = \frac{500 \times a}{g} + 190 + 200 \cos 36.87$$

$$\therefore \quad a = -6.86 \text{ m/s}^2$$

$$= 6.86 \text{ m/s}^2 \text{ (rightward)}$$

(c) As the body starts from rest the only applied force is 200 cos 36.87 = 160 N and frictional force is 190 N. As frictional force is greater than the applied force, body will not move.

Problem 46. : In Fig. assume A weight 890 N and B weight 445 N. Determine the acceleration of the bodies if the coefficient of kinetic friction is 0.1 between the cable and the drum.

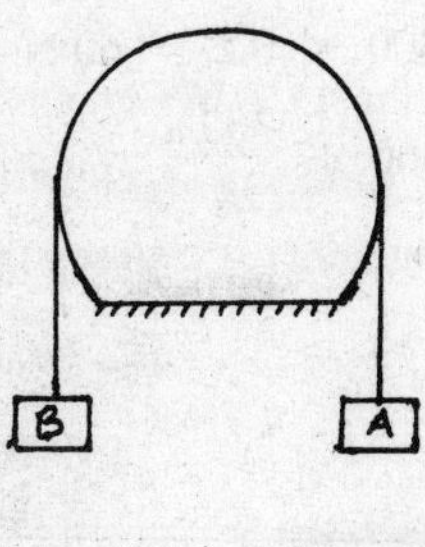

Solution : Let acceleration of A = Acceleration of B = a

From the free body diagram

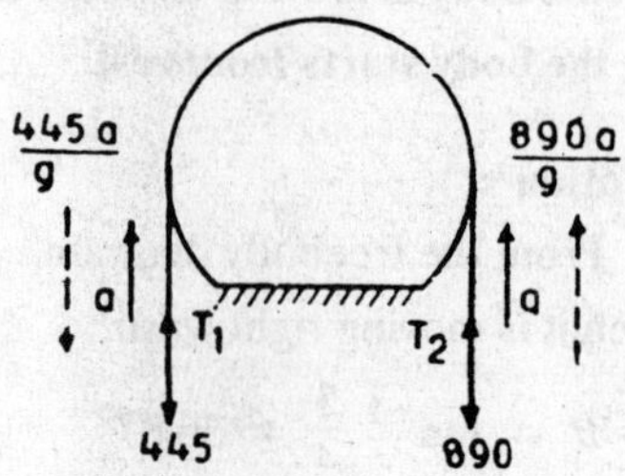

$$F_{yB} = T_1 - 445 - \frac{445}{g} a = 0$$

$$\therefore \; T_1 = 445 + \frac{445a}{g} \quad \ldots\ldots (1)$$

$$\Sigma F_{yA} = T_2 - 890 + \frac{890a}{g} = 0$$

$$\therefore \; T_2 = 890 - \frac{890 \cdot a}{g} \quad \ldots\ldots (2)$$

Now $\frac{T_2}{T_1} = e^{f\beta}$

$f = 0.1$ and $\beta = \pi$

Substituting the values from equation (1) and (2) in the above eqn.

$$\frac{890\,g - 890\,a}{445\,g + 445\,a} = e^{0.1 \times \pi} = 1.369$$

$$\Rightarrow 2g - 2a = 1.369\,(g + a)$$

$$\therefore \; 3.359\,a = 0.631\,g$$

$$\therefore \; a = 1.8355 \text{ m/s}^2$$

Problem 47. : Determine acceleration of bodies in fig., if coefficient of kinetic frcition is 0.2 at all contact surfaces. Body A weighs 200 N and B weighs 300 N.

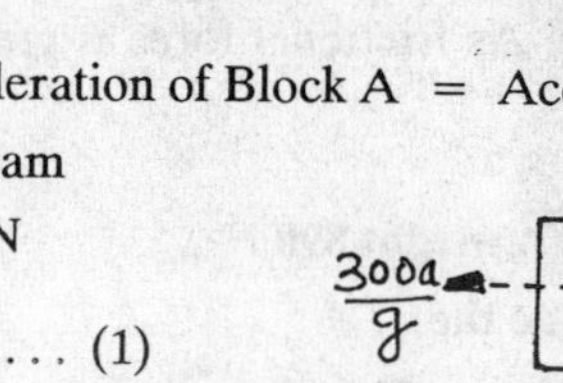

Solution : Let the acceleration of Block A = Acceleration of Block B = a

From the freebody diagram

$F = 300 \times 0.2 = 60$ N

$$T_1 = 60 + \frac{300\,a}{g} \quad \ldots\ldots (1)$$

Solution : From the figure $\tan\theta = \frac{3}{14}$

$\therefore \quad \theta = 36.86^o$

$$N_A = 400 \cos 36.86$$
$$= 400 \times 0.8 = 320 \text{ N}$$
$$N_B = 600 \cos 36.86$$
$$= 600 \times 0.8 = 480 \text{ N}$$
$$F_A = \mu N_A$$
$$= 0.3 \times 320 = 96 \text{ N}$$
$$F_B = \mu N_B$$
$$= 0.10 \times 480 = 48 \text{ N}$$

Consider Body A

$$400 \sin 36.86 + S_1 = \frac{W}{g} a_A + P + \mu \cdot N_A$$

$$240 + S_1 = \frac{400}{9.81} a_A + 200 + 96$$

$$\therefore \quad S_1 = 56 + 40.77\, a_A$$

Consider whole as a system

$$W_A \sin\theta + W_B \sin\theta - P - \mu N_A - \mu N_B = \frac{W_A + W_B}{g}$$

$$400 \sin 36.86 + 600 \sin 36.86 - 200 - 96 - 48 = \frac{400 + 600}{9.81} a$$

$$\therefore \quad a = 2.51 \text{ m/s}^2$$

Let S be the force in the spring which is in compression

$$S + W_A \sin 36.86 - P - \mu N_A = \frac{W_A}{g} a$$

$$S + 240 - 200 - 96 = \frac{400}{9.81} \times 2.511$$

$$\therefore \quad S = 158.40 \text{ N}$$

Problem 50. : Three bodies each of weight W are arranged as shown in figure. The coefficient of kinetic friction between B and C is 0.30 and 0.10 between C and ground. How far does C move when B has moved 's' m relative to C.

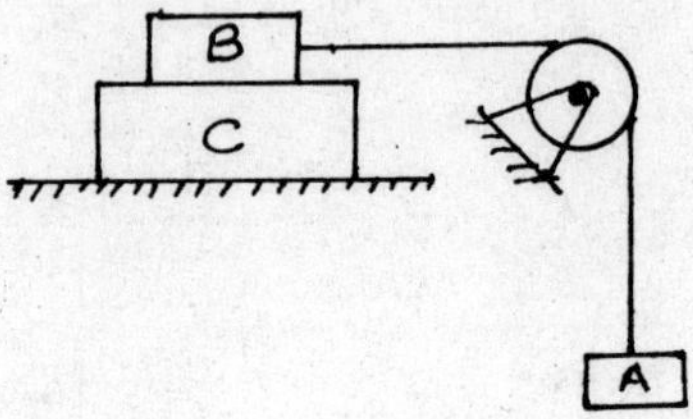

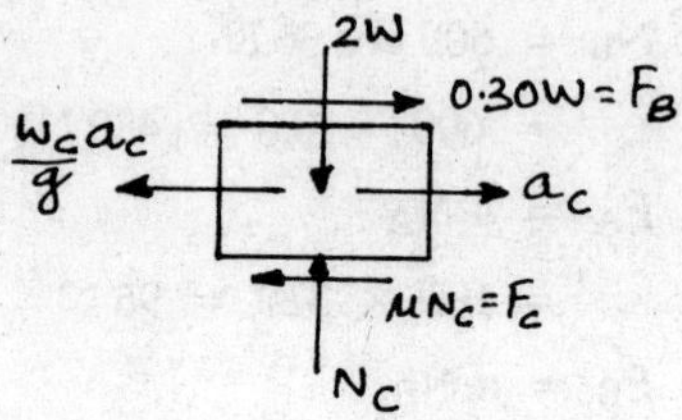

Solution : Consider FBD of C

$$N_C = 2W$$

$$F_C = \mu N_C = 0.1 \times 2W = 0.2\,W$$

$$F_B - F_C = \frac{W_C}{g} a_c$$

$$\therefore \quad 0.30\,W - 0.20\,W = \frac{W}{9.81} a_c$$

$$\therefore \quad a_c = 0.981\ \text{m/s}^2$$

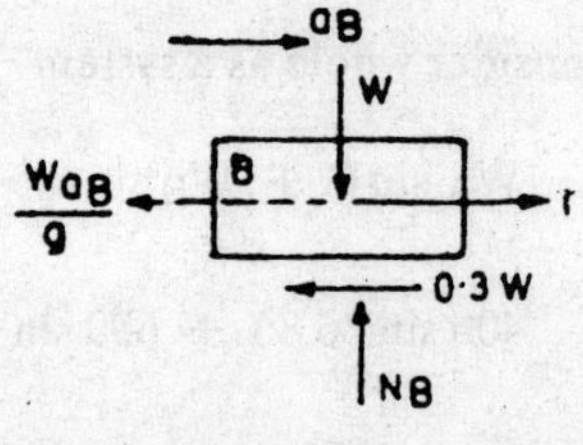

Consider FBD of Block B

$$T - F_B = \frac{W}{g} a_B$$

$$\therefore \quad T = \frac{W}{9.81} a_B + 0.30\,W \qquad \ldots\ldots (1)$$

Consider free body diagram of A

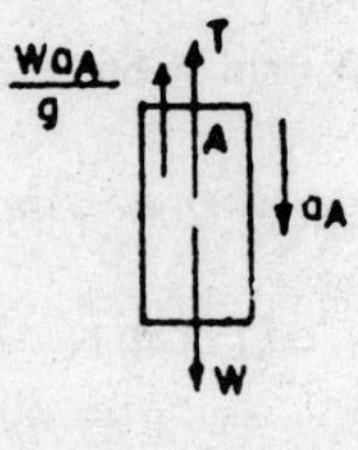

$$W - T = \frac{W}{g} a_A$$

$$= \frac{W}{g} a_B \quad (\text{As } a_A = a_B)$$

$$T = W - \frac{W}{g} a_B$$

$$T_2 = 200 - \frac{200\ a}{g} \quad \ldots\ldots (2)$$

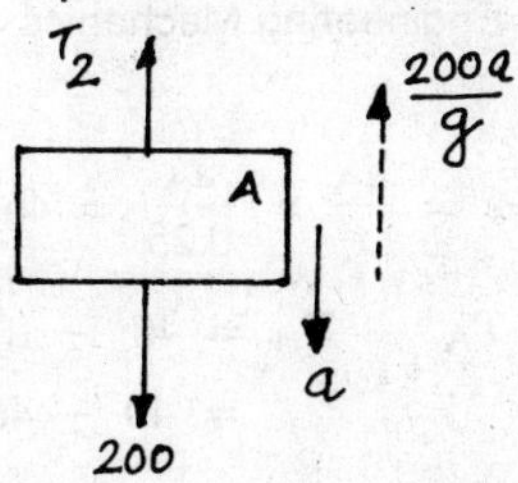

But $\frac{T_2}{T_1} = e^{f\beta}$

$f = 0.2\ ,\ \beta = \frac{\pi}{2}$

Substituting to values of T_1 and T_2 from equation (1) and (2) in above eqn.

$$\therefore \quad \frac{200\ g - 200\ a}{60g + 300a} = e^{0.2 \times \pi/2} = 1.369$$

$$200\ g - 200\ a = 1.369\ (60g + 300a)$$

On solving we get

$$a = 1.891\ m/s^2$$

Problem 48. : A block weighing 3000 N rest on an incline plane of 30°. It is connected by a pulley of weight 800 N and radius 25 cm at the other end of which is a 2000 N weight is attached. The system is released from rest. If the frictionless torque of the pulley is 10 N-m. Determine acceleration of the block. Coefficient of friction between the block A and plane is 0.10.

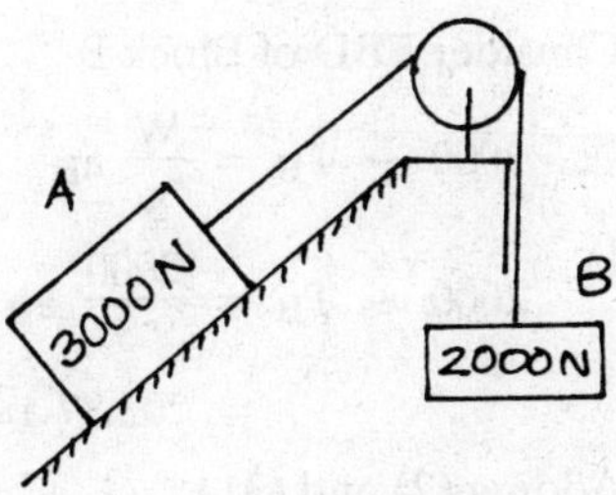

Solution : Consider FBD of the pulley r = 0.25 m

Inertia force for the pulley $I = \frac{800}{9.81} \times \frac{0.25^2}{2} = 2.548$

Total torque $T = I\alpha = 2.548\,\alpha$

$\therefore$ Frictional torque T – Frictionless torque

$$= 2.548\,\alpha - 10$$

$$\therefore \quad (T_B - T_A) \times r = 2.548\,\alpha - 10$$

$$(T_B - T_A) \times 0.25 = 2.548\,\alpha - 10$$

$$\therefore \quad T_A - T_B = 40 - 10.193\,\alpha$$

$$\alpha = \frac{a_A}{r} = \frac{a_A}{0.25} = 4a_A$$

$$\therefore \; T_A - T_B = 40 - 10.19 \times 4a_A$$

$$= 40 - 40.76\, a_A \qquad \ldots\ldots (1)$$

Consider FBD of Block A

$$N = 3000 \cos 30 = 2598.07 \text{ N}$$

$$T_A - 3000 \sin 30 - \mu N = \frac{W_A}{g} \times a_A$$

$$T_A - 1500 - 0.10 \times 2598.07 = \frac{3000}{9.81} a_A$$

$$T_A - 1759.80 = 305.81\, a_A \qquad \ldots\ldots (2)$$

Consider FBD of Block B

$$2000 - T_B = \frac{W}{g} a_B$$

$$2000 - T_B = \frac{2000}{9.81} a_B \quad \text{As } a_A = a_B$$

$$= 203.87\, a_A \qquad \ldots\ldots (3)$$

Adding (2) and (3)

$$T_A - 1759.80 + 2000 - T_B = 305.81\, a_A + 203.87\, a_A$$

$$(T_A - T_B) + 240.19 = 509.68\, a_A$$

Substuting the value from (1) in above equation

$$40 - 40.76\, a_A + 240.19 = 509.68\, a_A$$

$$550.44\, a_A = 280.19$$

$$\therefore \; a_A = 0.509 \text{ m/s}^2$$

Hence acceleration of Blocks $= 0.509 \text{ m/s}^2$

Problem 49. : Two bodies in figure are separated by a spring. Their motion down the incline is resisted by a force P = 200 N. The coefficient of friction is 0.30 under A and 0.10 under B. Determine the force in the spring.

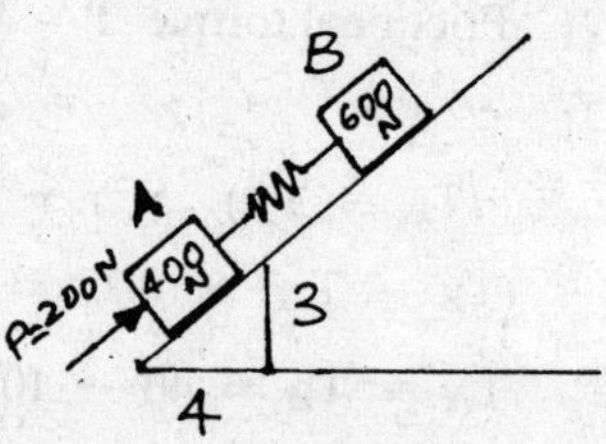

From equation (1)

$$W - \frac{W}{g} a_B = \frac{W}{g} a_B + 0.30\,W$$

$$\frac{2a_B}{g} = (1 - 0.3) = 0.70$$

$$\therefore \quad 2a_B = 0.70 \times 9.81$$

$$\therefore \quad a_B = 3.433 \ m/s^2$$

Relative acceleration of B with respect to C

$$= a_B - a_C$$

$$= 3.433 - 0.981 = 2.452$$

$$S_B - S_C = ut + \frac{1}{2}(a_B - a_C)\,t^2$$

$$= \frac{1}{2} \times 2.452 \times t^2$$

$$\therefore \quad S = 1.226\,t^2$$

$$\therefore \quad t = \left[\frac{S}{1.226}\right]^{1/2}$$

$$S_C = \frac{1}{2} a_C \times t^2$$

$$= \frac{1}{2} \times 0.981 \times \frac{S}{1.226}$$

$$\therefore \quad S_C = 0.40\,S$$

Problem 51. : If the coefficient of kinematic friction is 0.25 under each body in the system shown in figure. How far and in what direction will body B move in 5 sec. starting from rest ?

Solution : Considering FBD of A

$$\tan\theta = \frac{3}{4} \quad \therefore \quad \theta = 36.86^o$$

$$N_A = W_A \cos 36.86$$

$$= 2720 \times 0.8 = 2176 \text{ N}$$

$$\Sigma F = m \cdot a_A$$

$$T_1 - F_A - W_A \sin 36.86 = \frac{W}{g} a_A$$

$$T_1 - \mu N_A - W_A \sin 36.86 = \frac{W}{g} a_A$$

$$T_1 - 0.25 \times 2176 - 2720 \sin 36.86 = \frac{2720}{9.81} a_A$$

$$T_1 = 2176 + 277.268\, a_A \qquad \ldots\ldots (1)$$

Considering equilibrium of Block B

$$S_A = 2S_B$$

$$a_A = 2a_B$$

$$W_B - T_2 = \frac{W_B}{g} a_B$$

$$5440 - 2T_1 = \frac{5440}{9.81} a_B$$

$$\therefore \quad T_1 = 2720 - \frac{2720}{9.81} a_B$$

$$\therefore \quad 2176 + 277.268\, a_A = 2720 - \frac{2720}{9.81} a_B$$

$$2176 + 277.268 \times 2\, a_B = 2720 - \frac{2720}{9.81} a_B$$

$$\therefore \quad a_B = \frac{544}{831.804} = 0.654$$

$$S_B = ut + \frac{1}{2} a_B\, t^2$$

$$= 0 + \frac{1}{2} \times 0.654 \times 5^2 = 8.176 \text{ m}$$

Chapter Nine

WORK ENERGY METHOD

Work Energy Equation For Translation :

This is an approach in dynamics which bypasses acceleration and directly relates forces, displacement and velocity. This method is based on integrating the fundamental equations $\vec{R} = m\vec{a}$

The W-E equation for a particle is :

$$RW = \frac{W}{2g}(v^2 - v_o^2) = \Delta\,(KE)$$

This states that the resultant work done upon a particle is equal to change in kinetic energy.

When force and displacement are in the same direction then work done is positive, otherwise negative. While applying this method, only the work done by active external forces need to be considered. The work done by internal forces in cords connecting a system of bodies is not considered because it is automatically eliminated. Also forces perpendicular to the displacement do not work.

Work Energy Equation Applied to Connected System :

Steps :

(1) Determine direction of motion.
(2) Draw active force diagram for the system.
(3) Determine kinematic relations among the bodies of system.
(4) Use W-E equation.

Note : The work done in deforming a spring from its free length to an extension (or compression) of 'S' units is

$$RW = \frac{1}{2} kS^2$$

where k is stiffness of spring in lb/ft. Here average force is considered for computation of work.

Problem 1. : After the block in fig. has moved 10 ft. from rest, the constant force P is removed. Find the velocity of the block when it returns to its initial position.

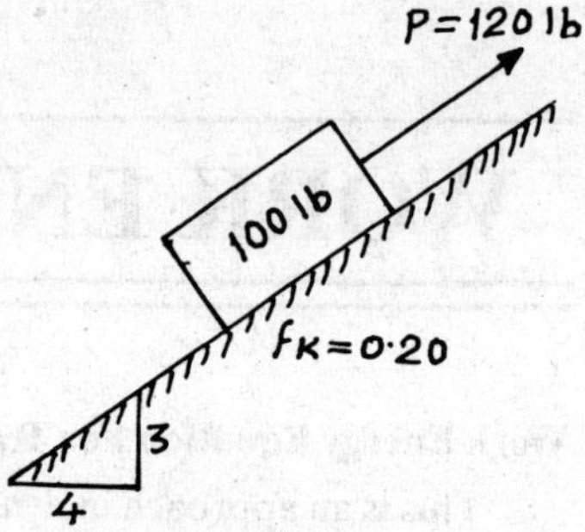

Solution :

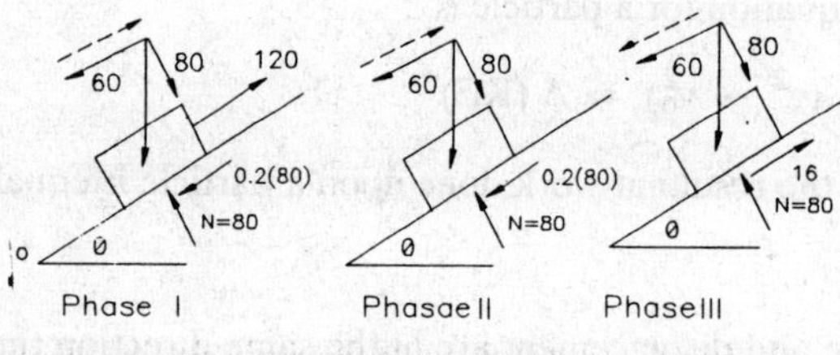

Dotted arrow show the direction of motion.

Phase I : Here body moves up by distance 10 ft. after which 120 lb force is removed.

Using Work–Energy equation we get,

$$RW = \Delta kE : (120 - 60 - 16)\ 10 = \frac{100}{2g} (V_1^2 - 0)$$

$$V_1 = 16.83 \text{ fps..}$$

Phase II : Here body moves along the incline by distance S before returning down. So final velocity for the phase is 0.

$$RW = \Delta KE : (-60 - 16) S = \frac{100}{2g} (0 - 16.83^2)$$

$$S = 5.79 \text{ ft.}$$

Phase III : Here body moves down by distance $(10 + s = 10 + 5.79 = 15.79)$ to pass initial position.

$$RW = \Delta KE : (60 - 16)(15.79) = \frac{100}{2g}(V_2^2 - 0)$$

$$V_2 = 21.15 \text{ fps}^2$$

Problem 2. : A weight of W lb is suspended from a vertical spring fig. whose modulus is k lb / ft. The weight is pulled down s ft from its equilibrium position and then released. Determine its velocity when it returns to its equilibrium position.

Solution :

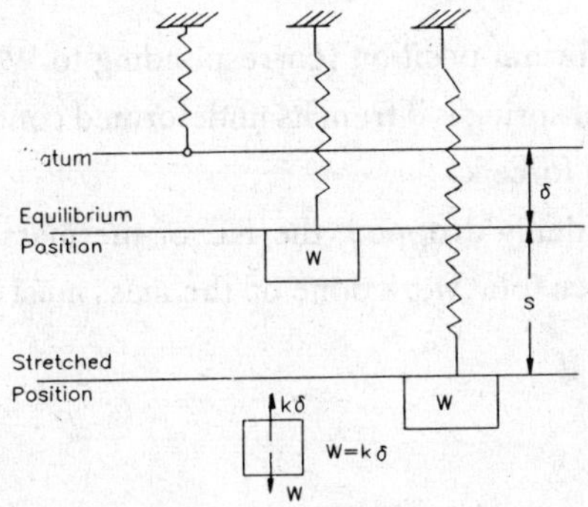

Let δ is extension of spring from free length when weight W is attached.

In equilibrium position : Thus W is balanced by $k\,\delta$

Here

$$W = k\,\delta \qquad \ldots\ldots (1)$$

Stretched Position : For the given problem body is first stretched to stretched position and then released. We have to find velocity V, corresponding to equilibrium position.

Between these 2 positions :

$$RW = \Delta KE :$$

$$\frac{1}{2}kS^2 = \frac{W}{2g}(V^2 - 0)$$

Note that since W is balanced by $k\,\delta$ its work is not included above.

$$V = S\left[\frac{kg}{W}\right]^{1/2}$$

Problem 3. : A weight is dropped from a position just above, but not touching a spring. Show that the maximum deformation produced will be twice that if the same weight were gradually lowered upon the spring.

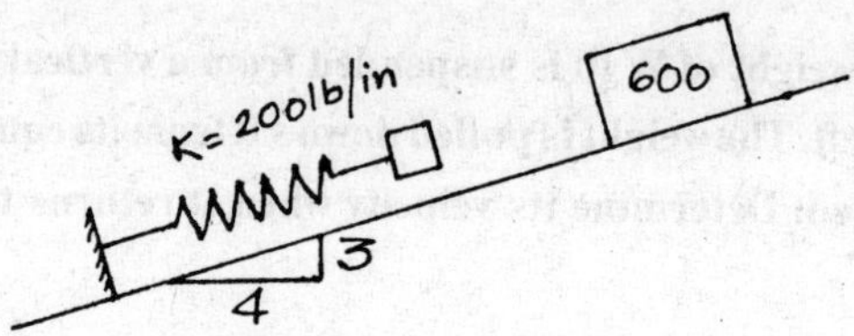

Solution : In equilibrium position (corresponding to W lowered gradually) let the deformation of spring is δ from its undeformed condition and weight W is balanced by spring force $k\delta$

When W is suddenly dropped, the KE of the mass at initial and final position is zero. Hence total work done on the mass must be zero.

i.e. $Wx - \frac{1}{2}kx^2 = 0$

But $W = k\delta$

$$k\delta x - \frac{1}{2}kx^2 = 0$$

$$x = 2\delta$$

Problem 4. : An elevator weighing 4000 lb is being lowered at a constant rate of 10 fps when the hoisting drum is suddenly stopped. If the elastic properties of the supporting cable are such that it is equivalent to a spring with a modulus of 2000 lb/in., determine by how many times the sudden stop of the hoisting drum momentarily increases the cable tension.

Solution : The given situation is modelled as shown in fig.

$$k = 2000 \text{ lb/in} = 24000 \text{ lb/ft}$$

In equilibrium position,

$$W = kS$$

$= 4000 = 24000\,S \qquad S = 0.167$ ft.

Between $V_o = 10$ to $V = 0$

$$RW = \Delta KE : -\frac{1}{2}(24000)\,x^2 = \frac{4000}{2g}(0 - 10^2)$$

$x = 0.7134$ ft.

T = Initial tension $= kS = 0.167\,(2400) = 4000$ lb

T_1 = Tension after sudden stop $= k\,(x + S)$

$= 24000\,(0.7134 + 0.167) = 21274.63$ lb

$$\therefore \frac{T_1}{T} = \frac{21274.63}{4000} = 5.32$$

Problem 5. : A 100–lb weight moves along the smooth rigid guide from A to B as shown in Fig. Find its velocity at B if it starts from rest at A. The free length of the spring is 2 ft and its modulus is 5 lb / in.

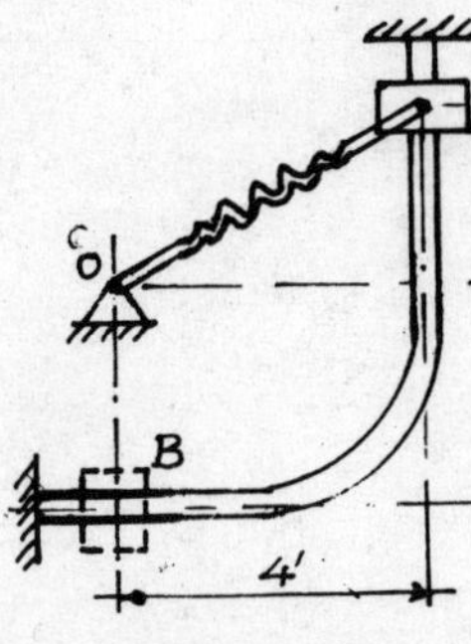

Solution : $K = 5$ lb/in $= 60$ lb/ft

Free length $= 2$ ft

$OA = [\,4^2 + 2^2\,]^{1/2} = 4.47$ ft

$OB = 3$ft.

S_1 = Extension at position A $= 4.47 - 2 = 2.47$ ft.

S_2 = Extension at position B $= 3 - 2 = 1$ ft.

Between A and B :

$$RW = \Delta KE : wh + \left(\frac{1}{2}kS_1^2 - \frac{1}{2}kS_2^2\right) = \frac{W}{2g}(V^2 - V_o^2)$$

$$\therefore \quad 100 \times 5 + \frac{1}{2} 60 (2.47)^2 - \frac{1}{2} \times 60 \times (1)^2 = \frac{100}{2(32.2)} (V^2 - 0)$$

$$V = 20.5 \text{ fps}$$

Problem 6. : A weight W is attached to one end of a stiff rod of length L and negligible weight that is hinged to a horizontal axis at the other end. The rod is released from rest in a horizontal position and allowed to swing freely in a vertical arc. Through what angle must it swing to cause a tension in it of 1.5 W ?

Solution : At any position B, the forces acting on body are shown. Considering dynamic equilibrium.

$$\Sigma F_n = 0$$

$$\frac{W}{g}\frac{V^2}{L} + W \sin\theta - T = 0$$

As $T = 1.5 \, W$

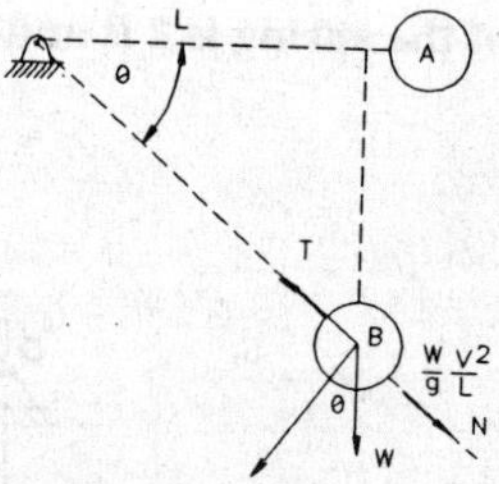

$$\frac{W}{g}\frac{V^2}{L} + W \sin\theta - 1.5\,W = 0$$

or $$\frac{V^2}{gL} = 1.5 - \sin\theta$$

$$V^2 = (1.5 - \sin\theta)\, gL$$

Between A and B :

$$RW = \Delta KE : Wh = \frac{W}{2g}(V^2 - 0)$$

$$W(L \sin\theta) = \frac{W}{2g}(1.5 - \sin\theta)\, gL$$

$$\sin\theta = 0.75 - 0.5 \sin\theta$$

$$\theta = 30^\circ$$

Problem 7. : In fig. by how much should the spring be compressed so that it will cause the 1–lb pellet to trace completely around the frictionless vertical loop ? What force is exerted by the track upon the pellet when it is at position B ?

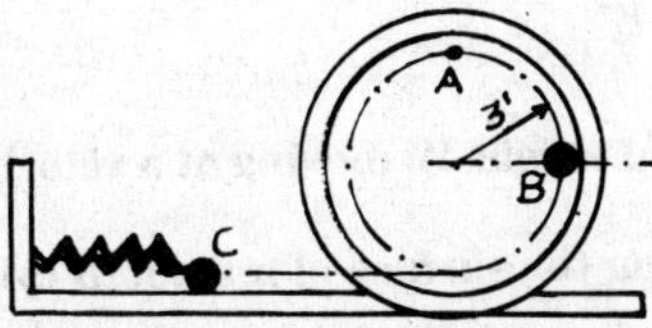

Solution : To travel 1 lb pellet completely, at top position 2, the normal reaction is zero.

Creating dynamic equilibrium at position 2 (top),

$$\Sigma F_n = 0 \; ; \; \frac{W}{g}\frac{V^2}{r} = W$$

$$V^2 = gr = 32.2\,(3) = 96.6$$

$$k = 5 \text{ lb/in} = 60 \text{ lb/ft.}$$

Between 1 and 2 :

Here spring is undeformed. So positive work is done by spring.

$$RW = \Delta KE \; ; \; \frac{1}{2}kS^2 - Wh = \frac{W}{2g}(V^2 - V_o^2)$$

$$\frac{1}{2} \times 60 \times S^2 - 1 \times 6 = \frac{1}{2g}(96.6)$$

or $S = 0.5$ ft

Between 1 and 3 :

$$RW = \Delta KE \; ; \; \frac{1}{2}\,60 \times 0.5^2 - 1 \times 3 = \frac{1}{2g}(V^2 - 0)$$

$$V = 17.02 \text{ fps}$$

At 3, creating dynamic equilibrium,

$$\Sigma F_x = : \; \frac{W}{g}\frac{V^2}{r} - N = 0$$

$$N = \frac{W}{g}\frac{V^2}{r} = \frac{1}{g}\frac{(17.02)^2}{3} = 3 \text{ lb}$$

Problem 8. : A particle of weight W, moving at a velocity of $\frac{1}{2}\sqrt{gr}$ fps at the top, slides vertically along the surface of a smooth cylinder of radius r. The axis of the cylinder is horizontal. Find the vertical distance the particle falls before it leaves the cylinder.

Solution : $V_o = \frac{1}{2}\sqrt{gr}$

$$V_o^2 = 0.25\ gr$$

When particle looses the contact with cylinder at position 2 . N = 0 for that position. Creating dynamic equilibrium at position 2.

$$\Sigma F_n = 0 : \frac{W}{g}\frac{V^2}{r} - W\cos\theta = 0$$

$$V^2 = gr\cos\theta$$

Also $h = r - r\cos\theta$

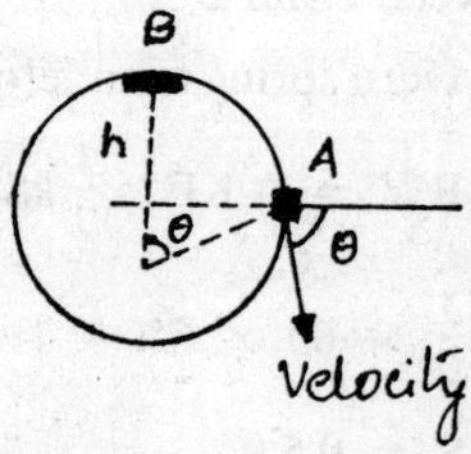

Between 1 and 2 :

$$RW = \Delta KE : Wh = \frac{W}{2g}(V^2 - V_o^2)$$

$$W(r - r\cos\theta) = \frac{W}{2g}(gr\cos\theta - 0.25\ gr)$$

$$1 - \cos\theta = 0.5\cos\theta - 0.125$$

$$\theta = 41.41^o$$

$$h = r - r\cos\theta = r - r\cos 41.41 = 0.25\ r$$

Problem 9. : A particle starts from rest at the top and slides in a vertical plane along the smooth surface of a cylinder of 6–ft radius. The axis of the cylinder is horizontal. Measuring from the center of the cylinder, how far does the particle travel horizontally before it strikes the ground on which the cylinder rests ?

Solution : $V_0 = 0$,

when particle looses contact with cylinder at position 2, N = 0.

Creating dynamic equilibrium at 2,

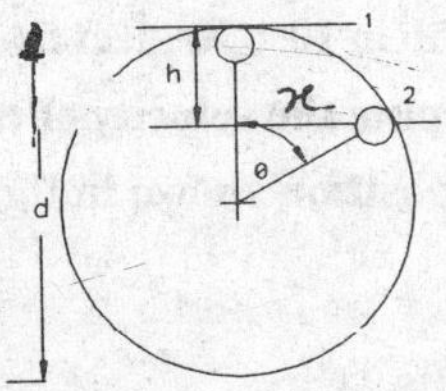

$$\Sigma F_n = 0 : \frac{W}{g}\frac{V^2}{r} - W\cos\theta = 0$$

$$\Rightarrow V^2 = g\, r\cos\theta \qquad \ldots\ldots (1)$$

Also $h = r - r\cos\theta$

Between 1 and 2 :

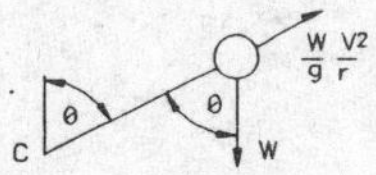

At position 2

$$RW = \Delta KE : Wh = \frac{W}{2g}(V^2 - V_o^2)$$

$$W(r - r\cos\theta) = \frac{W}{2g}(gr\cos\theta - 0)$$

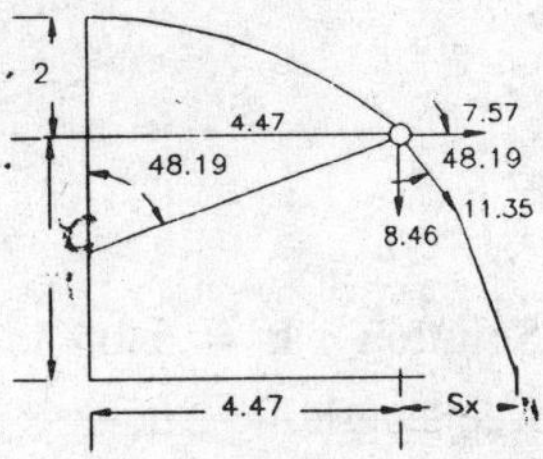

$$1 - \cos\theta = 0.5\cos\theta$$

$$\Rightarrow \theta = 48.19^o$$

$$h = r - r\cos\theta = 6 - 6\cos 48.19 = 2 \text{ ft}$$

$$d - h = 12 - 2 = 10 \text{ ft}$$

$$x = r\sin\theta = 6\sin 48.19 = 4.47 \text{ ft}$$

After leaving the contact with cylinder at point 2, its motion becomes that of projectile.

From (1), $V^2 = gr\cos\theta = 32.2\,(6)\cos 48.19$

$$V = 11.35 \text{ fps}$$

This V is perpendicular to the line joining C and 2

$$V_{ox} = 11.35\cos 48.19 = 7.57 \text{ fps}$$

$$V_{oy} = 11.35\sin 48.19 = 8.46 \text{ fps}$$

$$S_y = V_{oy}\, t + \frac{1}{2} g t^2$$

$$10 = 8.46\, t + \frac{1}{2}(32.2)\, t^2$$

Solving we get, t = 0.57 sec. other value is negative hence discarded.

$$S_x = V_{ox}\, t$$

$$= 7.57\,(0.57) = 4.3 \text{ ft}$$

Horizontal distance travelled from centre

$$= x + S_x = 4.47 + 4.3 = 8.77 \text{ ft}$$

Problem 10.: Neglect friction of the 60–lb collar against its vertical guide and compute the velocity of the collar after it has fallen 7 ft, starting from rest in the position shown in fig. The unstretched length of the spring is 2 ft.

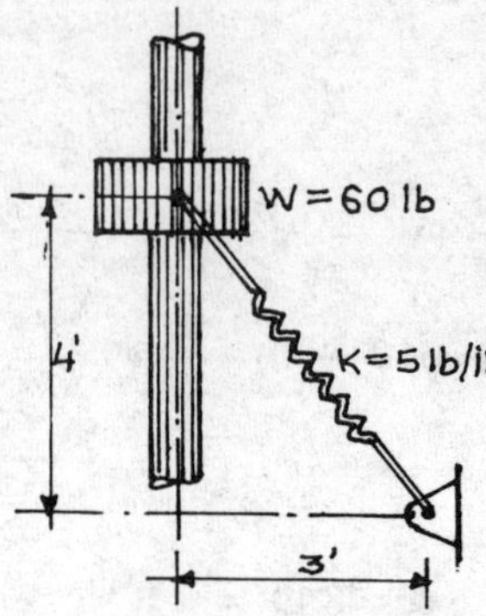

Solution : k = 5 lb / in = 60 lb/ft

free length = 2 ft.

$$OA = [3^2 + 4^2]^{1/2} = 5 \text{ ft}$$

$$OB = [3^2 + 3^2]^{1/2} = 4.24 \text{ ft}$$

Initial extension = $S_1 = 5 - 2 = 3$ ft

Final extension = $S_2 = 4.24 - 2 = 2.24$ ft

$$RW = \Delta KE : Wh + \frac{1}{2} kS_1^2 - \frac{1}{2} kS_2^2 = \frac{W}{2g}(V^2 - V_o^2)$$

$$60 \times 7 + \frac{1}{2} \times 60 \times 3^2 - \frac{1}{2} \times 60 \times 2.24^2 = \frac{60}{2g} V^2$$

V = 24.06 fps

Problem 11.: The car in fig. is moving toward the bumper spring and has a kinetic energy of 100,000 in–lb. The main bumper shield (a – a) is connected to the main spring which has a modulus of 1000 lb/in. The two auxiliary bumper shields (b) are 12 in. behind a – a and are attached to secondary springs, each of which has a modulus of 500 lb / in. Determine the greatest movement of a – a. What percentage of the energy will then be absorbed by the main spring ?

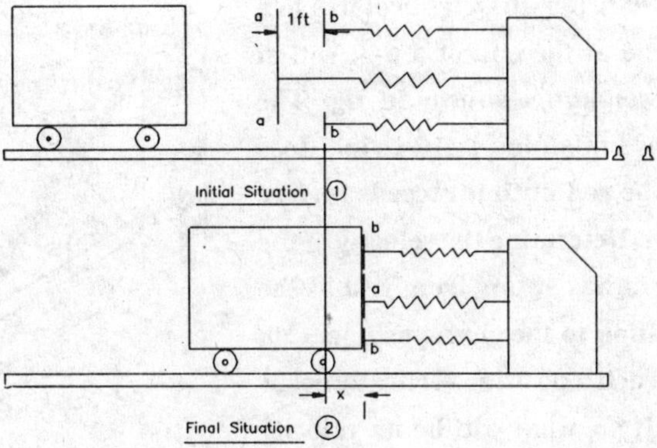

Solution : Δ KE $= -100000$ lb in

$$= -\frac{100000}{12} = 8333.33 \text{ lb/ft}$$

$K_{aa} = 1000$ lb/in $= 12000$ lb/ft

$K_{bb} = 500$ lb/in $= 6000$ lb/ft

When a – a is compressed by (1 + x) ft., bb is compressed by x ft.
Since both springs are deformed, work done will be negative work.

$$RW = \Delta KE : -\frac{1}{2} K_{aa} (1 + x)^2 - \frac{1}{2} K_{bb} x^2 = -8333.33$$

Δ KE is negative since car is brought to rest.

$$-\frac{1}{2}(12000)\ (1 + x)^2 - \frac{1}{2}(6000)\ (x^2) = -8333.33$$

Solving this, $x = 0.17$.

S = Total compression of spring a-a = 1 + 0.17

= 1.17 ft = 14.04 inch.

V_a = energy absorbed by aa

$$= \frac{1}{2} kS^2$$

$$= \frac{1}{2}(12000)\ (1.17)^2 = 8213.4 \text{ lb/ft}$$

$$\% \text{ energy absorbed by a – a} = \frac{\text{energy absorbed by a–a}}{\text{total KE}} \times 100$$

$$= \frac{8213.4}{8333.33} \times 100 = 98.56\ \%$$

Problem 12. : A 100–lb weight rotates in a vertical plane at the end of a 6–ft rod of negligible weight as shown in fig. The spring, whose modulus is 10 lb/in., does not act on the rod until it exceeds its free length of 3 ft. Determine the velocity of the weight after it has swung from rest at the dashed position to the given position. Can the weight reach a position vertically below the pivot ? If so, what will be its velocity then ?

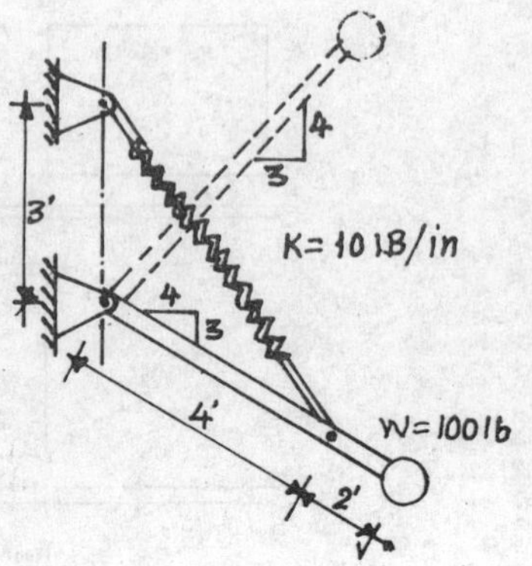

Solution : K = 10 lb / in = 120 lb / ft.

Free length = L = 3 ft.

Cosine Rule :

$$OA^2 = OB^2 + AB^2 - 2\,OB\,AB \cos 36.87$$

$$= 3^2 + 4^2 \;0\; 2\,(3)\,(4) \cos 36.87$$

$$OA = 2.41 \text{ ft}$$

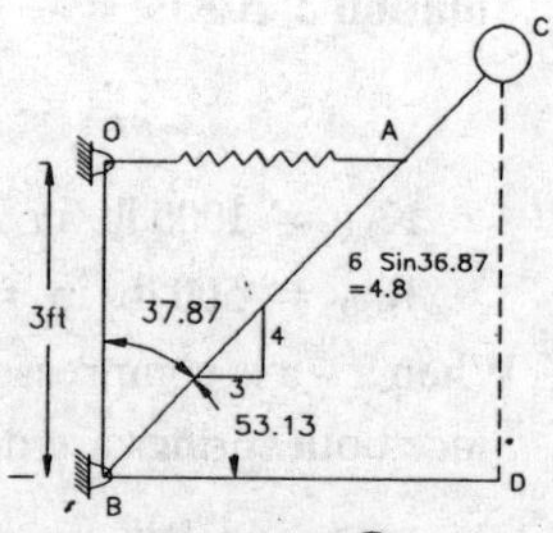

In?itial Position (1)

$$\text{Also } OE^2 = OB^2 + BE^2 - 2\,OB\,BE \cos (90 + 36.87)$$

$$= 3^2 + 4^2 - 2\,(3) \times (4) \cos 36.87^{\circ}$$

$$\therefore \quad OE = 6.28 \text{ ft}$$

$$S_1 = OA - L = 2.41 - 3 = -0.59$$

$$S_2 = OE - Ll = 6.28 - 3 = 3.28$$

Since spring does not act unless its length exceeds 3 ft as mentioned in problem, $S_1 = 0$

Between 1 and 2 :

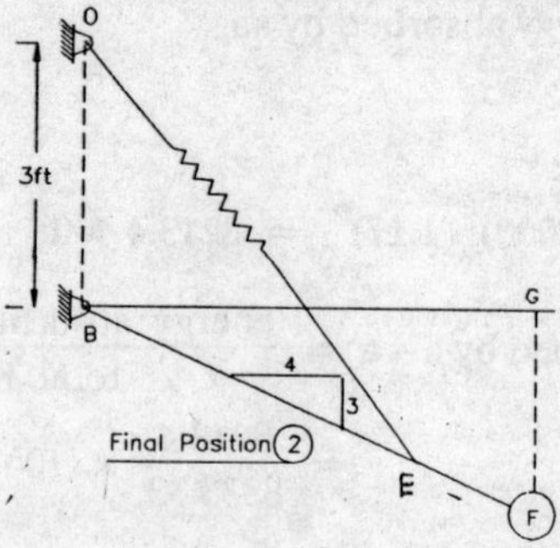

Final Position (2)

$$RW = \Delta KE \; ; \; \left(\frac{1}{2}KS_1^2 - \frac{1}{2}KS_2^2\right) + Wh = \frac{W}{2g}(V^2 - V_o^2)$$

$$h = CD + FG = 4.8 + 3.6 = 8.4 \text{ ft} \quad -\frac{1}{2}\, 120\, (3.28)^2 + 100\, (8.4)$$

$$= \frac{100}{2g}(V^2 - 0)$$

$$V = 11.19 \text{ fps}$$

When spring is stretched to position (3)

$OH = 7$ ft.

$S_3 = OH - L = 7 - 3 = 4$ ft

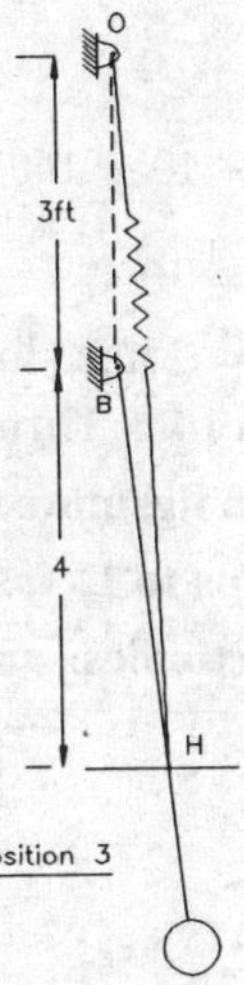

Position 3

Between 1 and 3 :

$$\left(\frac{1}{2}KS_1^2 - \frac{1}{2}KS_3^2\right) + Wh = \frac{W}{2g}(V^2 - V_o^2)$$

$$-\frac{1}{2} \times 120 \times 4^2 + 100 \times (4.8 + 6)$$

$$= \frac{100}{2g}(V^2 - 0)$$

$$V = 8.97 \text{ fps.}$$

So weight can reach to position (3)

So at position (3), $V = 8.79$ fps.

Problem 13. : Determine the constant force P that will give the system of bodies shown in fig. a velocity of 10 fps after moving 15 ft from rest.

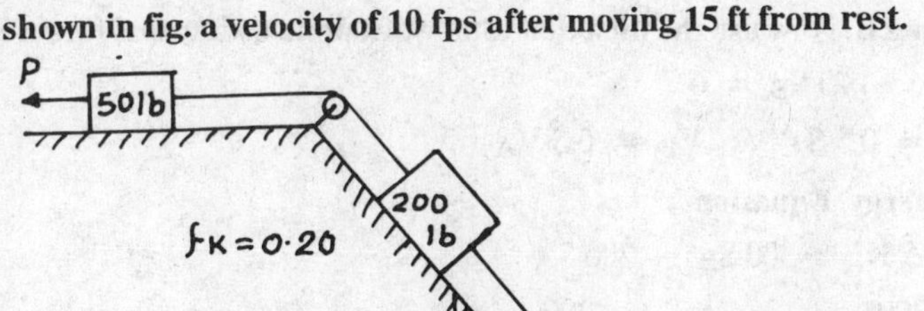

Solution : Direction of motion of system is shown by dotted line. All blocks move with the same velocity V.

$RW = \Delta KE$

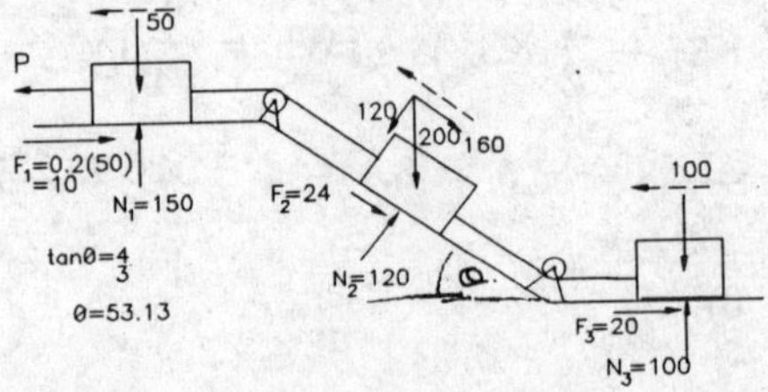

$$(P - 10 - 160 - 24 - 20)\ 15 = \frac{(50 + 200 + 100)}{2g}(V^2 - 0)$$

But $V = 10$

$P = 250.23$ lb.

Problem 14. : Through what distance will body A in fig. move in changing its velocity from 6 fps to 12 fps ? Assume the pulleys to be frictionless and of negligible weight.

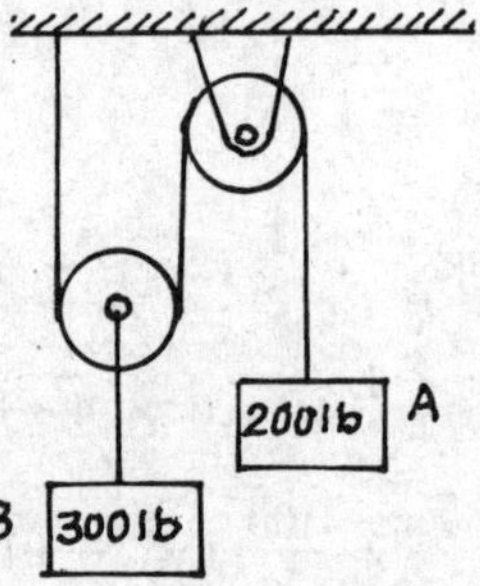

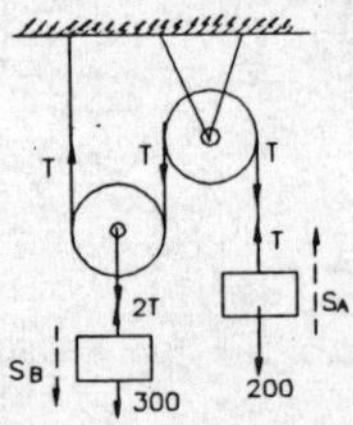

Solution : Let A moves up and B moves down.

Kinematics of Motion : Assuming A to be at rest we get tension 2T in cord supporting B. Now use the idea that total work done by internal forces is zero.

$$TS_A - 2TS_B = 0$$

or $S_B = 0.5\,S_A$ & $V_B = 0.5\,V_A$

Work Energy Equation :

$RW = \Delta KE : 300\,S_B - 200\,S_A$

$$= \frac{300}{2g}(V_B^2 - V_{OB}^2) + \frac{200}{2g}(V_A^2 - V_{OA}^2)$$

$300\,(0.5\,S_A) - 200\,S_A$

$$= \frac{300}{2g}[(0.5\,V_A)^2 - (0.5\,V_{OA})^2] + \frac{200}{2g}(V_A^2 - V_{OA}^2)$$

But $V_A = 12$, $V_{OA} = 6$

$$- 50 S_A = \frac{300}{2g}(6^2 - 3^2) + \frac{200}{2g}(12^2 - 6^2)$$

$$S_A = - 9.22 \text{ ft}$$

Negative sign indicates that A moves down.

Problem 15. : In what distance will body A of fig. attain a velocity of 10 fps, starting from rest ?

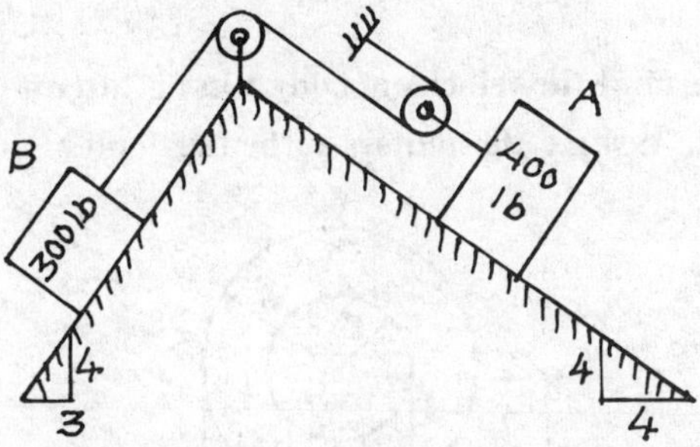

Solution :

Direction of Motion : Let A is at rest with no friction..

Then 2T = 240 and T = 120 lb. (for equilibrium)

Consider block B. The net force acting on B without friction is

(240 – 120) = 120 down the incline.

The friction available is 0.2 (180) = 36 lb which is not sufficient to stop the motion of block B. So bock B moves down the incline and A moves up the incline.

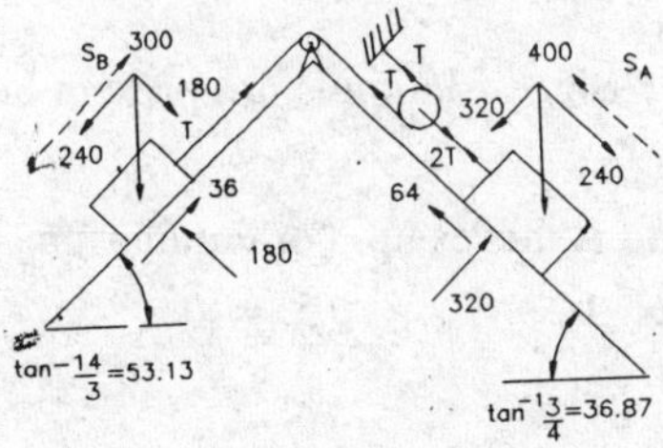

Kinematics of Motion : Using the idea that work done by internal connecting forces is zero.

$$2T\, S_A - T\, S_B = 0$$

$$S_B = 2\, S_A \,, \; V_B = 2V_A$$

Work Energy Equation :

$$RW = \Delta KE : (-240 + 64)\ S_A + (240 - 36)\ S_B$$

$$= \frac{400}{2g}(V_A^2 - V_{OA}^2) + \frac{300}{2g}(V_B^2 - V_{OB}^2)$$

$$V_A = 10 \text{ fps}, \quad V_{OA} = 0$$

$$(-176)\ S_A + (204)\ (2S_A) = \frac{400}{2g}(10^2 - 0) + \frac{300}{2g}[(2(10)^2 - 0]$$

$$S_A = 23.89 \text{ ft}$$

Problem 16. : Determine the velocity of body A in fig. after it has moved 12 ft, starting from rest. Assume the pulleys to be frictionless and of negligible weight.

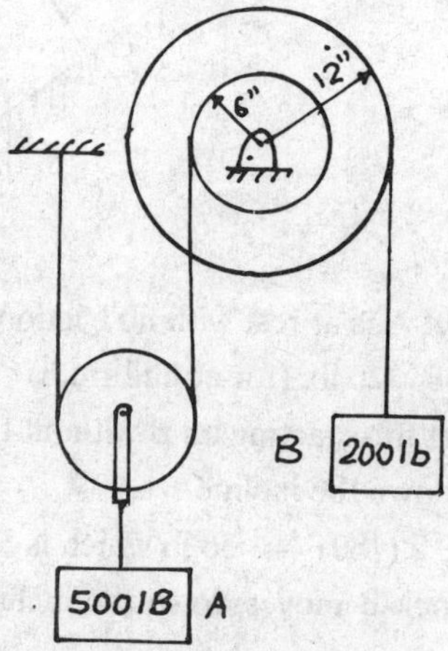

Solution :

Direction of Motion : Since friction is not present here, we can assume any direction of motion.

Let B moves down and A moves up. For equilibrium

$$\Sigma M_o = 0 : T \times 1 - T_1 \times 0.5 = 0$$

$$T_1 = 2T$$

All other values are marked in fig.

Kinematics of Motion : $-T S_A + S_B (4T) = 0$

$$S_B = 4S_A, \quad V_B = 4V_A$$

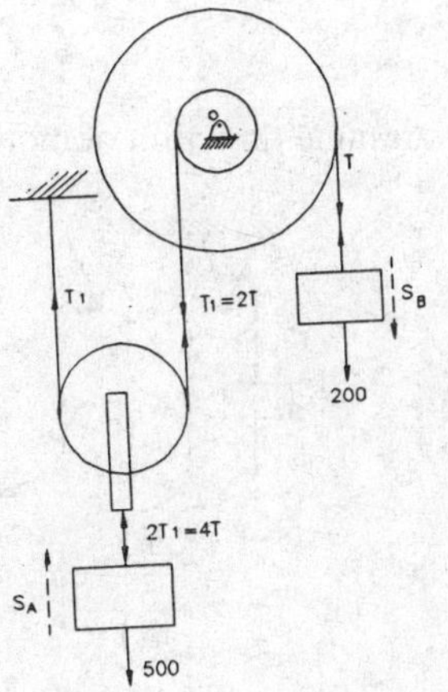

Work Energy Equation :

$$RW = \Delta KE : 200\, S_B - 500\, S_A = \frac{200}{2g}(V_B^2 - 0) + \frac{500}{2g}(V_A^2 - 0)$$

$$200\,(4S_A) - 500\, S_A = \frac{200}{2g}\left[(4V_A)^2\right] + \frac{500}{2g} V_A^2$$

$$300\, S_A = \frac{3700}{2g} V_A^2$$

$$300\,(12) = \frac{3700}{2g} V_A^2$$

$$V_A = 7.92 \text{ fps}$$

Problem 17. : Assume the pulleys in fig. to be frictionless and of negligible weight. Find the velocity of body B after it has moved 10 ft from rest. Then use this result to determine the acceleration of body B.

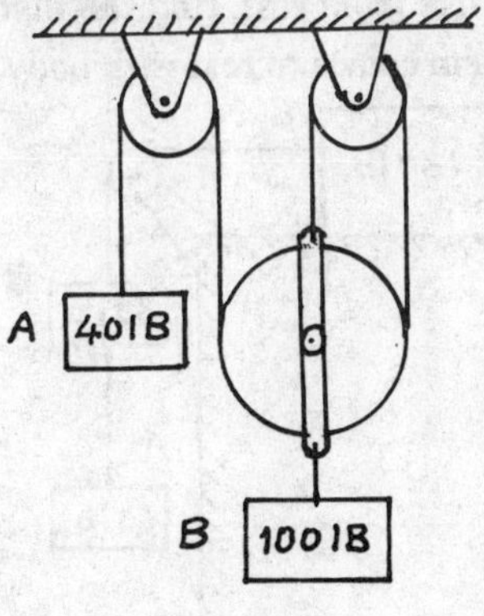

Solution :

Direction of Motion : Assume direction of motion as shown in fig.

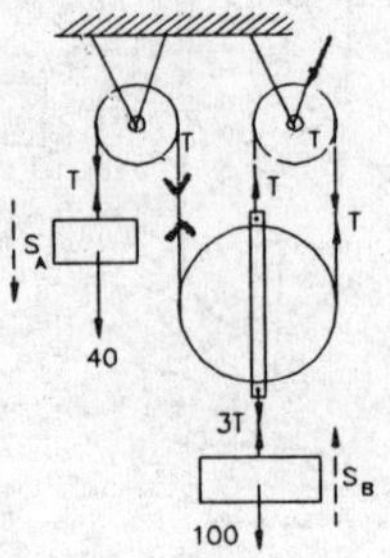

Kinematics of Motion : $3T\,S_B - T\,S_A = 0$

$S_A = 3S_B$, $V_A = 3V_B$

Work Energy Equation : RW $= \Delta$ KE :

$$40\,S_A - 100\,S_B = \frac{40}{2g}(V_A^2 - V_{OA}^2) + \frac{100}{2g}(V_B^2 - V_{OB}^2)$$

$$40\,(3\,S_B) - 100\,S_B = \frac{40}{2g}((3V_B)^2 - 0) + \frac{100}{2g}(V_B^2 - 0)$$

Putting $S_B = 10$

$$200 = \frac{460}{2g}V_B^2 \Rightarrow V_B = 5.29 \text{ fps.}$$

Also, $V_B^2 = V_{OB}^2 + 2\,a_B\,S_B$:

$$5.29^2 = 0 + 2a_B\,(10) \Rightarrow a_B = 1.4 \text{ fps}^2$$

Problem 18. : The system shown in fig. is connected by flexible, inextensible cords. If the system starts from rest, find the distance d between A and the ground so that the system comes to rest with body B just touching A.

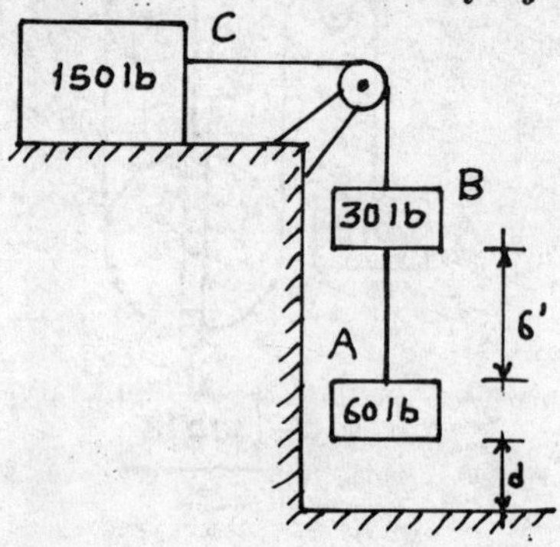

Solution : All blocks move with the same velocity V.

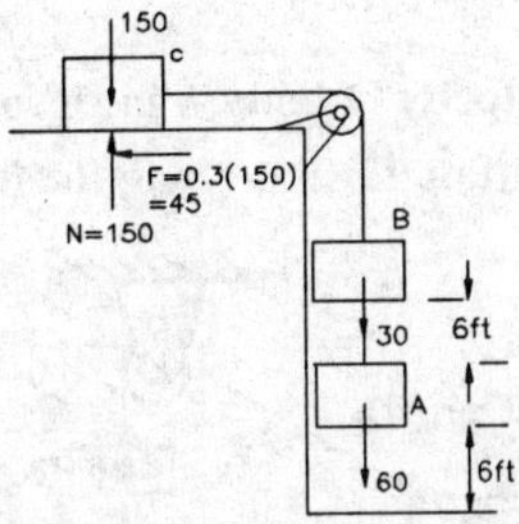

Phase I : All the blocks move by distance d starting from rest.

$$[60 + 30 - 45]\, d = \frac{(60 + 30 + 150)}{2g} (V^2 - 0) \quad \ldots\ldots (1)$$

Phase II : Here blocks B and C move by distance 6m so that B just touches A and system comes to rest.

$$(30 - 45)(6) = \frac{(30 + 150)}{2g} (0 - V^2) \Rightarrow V^2 = g \quad \ldots\ldots (2)$$

Substituting V^2 from (2) in (1) $\Rightarrow$ d = 2.67 ft

Problem 19. : If d = 5 ft in the preceding problem, determine the total distance moved by body C.

Solution : Refer to fig. of previous problem d = 5 ft.

Phase I : All the blocks move by distance 5 ft.

$$(60 + 30 - 45)(5) = \frac{(60 + 30 + 150)}{2g} (V^2 - 0)$$

$$V = 7.77 \text{ fps}$$

Phase II : Blocks B and C move by distance 6 ft and final velocity is V′.

$$(30 - 45)(6) = \left(\frac{30 + 150}{2g}\right) (V'^2 - 7.77^2)$$

$$V' = 5.31 \text{ fps.}$$

Phase III : Block C move by distance x and comes to rest

$$(-45)\, x = \frac{150}{2g} (0 - 5.31)^2$$

$\Rightarrow x = 1.46$ ft

Total distance moved by C = $d + 6 + x = 5 + 6 + 1.46 = 12.46$ ft .

Problem 20. : Find the velocity of body A in fig. after it has moved, starting from rest at the given position, for 9 ft along the frictionless surface.

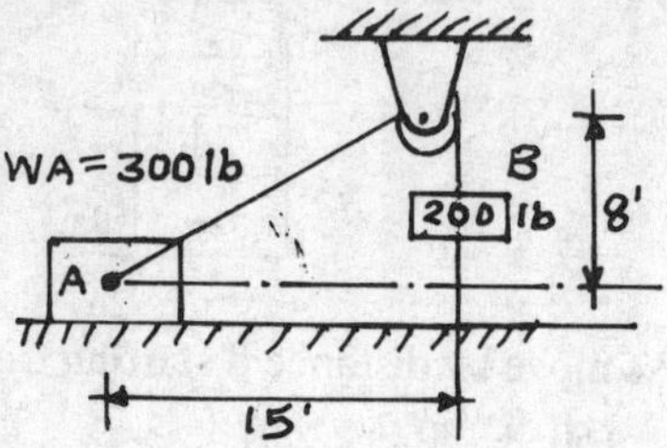

Solution :

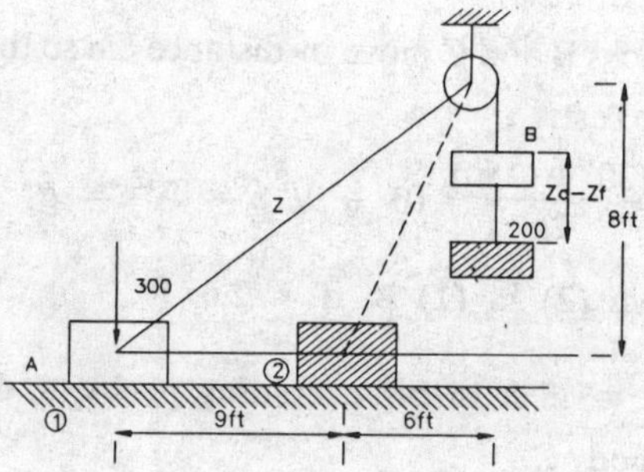

Let (1) be the initial position and (2) be the final position.

Corresponding to (1), $Z_o = [15^2 + 8^2]^{1/2} = 17$

Corresponding to (2), $Z_f = [8^2 + 6^2]^{1/2} = 10$

Distance moved by B $= Z_o - Z_f = 17 - 10 = 7$ft

$$Z^2 = x^2 + 8^2$$

Diff. w.r.t. t, $2Z \dfrac{dZ}{dt} = 2x \dfrac{dx}{dt}$

$$2Z\,V_B = 2x\,V_A$$

$$ZV_B = xV_A \Rightarrow V_B = \frac{x}{z} V_A$$

At (2), $V_B = \dfrac{x}{z} V_A = \dfrac{6}{10} V_A = 0.6\,V_A$

$RW = \Delta KE$:

$$200 \times 7 = \frac{200}{2g}(V_B^2 - 0) + \frac{300}{2g}(V_A^2 - 0)$$

$$1400 = \frac{200}{2g}[(0.6\, V_A)^2] + \frac{300}{2g} V_A^2$$

$V_A = 15.57$ fps

Problem 21. : Two sliders, connected by a light rigid link 10 ft long, move in the frictionless guides shown in fig. If B starts from rest when it is vertically below A, determine the velocity of B when x = 6 ft. Assume $W_A = W_B = 200$ lb and $W_C = 100$ lb .

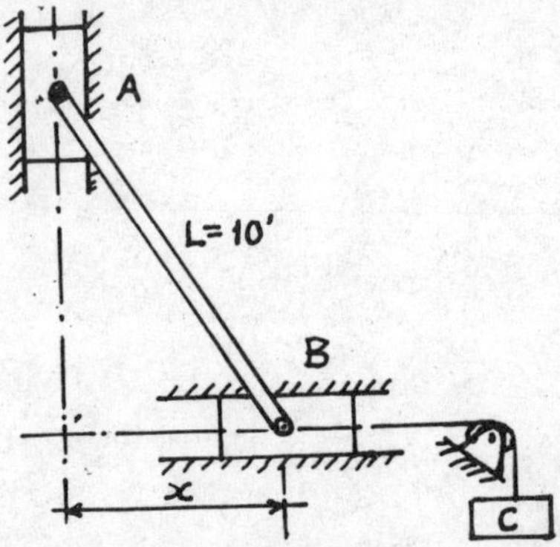

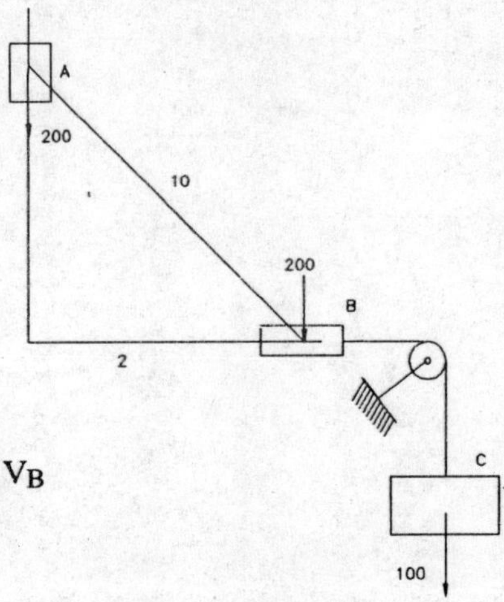

Solution : $V_B = V_C$ .... (1)

Also $x^2 + y^2 = 10^2$ ..... (2)

$2x \cdot \dot{x} + 2y \cdot \dot{y} = 0$

$xV_B + y \cdot V_A = 0$

$V_A = -\frac{x}{y} V_B$ ..... (3)

when $x = 6$, from (2) $y = 8$ ft.

from (3) $\Rightarrow V_A = -\frac{x}{y} V_B = -\frac{6}{8} V_B$

$V_A = -0.75\, V_B$ ..... (4)

at $x = 6$ ft.

Initially B is below A. For final configuration, distance moved by B and hence C is 6 ft. Distance moved by A is $(10 - 8) = 2$ ft.

$$RW = \Delta KE : (200 \times 2) + (100 \times 6)$$

$$= \frac{200}{2g}(V_A^2 - 0) + \frac{200}{2g}(V_B^2 - 0) + \frac{100}{2g}(V_C^2 - 0)$$

From (1) and (4),

$$1000 = \frac{200}{2g}(-0.75\,V_B)^2 + \frac{200}{2g}V_B^2 + \frac{100}{2g}V_B^2$$

$$V_B = 12.49 \text{ fps}$$

Chapter Ten

IMPULSE &MOMENTUM

Introduction :

The problem involving impact, variable mass etc. can be convieniently solved by Impulse Momentum method. Impulse is the produce of force and time i.e. $I = F. dt$. The momentum is product of mass and velocity i.e. $P = mv$. A significantly large force acting on a body for infinitesimally small time can considerably change the velocity of the body. The relationship between force, velocity, time for such problems can be established by Impulse Momentum method.

Impulse Momentum Equation for Translation :

This method is based on integrating the fundamental equation

$\vec{R} = m\vec{a}$ over time limits

The Impulse Momentum equation is

Impulse = Change in Momentum

i.e. $I = \Delta \rho$

The component of resultant linear impulse along any direction will be equal to the difference of components of final momentum and initial momentum in the direction of component of impulse.

Thus,

$$\int_0^t (\Sigma F_x) \cdot dt = \frac{W}{g} (v_x - v_{ox}) \qquad \ldots\ldots (A)$$

when ΣF_x = Sum of components of forces along X-axis.

$\frac{W}{g} (v_x - v_{ox})$ = Change in momentum along X-axis.

Problem 1. : A 4–oz baseball is thrown with a velocity of 60 fps toward a batter. After being struck by the bat B, the ball has a velocity of 160 fps directed as shown in fig. Find the average force exerted on the ball if the impact lasts for 0.02 sec.

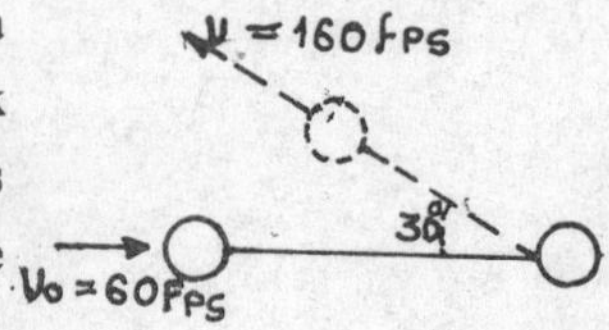

Solution : Weight of ball = 4–oz = 0.25 lb

$v_o = 60$ fps

$v = 160$ fps at 30°

$t = 0.02$ sec.

The x–direction component of velocity is

$- 160 \cos 30 = - 138.56$ fps

Thus From equation (A)

$$F\,(0.02) = \frac{0.25}{32.2}(-138.56 - 60)$$

$F = 77.1$ lb

Average force exerted on ball is 77.1lb.

Problem 2. : A 40–ton railroad car moving at 2 mph along a level track strikes and is coupled to a stationary 60–ton car in 0.50 sec. Determine the average force acting on each car during the coupling.

Solution : $W_1 = 40$ ton = 80000 lb

$$v_1 = 2 \text{ mph} = \frac{2 \times 5280}{60 \times 60} = 2.93 \text{ fps}$$

$W_2 = 60$ ton = 120000 lb

$v_2 = 0$ (since stationary)

$t = 0.5$ sec

let the common velocity be v

Thus $m_1 v_1 + m_2 v_2 = (m_1 + m_2)\, v$

$$\frac{80000}{g} \times 2.93 + \frac{120000 \times 0}{g} = \frac{(80000 + 120000)}{g} v$$

$v = 1.172$ fps

Let F be the average force acting.

From equation (A)

$$F \times 0.5 = \frac{120000}{32.2}(1.172 - 0)$$

$$F = 8735.40 \text{ lbs}$$

Problem 3. : During and Apollo mission to the moon, the lunar excursion module (LEM) is separated from the command module by an average thrust P lasting for 1 sec. As a result, the LEM has a velocity 20 fps slower than the common module. Determine P assuming the mass of the LEM and the command module to be respectively 120 lb–sec^2 / ft and 300 lb sec^2 / ft.

Solution : Suppose velocity of LEM and command module be v_1 and v_2 be final velocity of command module.

We know that

$$(W_1 + W_2)\frac{u}{g} = \frac{(W_1 V_1 + W_2 V_2)}{g}$$

or $(W_1 + W_2)\,u = W_1 V_1 + W_2 V_2$

$(120 + 300)\,x = 120\,(y - 20) + 300\,y$

It is given LEM has velocity 20 fps less than command module.

From above we have

$x = y - 5.71$ ..... (1)

Applying equation (A) we have

$P \times 1 = 120\,(y - 20 - x)$

$P = 120\,(5.71 - 20) = -1714.8$ lb

Negative sign indicates reverse direction.

Problem 4. : The system shown in fig. is moving rightward at a velocity of 15 fps when a constant horizontal force P is applied as shown. Determine the value of P that will give the system a leftward velocity of 30 fps in a time interval of 10 sec.

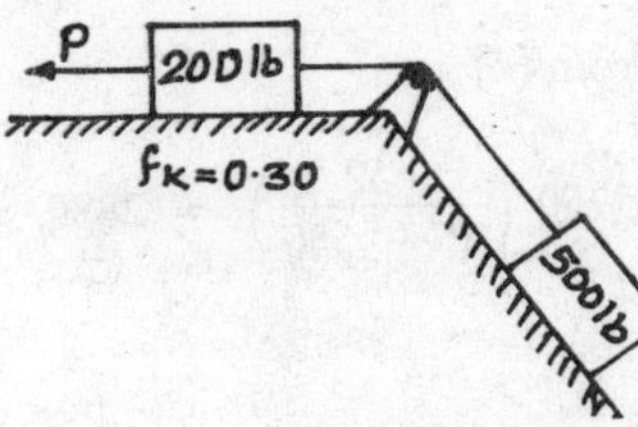

Solution :

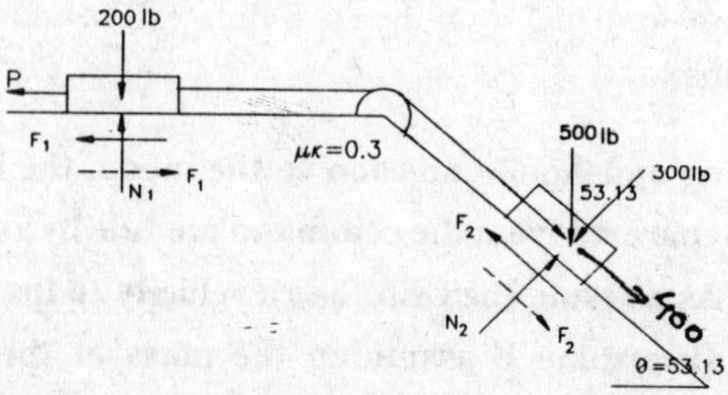

F_1 and F_2 shown by firm lines are when system moves towards rightward. Dashed lines are for when system moves leftward.

$$F_1 = \mu_k N_1 = 0.3 \times 200 = 60 \text{ lb}$$

$$F_2 = \mu_k N_2 = 0.3 \times 300 = 90 \text{ lb}.$$

For rightward motion

$$(400 - 90 - 60 - P)\, t_1 = \frac{200 + 500}{32.2}(0 - 15)$$

$$(P - 250)\, t_1 = 326 \qquad \ldots\ldots (1)$$

Also for leftward motion

$$(P - 60 - 90 - 400)\, t_2 = \frac{200 + 500}{32.2}(30 - 0)$$

$$(P - 550)\, t_2 = 652 \qquad \ldots\ldots (2)$$

given $t_1 + t_2 = 10$ sec. $\qquad \ldots\ldots (3)$

From (1) and (3) we get

$$(P - 250)(10 - t_2) = 326$$

or $10P - 2500 - Pt_2 + 250\, t_2 = 326$

or $10P - P\, t_2 + 250\, t_2 = 2886 \qquad \ldots\ldots (4)$

From (2) $Pt_2 - 550\, t_2 = 652$

or $t_2 = \dfrac{652}{P - 550}$ from (4)

$$10P - \frac{652P}{P - 550} + 250\left(\frac{652}{P - 550}\right) = 2886, \text{ solving we get}$$

$P = 622.5$ lb.

Problem 5. : Oil with a specific gravity of 0.9 is flowing at 10 fps in a straight pipe 2 ft. in diameter and 2000 ft. long. How long will it take to close a valve at a discharge end without raising the average pressure more than 40 psi ?

Solution : Pressure is $40 \dfrac{\text{lb}}{\text{in}^2}$

Diameter of pipe = 2ft = 24 in

$$F = PA = 40 \times \frac{\pi}{4} (24)^2$$

$$\text{or } F = 18095.57 \text{ lb}$$

From equation of impulse momentum.

$$18095.57\,(t) = \frac{W}{g}(v_2 - v_1)$$

W = Area × length × Density of water × Sp. gravity of oil

$$= \frac{\pi}{4}(2)^2 \times 2000 \times 62.5 \times 0.9 = 353429.17 \text{ lb (weight oil)}$$

Placing in above equation

$$18095.57\,(t) = \frac{353429.17}{32.2}(10 - 0)$$

$$t = 6.066 \text{ sec.}$$

Conservation of Linear Momentum :

It is applied to system as a whole and never to its component parts. It can be stated as

$$m_1 v_1 + m_2 v_2 + \ldots = m_1 v_1' + m_2 v_2' + \ldots.$$

where m_1 & m_2 is the mass of colliding bodies

and v_1 & v_2 are velocities before impact

v_1' & v_2' are velocities after impact.

it may also be write as

$$\frac{W_1}{g} v_1 + \frac{W_2}{g} v_2 + \ldots. = \frac{W_1}{g} v_1' + \frac{W_2}{g} v_2' + \ldots. \qquad \text{(B)}$$

Problem 6. : A 1000-lb shell is fired from a 200,000-lb cannon with a velocity of 2000 fps. Find the modulus of a nest of springs that will limit the recoil of the

Solution : $W_1 = 1000$ lb , $W_2 = 200000$ lb , $v_1 = 0$ & $v_2 = 0$

$v_1' = 2000$ fps

From equation (B) we have

$$0 + 0 = \frac{1000}{g} \times 2000 + \frac{200000}{g} \times v_2'$$

$v_2' = -10$ fps

(negative indicates v_2' is opposite to v_1')

$$\text{K.E. stored in spring} = \frac{1}{2} k \delta^2 = \frac{1}{2} \frac{W_2}{g} v_2'$$

As $\delta = 3$ ft

$$\therefore \frac{1}{2} k (3)^2 = \frac{1}{2} \frac{200000}{32.2} \times (10)^2$$

$K = 69013.11$ lb/ft $= 5751.1$ lb/inch .

Problem 7. : Just before they collide, two disks on a horizontal surface have the velocities shown in fig. Knowing that the 20-lb disk A rebounds to the left with a velocity of 6 fps, determine the rebound velocity of the 30-lb disk B. Assume the colliding surfaces are smooth.

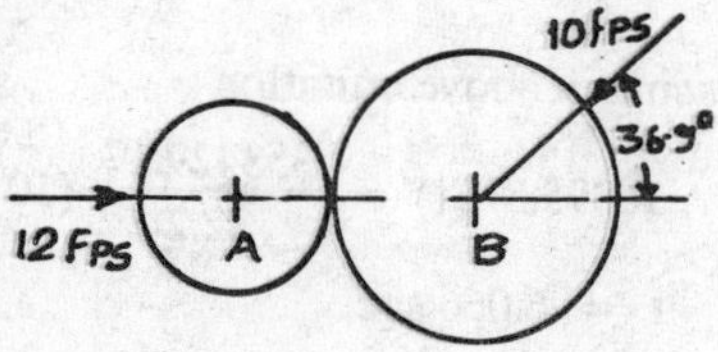

Solution : From equation (B) we have velocity towards right positive

$$\frac{20}{g} \times 12 - \frac{30}{g} 10 \cos 36.9 = \frac{-20 \times 6}{g} + \frac{30\, v_{2H}'}{g}$$

$v_{2H}' = 4$ fps (v_{2H}' is horizontal component of velocity)

$v_{2v}' =$ vertical component will remain unchanged i.e.

$v_{2v}' = 10 \sin 36.9^\circ = 6$ fps

$$\theta = \tan^{-1} \frac{6}{4} = 56.3^\circ$$

$v_2' = [6^2 + 4^2]^{1/2} = 7.21$ fps

$v_2' = 7.21$ fps

Problem 8. : Two moving balls of putty collide and cling together. Determine the magnitude of their common final velocity and the loss in kinetic energy. The first ball A weighs 1 -lb and has a velocity $v_A = 10\hat{k} + 15\hat{j}$ fps . The second ball B weighs 2-lb and has a velocity $v_B = 4\hat{k} + 3\hat{j} + 5\hat{k}$ fps .

Solution : $W_A = 1$ lb , $W_B = 2$ lb

$$v_A = 10\hat{i} + 15\hat{j},\ v_B = 4\hat{i} - 3\hat{j} + 6\hat{k}$$

Let combined velocity be V_c

from equation (B) we have

$$1 \times (10i + 15j) + 2(4i - 3j + 6k) = (1 + 2)V_C$$

$$\text{or } V_C = 6i + 3j + 4k$$

$$|V_C| = [6^2 + 3^2 + 4^2]^{1/2} = 7.81 \text{ fps}$$

$$|V_A| = [10^2 + 15^2]^{1/2} = 18.03 \text{ fps}$$

$$|V_B| = [4^2 + 3^2 + 6^2]^{1/2} = 7.81 \text{ fps}$$

$$\text{Initial Energy} = \frac{1}{2}\frac{W_A}{g}V_A^2 + \frac{1}{2}\frac{W_B}{g}V_B^2$$

$$= \frac{1}{2} \times \frac{1}{32.2} \times (18.03)^2 + \frac{1}{2} \times \frac{2}{32.2}(7.81)^2$$

$$= 6.942 \text{ ft} - \text{lb}$$

$$\text{Final Energy} = \frac{1}{2}\frac{(W_A + W_B)}{g}(V_C)^2$$

$$= \frac{1}{2}\frac{(1 + 2)}{32.2}(7.81)^2 = 2.841 \text{ ft} - \text{lb}$$

$$\text{Less of Energy} = 6.942 - 2.841 = 4.1 \text{ ft} - \text{lb}$$

Problem 9. : The spring shown in fig. has a free length of 12 in. It is compressed to half its length and the blocks are suddenly released from rest. Determine the velocity of each block when the spring is agian 12 in. long.

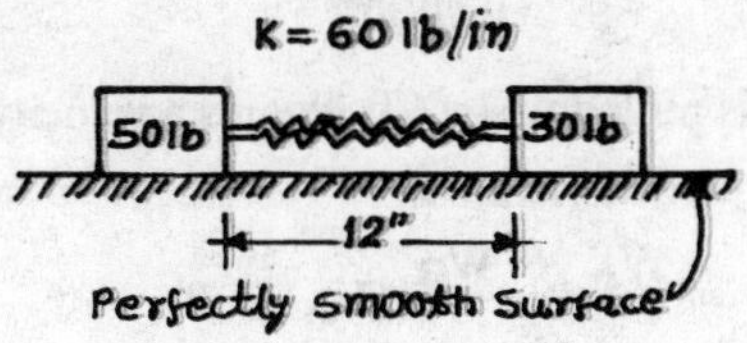

Solution : Energy stored in spring when compressed $= \frac{1}{2} kS^2$

$$E_S = \frac{1}{2} \times 60 \times 6^2 = 1080 \text{ lb-in}$$

$$\text{or } E_S = 90 \text{ lb-ft}$$

$$\text{K.E. gained by the blocks} = \frac{1}{2}\frac{W_1}{g} V_1^2 + \frac{1}{2}\frac{W_2}{g} V_2^2$$

$$= \frac{1}{2g}(50 V_1^2 + 30 V_2^2)$$

Equating the two energies (law of conservation of energy)

$$\frac{1}{2 \times 32.2}(50 V_1^2 + 30 V_2^2) = 90 \qquad \ldots\ldots (1)$$

We also have from law of conservation of momentum (equation B)

$$\frac{50}{g} V_1 - \frac{30}{g} V_2 = 0$$

$$\text{or } \frac{V_1}{V_2} = \frac{3}{5} \text{ or } V_1 = 0.6 V_2$$

from eqn. (1) $V_2 = 10.99$ fps & $V_1 = 6.59$ fps

Problem 10. : After body A in fig. moves 6 ft. from rest, it picks up body C. How much further does A move before reversing its direction ? Assume $W_A = 10$ lb, $W_B = 20$ lb, $W_C = 15$ lb.

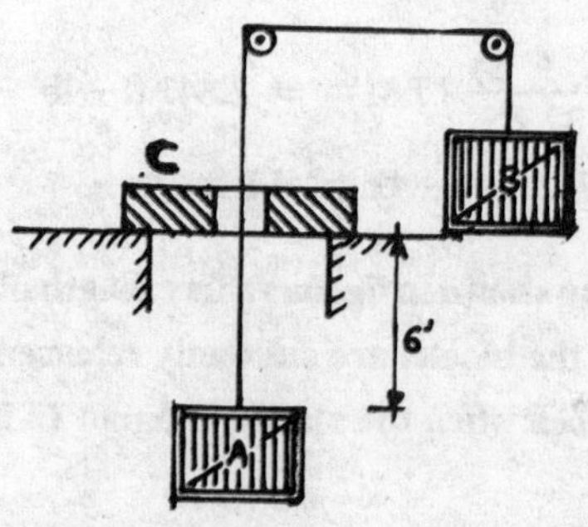

Solution : When A is pulled up by 6 ft it gains a velocity which is same as B. From work energy equation we have

$$(W_B - W_A)\,x = \frac{W_A + W_B}{2g}(V_A^2 - 0)$$

$$\text{or } (20 - 10)\, 6 = \frac{20 + 10}{2 \times 32.2} V_A^2$$

We get $V_A = 11.35 \text{ fps} = V_B$

After A collides with C. Both A and C move with same velocity as B

$$\frac{10}{g} V_A + \frac{20}{g} V_B = \frac{(10 + 15 + 20)}{g} V_C$$

Solving we have

$V_C = 7.57$ fps

We also have from work energy

$$(W_B - W_A - W_C)\, x = \frac{(W_A + W_B + W_C)}{2g} (0 - V_C^2)$$

$$(5)\, x = \frac{-45 \times 7.57^2}{2 \times 32.2}$$

or $x = 8$ ft

Problem 11.: A wooden pile that weighs 500 lb is driven vertically into the ground by successive blows from a piledriver whose 1000-lb hammer falls freely through a height of 6 ft upon the head of the pile. How deep does a single blow of the hammer drive the pile when the ground has developed an average resitance to penetration of 6500 lb ? Assume that the hammer and pile cling together after impact.

[Hing : Resultant work done equals kinetic energy lost in impact.]

Solution : The potential energy of hammer is converted to K.E.

Thus

$$1000 \times 6 = \frac{1}{2} \times \frac{1000}{g} v^2$$

or $v = 19.66$ fps

v is the velocity with which hammer strikes the pile. After striking hammer and pile move together.

Hence

$$\frac{1000}{g} \times 19.66 + 0 = \frac{(1000 + 500)}{g} V_C$$

$V_C = 13.1$ fps V_C is the combined velocity of pile and hammer.

Loss in K.E. = Initial K.E. − Final K.E.

$$= 1000 \times 6 - \frac{1}{2} \times \frac{1500}{g} (13.10)^2$$

$$= 2000 \text{ ft-lb}$$

Suppose pile penetrates by distance y in single blow then

Work done = Loss in K.E.

$$\text{or } \frac{1}{2} (6500)\ y^2 = 2000$$

$$\text{or } y = 0.785 \text{ ft}$$

ELASTIC IMPACT :

When two elastic bodies collide, they are deformed at first. Then they spring apart because of restoring elastic forces. The resultant force on the system is zero and law of conservation of momentum may be applied. There are four types of impact as given below

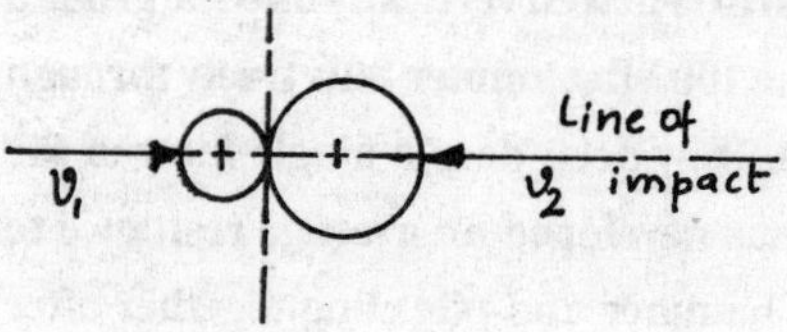

(i) Direct Central impact

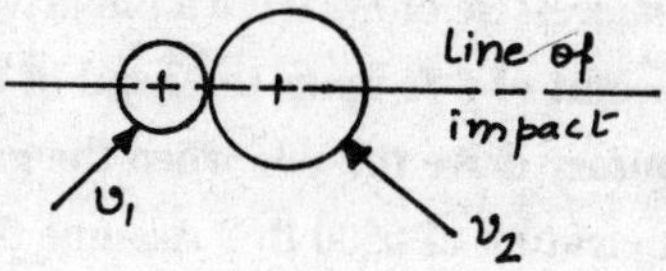

(ii) Oblique central impact.

Other two types of impact are :

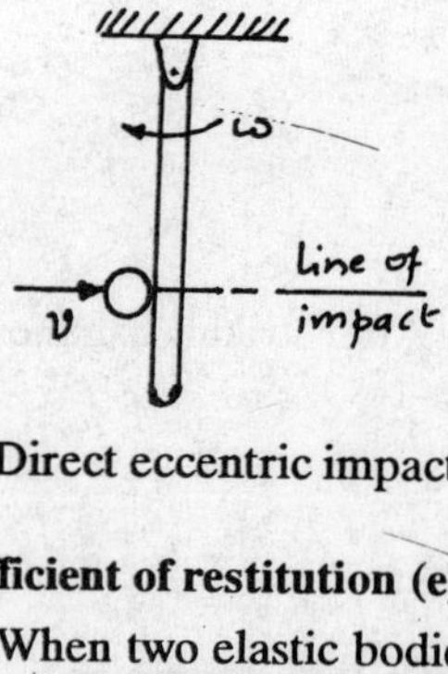

(iii) Direct eccentric impact

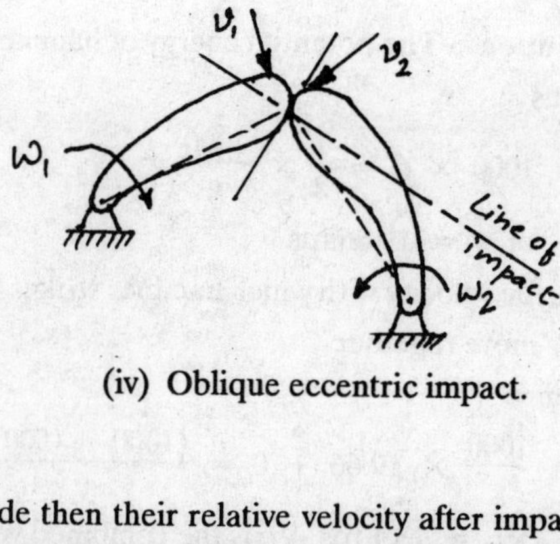

(iv) Oblique eccentric impact.

Coefficient of restitution (e) :

When two elastic bodies collide then their relative velocity after impact bears a constant ratio with their relative velocity before impact and it is in opposite direction and proportionality constant between two terms is called coefficient of restitution.

$$e = \frac{\text{Relative velocity after impact}}{\text{Relative velocity before impact}}$$

$$= \frac{v_2' - v_1'}{v_1 - v_2}$$

It is preferred in the form $e(v_1 - v_2) = v_2' - v_1'$

or $v_2' - v_1' = -e(v_2 - v_1)$

Problem 12. : Direct central impact occurs between a 60-lb body moving rightward at 10 fps and 20-lb body moving leftward at 20 fps. If the coefficient of restitution is e = 0.6, determine the average impact force for a time of impact lasting 002 sec.

Solution : $W_1 = 60$ lb , $W_2 = 30$ lb , $v_1 = 10$ fps ,

$v_2 = -20$ fps , $e = 0.6$, $t = 0.02$ sec.

Let the velocities after impact be v_2' & v_1'

$$e(v_1 - v_2) = v_2' - v_1'$$

or $0.6\left(10 - (-20)\right) = v_2' - v_1'$ or $v_2' - v_1' = 18$ .... (1)

From conservation of momentum

$$m_1 v_1 + m_2 v_2 = m_1 v_1' + m_2 v_2'$$

$$60 \times 10 + 30(-20) = 60 v_1' + 30 v_2'$$

$2 v_1' + v_2' = 0$ or $v_2' + 2v_1' = 0$ ..... (2)

Solving simultaneously (1) & (2), we get

$v_1' = -6$ fps , $v_2' = +12$ fps

Negative sign indicates v_1' is leftwards

Average force of impact

$$F(t) = \frac{W_2}{g}(v_2' - v_2)$$

$$F(0.02) = \frac{30}{32.2}(12 - (-20))$$

$$F = 1490.68 \text{ lb}$$

Problem 13. : Direct central impact occurs between a 30-lb body moving to the right at 6 fps and a body of weight W moving to the left at 4 fps. The coeffcient of restitution is e = 0.5. After impact the 30-lb body has a leftward velocity of 3 fps. (a) Find the value of W. (b) If the impact lasts for 0.02 sec, find the average impact force.

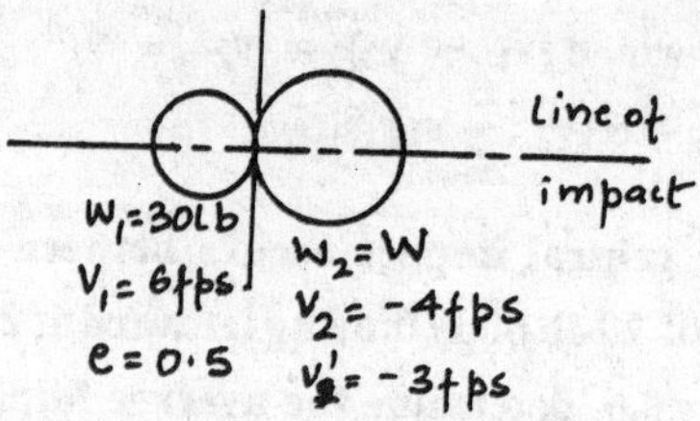

Solution : $W_1 = 30$ lb , $W_2 = W$ (say) , $v_1 = 6$ fps ,

$v_2 = -4$ fps , e = 0.5 , $v_1' = -3$ fps (leftwards)

(a) $e = \dfrac{v_2' - v_1'}{v_1 - v_2}$ or $0.5(6 + 4) = v_2' + 3$

or $v_2' = 2$ fps

From law of conservation of momentum

$$30 \times 6 + W(-4) = 30-(-3) + W(2)$$

or W = 45 lb

(b) t = 0.02 sec find F

$$F(t) = \frac{W_2}{g}(v_2' - v_2)$$

$$F(0.02) = \frac{45}{32.2}(2 + 4)$$

or F = 419.26 lb

Problem 14. : A golf ball is dropped from a height of 10 ft. upon a concrete floor. The coefficient of restitution is 0.984. Find the height to which the ball rebounds on the first, second, and third bounces.

Solution : When ball reaches the floor the velocities

$$v^2 = u^2 + 2gh \text{ as } u = 0 \text{ ; } v = \sqrt{2gh}$$

$$v_1 = [2 \times 32.2 \times 10]^{1/2} = 25.38 \text{ fps}$$

$$e = \frac{v_2' - v_1'}{v_1 - v_2}$$

$$0.984 = \frac{0 - v_1'}{25.38 - 0} \quad \text{or} \quad v_1' = -24.97 \text{ fps (upwards)}$$

$$\text{Height reached } h_1 = \frac{v_1^2}{2g} = \frac{(24.97)^2}{2 \times 32.2} = 9.68 \text{ ft}$$

For Second rebound :

$$v_1 = [2 \times 32.2 \times 9.68]^{1/2} = 24.97 \text{ fps}$$

$$0.984 = \frac{0 - v_1'}{24.97 - 0} \quad \text{or} \quad v_1' = -24.57 \text{ fps (upward)}$$

$$h_2 = \frac{24.57^2}{2 \times 32.2} = 9.375 \text{ ft}$$

For Third : v = 24.57 fps

$$0.984 = \frac{0 - v_1'}{24.57 - 0} \quad \text{or} \quad v_1' = 24.1768 \text{ fps}$$

$$h_3 = \frac{24.2768^2}{2 \times 32.2} = 9.076 \text{ ft}.$$

Problem 15. : The balls A and B in fig. are attached to stiff rods of neglibible wright. Ball A is released from rest and allowed to strike B. If e = 0.6, determine the maximum angle θ through which ball B will swing. What is the maximum and minimum tension in the rod attached to B ? If the impact lasts for 0.01 sec, also find the average impact force.

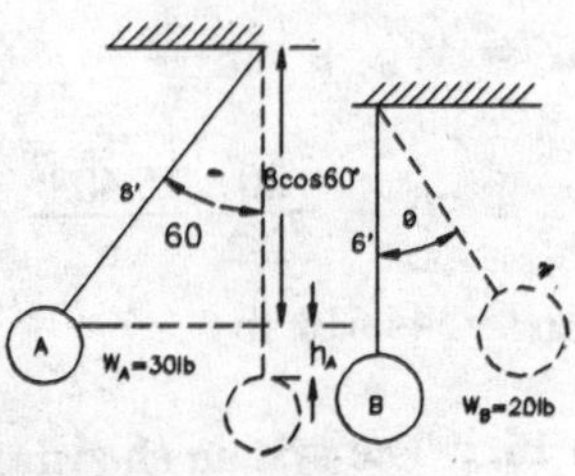

Solution : Ball A is released from rest

$h_A = 8 - 8\cos 60 = 4$ ft (refer fig.)

P.E. = K.E. gained

$$30 \times 4 = \frac{1}{2} \times \frac{30}{3.2.2} \times v_A^2$$

or $V_A = 16.05$ fps

V_A is the velocity with which A strikes B.

Applying law of conservation of momentum

$$30 \times 16.05 + 0 = 30\,v_A' + 20\,v_B'$$

$$\text{or } 3v_A' + 2v_B' = 48.15 \qquad \ldots..(1)$$

$$e = \frac{v_B' - v_A'}{v_A - v_B} \text{ or } 0.6 = \frac{v_B' - v_A'}{16.05 - 0}$$

$$\text{or } v_A' - v_B' = 9.63 \qquad \ldots..(2)$$

Solving eqn. (1) & (2) $v_A' = 5.778$ fps

$v_B' = 15.408$ fps

Height by which B rise $= 6 - 6\cos\theta$

$$20\,(6 - 6\cos\theta) = \frac{1}{2} \times \frac{20}{32.2} \times 15.408^2 \quad (\text{P.E.} = \text{K.E.})$$

Solving $\theta = 67.32^\circ$

Impact takes for 0.01 sec hence

$$F\,(0.01) = \frac{20}{32.2}\,(15.408 - 0)$$

or $F = 957$ lb

$$T_{min} = 20\cos\theta = 20\cos 67.32 = 7.71 \text{ lb}$$

$$T_{max} = W_B + m_B \frac{v_B^2}{r}$$

$$= 20 + \frac{20}{32.2.}\,\frac{15.408^2}{6}$$

$$T_{max} = 44.616 \text{ lb}$$

Problem 16. : The system shown in fig. of problem 15 is used to determine experimentally the coefficient of restitution. If ball A is released from rest and ball B swings through $\theta = 53.1^\circ$ after being strick, determine e.

Solution : From the previous problem we have

$h_A = 4$ft & $v_A = 16.05$ fps

$$e = \frac{v_B' - v_A'}{v_A - v_B} = \frac{v_B' - v_A'}{16.05 - 0} = \frac{v_B' - v_A}{16.05} \qquad \ldots...(1)$$